COMPUTER ANALYSIS
OF POWER SYSTEMS

COMPUTER ANALYSIS
OF POWER SYSTEMS

J. Arrillaga

and

C. P. Arnold

University of Canterbury, Christchurch, New Zealand

JOHN WILEY & SONS
Chichester · New York · Brisbane · Toronto · Singapore

Other Wiley Editorial Offices

John Wiley & Sons, Inc., 605 Third Avenue,
New York, NY 10158–0012, USA

Jacaranda Wiley Ltd, G.P.O. Box 859, Brisbane,
Queensland 4001, Australia

John Wiley & Sons (Canada) Ltd, 22 Worcester Road,
Rexdale, Ontario M9W 1L1, Canada

John Wiley & Sons (SEA) Ptd Ltd, 37 Jalan Pemimpin 05-04.
Block B, Union Industrial Building, Singapore 2057

Library of Congress Cataloging-in-Publication Data:

Arrillaga, J.
    Computer analysis of power systems / J. Arrillaga and C. P. Arnold.
        p.   cm.
    Includes bibliographical references and index.
    ISBN   0 471 92760 0
        1.   Electric power systems—Data processing.   I.   Arnold, C. P.
    II.   Title.
    TK1005.A757   1990   90-39424
    621.31—dc20   CIP

British Library Cataloguing in Publication Data:

Arrillaga, J.
    Computer analysis of power systems.
    1.   Electricity transmission systems.   Mathematical models.
    Applications of computer systems
    I.   Title   II.   Arnold, C. P.
    621.31910113

    ISBN   0 471 92760 0

Typeset by Thomson Press (India) Limited, New Delhi
Printed in Great Britain by Courier International, Tiptree, Essex

CONTENTS

7   POWER SYSTEM STABILITY—ADVANCED COMPONENT MODELLING   197

PREFACE

In an earlier book entitled *Computer Modelling of Electrical Power Systems* the authors described some of the component models and numerical techniques that have established the digital computer as the primary tool in Power System Analysis. That book also included, for the first time, the incorporation of h.v.d.c. convertor and systems within conventional a.c. power system models. From an educational viewpoint some of that material can be considered of a specialised nature and can be substantially reduced to make room for several other basic and important topics of more general interest.

After three decades of computer-aided power system analysis the basic algorithms in current use have reached high levels of efficiency and sophistication.

In this new book the authors describe the main computer modelling techniques that, having gained universal acceptance, constitute the basic framework of modern power system analysis.

Some basic knowledge of power system theory, matrix analysis and numerical techniques is presumed, although several appendices and many references have been included to help the uninitiated to pick up the relevant background.

An introductory chapter describes the main computational and transmission system developments which influence modern power system analysis. This is followed by three chapters (2, 3 and 4) on the subject of load or power flow with emphasis on the Newton–Raphson fast-decoupled algorithm. Chapter 5 describes the subject of a.c. system faults.

The next two chapters (6 and 7) deal with the electromechanical behaviour of power systems. Chapter 6 describes the basic dynamic models of power system plant and their use in multi-machine transient stability analysis. More advanced dynamic models and a quasi-steady-state representation of large converter plant and h.v.d.c. transmission are developed in Chapter 7.

A description of the Electromagnetic Transients Program with the marriage between 'Bergeron's and Trapezoidal' methods is presented in Chapter 8.

A generalisation of the multi-phase models described in Chapter 3 is used in Chapter 9 as the framework for harmonic flow analysis.

Chapter 10 describes the state of the art in power system security and optimisation analysis.

Finally, Chapter 11 deals with recent advances made on the subject of interactive power system analysis and developments in computer graphics with emphasis on the use of personal computers.

The authors should like to acknowledge the considerable help received from so many of their present and earlier colleagues and in particular from P. S. Bodger, A. Brameller, T. J. Densem, H. W. Dommel, B. J. Harker, M. D. Heffernan, N. C. Pahalawaththa, M. Shurety, B. Stott, K. S. Turner and N. R. Watson.

1. INTRODUCTION

1.1 COMPUTERS IN POWER SYSTEMS

The appearance of large digital computers in the 1960s paved the way for unprecedented developments in power system analysis and with them the availability of a more reliable and economic supply of electrical energy with tighter control of the system frequency and voltage levels.

In the early years of this development the mismatch between the size of the problems to be analysed and the limited capability of the computer technology encouraged research into algorithmic efficiency. Such efforts have proved invaluable to the development of real time power system control at a time when the utilities are finding it increasingly difficult to maintain high levels of reliability at competitive cost.

Fortunately the cost of processing information and computer memory is declining rapidly. By way of example, in less than two decades the cost of computer hardware of similar processing power has reduced by about three hundred times.

The emphasis in modern power systems has turned from resource creation to resource management. The two primary functions of an energy management system are security and economy of operation and these tasks are achieved in main control centres. In the present state of the art the results derived by the centre computers are normally presented to the operator who can then accept, modify or ignore the advice received. However, in the longer term the operating commands should be dispatched automatically without human intervention, thus making the task of the computer far more responsible.

1.2 COMPUTER TASKS

The basic power system functions involve very many computer studies requiring processing power capabilities in millions of instructions per second (MIPS). The most demanding in this respect are the network solutions, the specific task of electrical power system analysis.

In order of increasing processing requirements the main computer tasks involved in the management of electrical energy systems are as follows.

- Automatic generation control (AGC).
- Supervisory control and data acquisition (SCADA).
- Generation scheduling.
- Network analysis.

The subject of this book is power system analysis and it is therefore important to consider the above computing tasks in relation to network analysis.

1.2.1 Automatic Generation Control [1]

During normal operation the following four tasks can be identified with the purpose of AGC:

- Matching of system generation and system load.
- Reducing the system frequency deviations to zero.
- Distributing the total system generation among the various control areas to comply with the scheduled tie flows.
- Distributing the individual area generation among its generating sources so as to minimise operating costs.

The first task is met by governor speed control. The other tasks require supplementary controls coming from the other control centres. The second and third tasks are associated with the regulation function, or load-frequency control and the last one with the economic dispatch function of AGC.

The above requirements are met with modest computer processing power (of the order of 0.1 MIPS).

1.2.2 Supervisory Control and Data Acquisition [2]

The modern utility control system relies heavily on the operator control of remote plant. In this task the operator relies on SCADA for the following tasks:

- Data acquisition
- Information display
- Supervisory control
- Alarm processing
- Information storage and reports
- Sequence of events acquisition
- Data calculations
- Remote terminal unit processing

Typical computer processing requirements of SCADA systems are 1–2 MIPS.

1.2.3 Generation Scheduling [3]

The operation scheduling problem is to determine which generating units should be committed and available for generation, the units' nominal generation or dispatch and in some cases even the type of fuel to use.

In general, utilities may have several sources of power such as thermal plant (steam and gas), hydro and pumped storage plants, dispersed generation (such as wind power or photovoltaic), interconnections with other national or international companies, etc. Also many utilities use load management control to influence the loading factor, thus affecting the amount of generation required.

The economic effect of operations scheduling is very important when fuel is a major component of the cost. The time span for scheduling studies depends on a number of factors. Large steam turbines take several hours to start up and bring on-line; moreover they have costs associated with up- and down-time constraints and start-ups. Other factors to be considered are maintenance schedules, nuclear refuelling schedules and long-term fuel contracts which involve making decisions for one or more years ahead. Hydro scheduling also involves long time frames due to the large capacity of the reservoirs. However many hydro and pump storage reservoirs have daily or weekly cycles.

Scheduling computer requirements will normally be within 2 MIPS.

1.3 NETWORK ANALYSIS

This is by far the more demanding task, since it develops basic information for all the others and needs to be continuously updated. Typical computer requirements will be of the order of 5 MIPS.

The primary subject of power system analysis is the load-flow or power-flow problem which forms the basis for so many modern power system aids such as state estimation, unit commitment, security assessment and optimal system operation. It is also needed to determine the state of the network prior to other basic studies like fault analysis and stability.

The methodology of load-flow calculations has been well established for many years, and the primary advances today are in size and modelling detail. Simulation of networks with more than 4000 buses and 8000 branches is now common in power system analysis.

While the basic load-flow algorithm only deals with the solution of a system of continuously differentiable equations, there is probably not a single routine program in use anywhere that does not model other features. Such features often have more influence on convergence than the performance of the basic algorithm.

The most successful contribution to the load-flow problem has been the application of Newton–Raphson and derived algorithms. These were finally established with the development of programming techniques for the efficient handling of large matrices and in particular the sparsity-oriented ordered elimination methods. The Newton algorithm was first enhanced by taking advantage of the decoupling characteristics of load flow and finally by the use of reasonable approximations directed towards the use of constant Jacobian matrices.

In transient stability studies the most significant modelling development has probably been the application of implicit integration techniques which allow the differential equations to be algebraised and then incorporated with the network's algebraic equations to be solved simultaneously. The use of implicit trapezoidal integration has proved to be very stable, permitting step lengths greater than the smallest time constant of the system. This technique allows detailed representation of synchronous machines with their voltage regulators and governors, induction motors and nonimpedance loads.

The trapezoidal method has also found application in the area of electromagnetic transients and, combined with Bergeron's method of characteristics, has resulted in a versatile and reliable algorithm known as the EMTP, which has found universal acceptance.

1.3.1 Security Assessment

The overall aim of the economy–security process is to operate the system at lowest cost with a guarantee of continued prespecified energy supply during emergency conditions. An emergency situation results from the violation of the operating limits and the most severe violations result from contingencies. A given operating state can be judged secure only with reference to one or several contingency cases [4].

The security functions include security assessment and control. These are carried out either in the 'real time' or 'study' modes.

The real time mode derives information from state estimates and upon detection of any violations, security control calculations are needed for immediate implementation. Thus computing speed and reliability are of primary importance.

The study mode represents a forecast operating condition. It is derived from stored information and its main purpose is to ensure future security and optimality of power system operation. The difficulty is that carrying load-flow solutions for large numbers of contingency cases involves massive computational requirements.

Modern energy management systems are using more open architectures permitting the connection of auxiliary computing devices on to which self-contained but computation-intensive calculations can be down-loaded. Contingency analysis is ideally suited to distributed processing. The separate cases in the contingency list can be shared between multiple inexpensive processors.

1.3.2 Optimal Power Flow

The computational need becomes even more critical when it is realised that contingency-constrained optimal power flow (OPF) usually needs to iterate with contingency analysis.

The purpose of an on-line function is to schedule the power system controls to achieve operation at a desired security level while optimising an objective function such as cost of operation. The new schedule may take system operation from one security level to another, or it may restore optimality at an already achieved security level. In the real time mode, the calculated schedule, once accepted, may be implemented manually or automatically. The ultimate goal is to have the security-constrained scheduling calculation initiated, completed and dispatched to the power system entirely automatically without human intervention.

1.4 TRANSMISSION SYSTEM DEVELOPMENT

The basic algorithms developed by power system analysts are built around conventional power transmission plant with linear characteristics. However, the advances made in power electronic control, the longer transmission distances and the justification for more interconnections (national and international) have resulted in more sophisticated means of active and reactive power control and the use of h.v.d.c. transmission.

Although the number of h.v.d.c. schemes in existence is still relatively low, most of the world's large power systems already have or plan to have such links. Moreover, considering the large power ratings of the h.v.d.c. schemes, their presence influences considerably the behaviour of the interconnected systems and they must be properly represented in power system analysis.

Whenever possible, any equivalent models used to simulate the convertor behaviour should involve traditional power-system concepts, for easy incorporation within existing programs. However, the number of degrees of freedom of d.c. power transmission is higher and any attempt to model its behaviour in the more restricted a.c. framework will have limited application. The integration of h.v.d.c. transmission with conventional a.c. load-flow and stability models has been given sufficient coverage in recent years and is now well understood.

1.5 INTERACTIVE POWER SYSTEM ANALYSIS

Probably the main development of the decade in power system analysis has been the change of emphasis from mainframe-based to interactive analysis software.

Until IBM introduced the PC/AT in 1984 it was out of the question to use a PC to perform power system analyses. At the time of writing, the 32-bit architecture and speed of the Intel 80286 chip combined with the highly increased storage capablity and speed of hard disks has made it possible for power system analysts to perform most of their studies on the PC. Moreover FORTRAN compilers have become available which are capable of handling the memory and code requirements of most existing power system programs.

Recent advances in graphics devices in terms of speed, resolution, colour, reduced costs and improved reliability have enhanced the interactive capabilities and made the designer's task more effective and attractive. The full potential of interactive analysis on the PC is still somehow limited by the resolution of typical displays available on the PC today, though this problem can be overcome to some extent by the use of zooming and panning techniques.

In parallel with the improvements in PCs there has been an equally impressive development in workstations, with sizes and prices sufficiently attractive to compete with PCs and without their limitation in graphic displays. Practically all large system study programs can now be run efficiently in such workstations.

These capabilities are beginning to have an impact in the educational scene too where, for a fraction of the cost of earlier computers, complete classes of students can now perform interactive power system studies individually and simultaneously in CAE laboratories.

Many commercial packages have already appeared offering power system software for the AT and PC market and their capabilities are expanding all the time. Early packages were restricted to basis load-flow, faults and stability studies, whereas more recent ones include more advanced programs and specialised features such as electromagnetic transients and harmonic propagation.

1.6 REFERENCES

[1] T. M. Athay, 1987. Generation scheduling and control, *Proc. IEEE* **75** 1592–1606.
[2] D. J. Gaushell and H. T. Darlington, 1987. Supervisory control and data acquisition, *Proc. IEEE* **75** 1645–1658.
[3] A. J. Cohen and V. R. Sherkat, 1987. Optimization-based methods for operations scheduling, *Proc. IEEE* **75** 1574–1591.
[4] B. Stott, O. Alsac and A. J. Monticelli, 1987. Security analysis and optimization, *Proc. IEEE* **75** 1623–1644.
[5] W. F. Tinney, 1972. Compensation methods with ordered triangular factorization, *IEEE Trans.* **PAS-91** 123–127.

2. LOAD FLOW

2.1 INTRODUCTION

Under normal conditions electrical transmission systems operate in their steady-state mode and the basic calculation required to determine the characteristics of this state is termed load flow (or power flow).

The object of load-flow calculations is to determine the steady-state operating characteristics of the power generation/transmission system for a given set of busbar loads. Active power generation is normally specified according to economic-dispatching practice and the generator voltage magnitude is normally maintained at a specified level by the automatic voltage regulator acting on the machine excitation. Loads are normally specified by their constant active and reactive power requirement, assumed unaffected by the small variations of voltage and frequency expected during normal steady-state operation.

The solution is expected to provide information of voltage magnitudes and angles, active and reactive power flows in the individual transmission units, losses and the reactive power generated or absorbed at voltage-controlled buses.

The load-flow problem is formulated in its basic analytical form in this chapter with the network represented by linear, bilateral and balanced lumped parameters. However the power and voltage constraints make the problem nonlinear and the numerical solution must therefore be iterative in nature.

The current problems faced in the development of load flow are an ever increasing size of systems to be solved, on-line applications for automatic control, and system optimization. Hundreds of contributions have been offered in the literature to overcome these problems [1].

Five main properties are required of a load-flow solution method.

(i) High computational speed. This is especially important when dealing with large systems, real time applications (on-line), multiple case load flow such as in system security assessment, and also in interactive applications.

(ii) Low computer storage. This is important for large systems and in the use of computers with small core storage availability, e.g. mini-computers for on-line application.

(iii) Reliability of solution. It is necessary that a solution be obtained for ill-conditioned problems, in outage studies and for real time applications.

(iv) Versatility. An ability on the part of load flow to handle conventional and special features (e.g. the adjustment of tap ratios on transformers; different representations

7

of power system apparatus), and its suitability for incorporation into more complicated processes.

(v) Simplicity. The ease of coding a computer program of the load-flow algorithm.

The type of solution required for a load flow also determines the method used:

accurate	or	approximate
unadjusted	or	adjusted
off-line	or	on-line
single case	or	multiple cases

The first column are requirements needed for considering optimal load-flow and stability studies, and the second column those needed for assessing security of a system. Obviously, solutions may have a mixture of the properties from either column.

The first practical digital solution methods for load flow were the Y matrix-iterative methods [2]. These were suitable because of the low storage requirements, but had the disadvantage of converging slowly or not at all. Z matrix methods [3] were developed which overcame the reliability problem but storage and speed were sacrificed with large systems.

The Newton–Raphson method [4, 5] was developed at this time and was found to have very strong convergence. It was not, however, made competitive until sparsity programming and optimally ordered Gaussian-elimination [6–8] were introduced, which reduced both storage and solution time.

Nonlinear programming and hybrid methods have also been developed, but these have created only academic interest and have not been accepted by industrial users of load flow. The Newton–Raphson method and techniques derived from this algorithm satisfy the requirements of solution-type and programming properties better than previously used techniques and are gradually replacing them.

2.2   NETWORK MODELLING

Transmission plant components are modelled by their equivalent circuits in terms of inductance, capacitance and resistance. Each unit constitutes an electric network in its own right and their interconnection constitutes the transmission system.

Among the many alternative ways of describing transmission systems to comply with Kirchhoff's laws, two methods—mesh and nodal analysis—are normally used. Nodal analysis has been found to be particularly suitable for digital computer work, and is almost exclusively used for routine network calculations.

The nodal approach has the following advantages.

- The numbering of nodes, performed directly from a system diagram, is very simple.
- Data preparation is easy.
- The number of variables and equations is usually less than with the mesh method for power networks.
- Network crossover branches present no difficulty.
- Parallel branches do not increase the number of variables or equations.

- Node voltages are available directly from the solution, and branch currents are easily calculated.
- Off-nominal transformer taps can easily be represented.

2.2.1 Transmission Lines

In the case of a transmission line the total resistance and inductive reactance of the line is included in the series arm of the equivalent-π and the total capacitance to neutral is divided between its shunt arms.

2.2.2 Transformer on Nominal Ratio

The equivalent-π model of a transformer is illustrated in Fig.2.1, where y_{oc} is the reciprocal of z_{oc} (magnetising impedance) and y_{sc} is the reciprocal of z_{sc} (leakage impedance). z_{sc} and z_{oc} are obtained from the standard short-circuit and open-circuit tests.

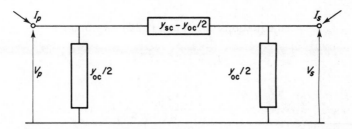

Figure 2.1
Transformer equivalent circuit

This yields the following matrix equation:

$$
\begin{bmatrix} I_p \\ I_s \end{bmatrix} = \begin{bmatrix} y_{sc} & -y_{sc} + y_{oc}/2 \\ -y_{sc} + y_{oc}/2 & y_{sc} \end{bmatrix} \cdot \begin{bmatrix} V_p \\ V_s \end{bmatrix} \tag{2.2.1}
$$

where y_{sc} is the short-circuit or leakage admittance and y_{oc} is the open-cicuit or magnetising admittance.

The use of a three-terminal network is restricted to the single-phase representation and cannot be used as a building block for modelling three-phase transformer banks.

The magnetising admittances are usually removed from the transformer model and added later as small shunt-connected admittances at the transformer terminals. In the per unit system the model of the single-phase transformer can then be reduced to a lumped leakage admittance between the primary and secondary busbars.

2.2.3 Off-nominal Transformer Tap Settings

A transformer with turns ratio a interconnecting two nodes i, k can be represented by an ideal transformer in series with the nominal transformer leakage admittance as shown in Fig. 2.2(a).

If the transformer is on nominal tap ($a = 1$), the nodal equations for the network branch in the per unit system are

$$I_{ik} = y_{ik} V_i - y_{ik} V_k \tag{2.2.2}$$

$$I_{ki} = y_{ik} V_k - y_{ik} V_i. \tag{2.2.3}$$

In this case $I_{ik} = - I_{ki}$.

For an off-nominal tap setting and letting the voltage on the k side of the ideal transformer be V_t we can write

$$V_t = \frac{V_i}{a} \tag{2.2.4}$$

$$I_{ki} = y_{ik}(V_k - V_t) \tag{2.2.5}$$

$$I_{ik} = - \frac{I_{ki}}{a}. \tag{2.2.6}$$

Eliminating V_t between equations (2.2.4) and (2.2.5) we obtain

$$I_{ki} = y_{ik} V_k - \frac{y_{ik}}{a} V_i \tag{2.2.7}$$

$$I_{ik} = - \frac{y_{ik}}{a} V_k + \frac{y_{ik}}{a^2} V_i. \tag{2.2.8}$$

A simple equivalent-π circuit can be deduced from equations (2.2.7) and (2.2.8) the elements of which can be incorporated into the admittance matrix. This circuit is illustrated in Fig. 2.2(b).

The equivalent cicuit of Fig. 2.2(b) has to be used with care in banks containing delta-connected windings. In a star–delta bank of single-phase transformer units, for example, with nominal turns ratio, a value of 1.0 per unit voltage on each leg of the star winding produces under balanced conditions 1.732 per unit voltage on each leg of the delta winding (rated line to neutral voltage as base). The structure of the bank

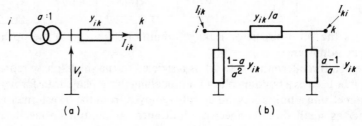

(a)   (b)

Figure 2.2
Transformer with off-nominal tap setting

requires in the per unit representation an effective tapping at $\sqrt{3}$ nominal turns ratio on the delta side, i.e. $a = 1.732$.

For a delta–delta or star–delta transformer with taps on the star winding, the equivalent circuit of Fig. 2.2(b) would have to be modified to allow for effective taps to be represented on each side. The equivalent-circuit model of the single-phase unit can be derived by considering a delta–delta transformer as comprising a delta–star transformer connected in series (back to back) via a zero-impedance link to a star–delta transformer, i.e. star windings in series. Both neutrals are solidly earthed. The leakage impedance of each transformer would be half the impedance of the equivalent delta–delta transformer. An equivalent per unit representation of this coupling is shown in Fig. 2.3. Solving this circuit for terminal currents

$$I_p = \frac{I'}{\alpha} = \frac{(V' - V'')y}{\alpha}$$

$$= \frac{(V_p/\alpha - V_s/\beta)y}{\alpha} = \frac{y}{\alpha^2} V_p - \frac{y}{\alpha\beta} V_s \qquad (2.2.9)$$

$$-I_s = \frac{I'}{\beta} = \frac{y}{\alpha\beta} V_p - \frac{y}{\beta^2} V_s \qquad (2.2.10)$$

or in matrix form

$$\begin{vmatrix} I_p \\ I_s \end{vmatrix} = \begin{vmatrix} y/\alpha^2 & -y/\alpha\beta \\ -y/\alpha\beta & y/\beta^2 \end{vmatrix} \cdot \begin{vmatrix} V_p \\ V_s \end{vmatrix} \qquad (2.2.11)$$

These admittance parameters form the primitive network for the coupling between a primary and secondary coil.

2.2.4 Phase-shifting Transformers

To cope with phase shifting, the transformer of Fig. 2.3 has to be provided with a complex turns ratio. Moreover, the invariance of the product VI^* across the ideal transformer requires a distinction to be made between the turns ratios for current

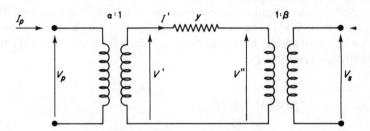

Figure 2.3
Basic equivalent circuit in p.u. for coupling between primary and secondary coils with both primary and secondary off-nominal tap ratios of α and β

and voltage, i.e.

$$V_p I_p^* = - V' I'^*$$

or

$$V_p = (a + jb)V' = \alpha V'$$

$$I_p^* = - \frac{I'^*}{a + jb}$$

$$I_p = - \frac{I'}{a - jb} = - \frac{I'}{\alpha^*}.$$

Thus the circuit of Fig. 2.3 has two different turns ratios, i.e.

$$\alpha_v = a + jb \quad \text{for the voltages}$$

and

$$\alpha_i = a - jb \quad \text{for the currents.}$$

Solving the modified circuit for terminal currents:

$$I_p = \frac{I'}{\alpha_i} = \frac{(V' - V'')y}{\alpha_i}$$

$$= \frac{(V_p/\alpha_v - V_s/\beta)y}{\alpha_i} = \frac{y}{\alpha_v \alpha_i} V_p - \frac{y}{\alpha_i \beta} V_s \qquad (2.2.12)$$

$$- I_s = \frac{I'}{\beta} = \frac{y}{\alpha_v \beta} V_p - \frac{y}{\beta^2} V_s. \qquad (2.2.13)$$

Thus, the general single-phase admittance of a transformer including phase shifting is

$$[y] = \begin{array}{|c|c|} \hline \dfrac{y}{\alpha_i \alpha_v} & -\dfrac{y}{\alpha_i \beta} \\ \hline -\dfrac{y}{\alpha_v \beta} & \dfrac{y}{\beta^2} \\ \hline \end{array} \qquad (2.2.14)$$

Note that, although an equivalent lattice network similar to that in Fig. 2.3 could be constructed, it is no longer a bilinear network as can be seen from the asymmetry of y in equation (2.2.14). The equivalent circuit of a single-phase phase-shifting transformer is thus of limited value and the transformer is best represented analytically by its admittance matrix.

2.3 BASIC NODAL METHOD

In the nodal method as applied to power system networks, the variables are the complex node (busbar) voltages and currents, for which some reference must be designated. In fact, two different references are normally chosen: for voltage

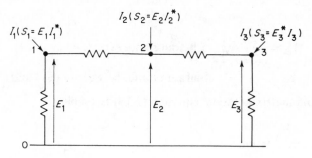

Figure 2.4
Simple network showing nodal quantities

magnitudes the reference is ground, and for voltage angles the reference is chosen as one of the busbar voltage angles, which is fixed at the value zero (usually). A nodal current is the net current entering (injected into) the network at a given node, from a source and/or load external to the network. From this definition, a current entering the network (from a source) is positive in sign, while a current leaving the network (to a load) is negative, and the net nodal injected current is the algebraic sum of these. One may also speak in the same way of nodal injected powers $S = P + jQ$.

Figure 2.4 gives a simple network showing the nodal currents, voltages and powers.

In the nodal method, it is convenient to use branch admittances rather than impedances. Denoting the voltages of nodes k and i as E_k and E_i respectively, and the admittance of the branch between them as y_{ki}, then the current flowing in this branch from node k to node i is given by

$$I_{ki} = y_{ki}(E_k - E_i). \tag{2.3.1}$$

Let the nodes in the network be numbered $0, 1, \ldots, n$, where 0 designates the reference node (ground). By Kirchhoff's current law, the injected current I_k must be equal to the sum of the currents leaving node k, hence

$$I_k = \sum_{i=0}^{n} I_{ki} = \sum_{i=0}^{n} y_{ki}(E_k - E_i). \tag{2.3.2}$$

Since $E_0 = 0$, and if the system is linear,

$$I_k = \sum_{i=0 \neq k}^{n} y_{ki} E_k - \sum_{i=1 \neq k}^{n} y_{ki} E_i. \tag{2.3.3}$$

If this equation is written for all the nodes except the reference, i.e. for all busbar in the case of a power system network, then a complete set of equations defining the network is obtained in matrix form as

$$
\begin{array}{|c|}
\hline
I_1 \\ \hline
I_2 \\ \hline
\\ \hline
I_n \\ \hline
\end{array}
=
\begin{array}{|c|c|c|c|}
\hline
Y_{11} & Y_{12} & & Y_{1n} \\ \hline
Y_{21} & Y_{22} & & Y_{2n} \\ \hline
& & & \\ \hline
Y_{n1} & Y_{n2} & & Y_{nn} \\ \hline
\end{array}
\cdot
\begin{array}{|c|}
\hline
E_1 \\ \hline
E_2 \\ \hline
\\ \hline
E_n \\ \hline
\end{array}
\tag{2.3.4}
$$

where

$$Y_{kk} = \sum_{i=0 \neq k}^{n} y_{ki} = \text{self-admittance of node } k$$

$$Y_{ki} = - y_{ki} = \text{mutual admittance between nodes } k \text{ and } i.$$

In shorthand matrix notation, equation (2.3.4) is simply

$$I = Y \cdot E \qquad (2.3.5)$$

or in summation notation

$$I_k = \sum_{i=1}^{n} Y_{ki} E_i \qquad \text{for } i = 1, \ldots, n. \qquad (2.3.6)$$

The nodal admittance matrix in equations (2.3.4) or (2.3.5) has a well-defined structure, which makes it easy to construct automatically. Its properties are as follows.

- Square of order $n \times n$.
- Symmetrical, since $y_{ki} = y_{ik}$.
- Complex.
- Each off-diagonal element y_{ki} is the negative of the branch admittance between nodes k and i, and is frequently of value zero.
- Each diagonal element y_{kk} is the sum of the admittance of the branches which terminate on node k, including branches to ground.
- Because in all but the smallest practical networks very few nonzero mutual admittances exist, matrix Y is highly sparse.

2.4 CONDITIONING OF Y MATRIX

The set of equations $I = Y \cdot E$ may or may not have a solution. If not, a simple physical explanation exists, concerning the formulation of the network problem. Any numerical attempt to solve such equations is found to break down at some stage of the process. (What happens in practice is usually that a finite number is divided by zero.)

The commonest case of this is illustrated in the example of Fig. 2.5. The nodal equations are constructed in the usual way as

$$
\begin{array}{|c|}
\hline I_1 \\ \hline I_2 \\ \hline I_3 \\ \hline
\end{array}
=
\begin{array}{|c|c|c|}
\hline y_{12} + y_{13} & -y_{12} & -y_{13} \\ \hline
-y_{12} & y_{12} + y_{23} & -y_{23} \\ \hline
-y_{13} & -y_{23} & y_{13} + y_{23} \\ \hline
\end{array}
\cdot
\begin{array}{|c|}
\hline E_1 \\ \hline E_2 \\ \hline E_3 \\ \hline
\end{array}
\qquad (2.4.1)
$$

Suppose that the injected currents are known, and nodal voltages are unknown. In this case no solution for the latter is possible. The Y matrix is described as being singular, i.e. it has no inverse, and this is easily detected in this example by noting that the sum of the elements in each row and column is zero, which is a sufficient condition for singularity, mathematically speaking. Hence, if it is not possible to

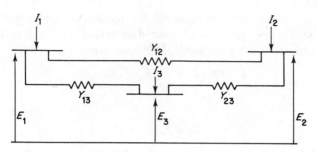

Figure 2.5
Example of singular network

express the voltages in the form $E = Y^{-1} \cdot I$, it is clearly impossible to solve equation (2.4.1) by any method, whether it involves inversion of Y or otherwise.

The reason for this is obvious—we are attempting to solve a network whose reference node is disconnected from the remainder, i.e. there is no effective reference node, and an infinite number of voltage solutions will satisfy the given injected current values.

When, however, a shunt admittance from at least one of the busbars in the network of Fig. 2.5. is present, the problem of insolubility immediately vanishes in theory, but not necessarily in practice. Practical computation cannot be performed with absolute accuracy, and during a sequence of arithmetic operations, rounding errors due to working with a finite number of decimal places accumulate. If the problem is well conditioned and the numerical solution technique is suitable, these errors remain small and do not mask the eventual results. If the problem is ill-conditioned, and this usually depends upon the properties of the system being analysed, any computational errors introduced arc likely to become large with respect to the true solution.

It is easy to see intuitively that if a network having zero shunt admittances cannot be solved even when working with absolute computational accuracy, then a network having very small shunt admittances may well present difficulties when working with limited computational accuracy. This reasoning provides a key to the practical problems of network, i.e. Y matrix, conditioning. A network with shunt admittances which are small with respect to the other branch admittances is likely to be ill-conditioned, and the conditioning tends to improve with the size of the shunt admittances, i.e. with the electrical connection between the network busbars and the reference node.

2.5   THE CASE WHERE ONE VOLTAGE IS KNOWN

In load-flow studies, it usually happens that one of the voltages in the network is specified, and instead the current at that busbar is unknown. This immediately alleviates the problem of needing at least one good connection with ground, because the fixed busbar voltage can be interpreted as an infinitely strong ground tie. If it is represented as a voltage source with a series impedance of zero value, and then converted to the Norton equivalent, the fictitious shunt admittance is infinite, as is the injected current. This approach is not computationally feasible, however.

The usual way to deal with a voltage which is fixed is to eliminate it as a variable from the nodal equations. Purely for the sake of analytical convenience, let this busbar be numbered 1 in an n busbar network. The nodal equations are then

$$
\begin{aligned}
I_1 &= Y_{11}E_1 + Y_{12}E_2 + \ldots Y_{1n}E_n \\
I_2 &= Y_{21}E_1 + Y_{22}E_2 + \ldots Y_{2n}E_n \\
&\vdots \\
I_n &= Y_{n1}E_1 + Y_{n2}E_2 + \ldots Y_{nn}E_n.
\end{aligned}
$$ (2.5.1)

The terms in E_1 on the right-hand side of equations (2.5.1) are known quantities, and as such are transferred to the left-hand side.

$$
\begin{aligned}
I_1 - Y_{11}E_1 &= Y_{12}E_2 + \ldots Y_{1n}E_n \\
I_2 - Y_{21}E_1 &= Y_{22}E_2 + \ldots Y_{2n}E_n \\
&\vdots \\
I_n - Y_{n1}E_1 &= Y_{n2}E_2 + \ldots Y_{nn}E_n.
\end{aligned}
$$ (2.5.2)

The first row of this set may now be eliminated, leaving $(n-1)$ equations in $(n-1)$ unknowns, $E_2 \ldots E_n$. In matrix form, this becomes

$$
\begin{bmatrix} I_2 - Y_{21}E_1 \\ \vdots \\ I_n - Y_{n1}E_1 \end{bmatrix} = \begin{bmatrix} Y_{22} & & Y_{2n} \\ & & \\ Y_{n2} & & Y_{nn} \end{bmatrix} \cdot \begin{bmatrix} E_2 \\ \vdots \\ E_n \end{bmatrix}
$$ (2.5.3)

or

$$
I = Y \cdot E.
$$ (2.5.4)

The new matrix Y is obtained from the full admittance matrix Y merely by removing the row and column corresponding to the fixed-voltage busbar, both in the present case where it is numbered 1 or in general.

In summation notation, the new equations are

$$
I_k - Y_{k1}E_1 = \sum_{i=2}^{n} Y_{ki}E_i \qquad \text{for } k = 2, \ldots, n
$$ (2.5.5)

which is an $(n-1)$ set in $(n-1)$ unknowns. The equations are then solved by any of the available techniques for the unknown voltages. It is noted that the problem of singularity when there are no ground ties disappears if one row and column are removed from the original Y matrix.

Eliminating the unknown current I_1 and the equation in which it appears is the simplest way of dealing with the problem, and reduces the order of the equations by one. I_1 is evaluated after the solution of the first equation in equation (2.5.1).

2.6 ANALYTICAL DEFINITION OF THE PROBLEM

The complete definition of power flow requires knowledge of four variables at each bus k in the system:

- P_k—real or active power
- Q_k—reactive or quadrature power
- V_k—voltage magnitude
- θ_k—voltage phase angle.

Only two are known *a priori* to solve the problem, and the aim of the load flow is to solve the remaining two variables at a bus.

We define three different bus conditions based on the steady-state assumptions of constant system frequency and constant voltages, where these are controlled.

(i) Voltage-controlled bus. The total injected active power P_k is specified, and the voltage magnitude V_k is maintained at a specified value by reactive power injection. This type of bus generally corresponds to either a generator where P_k is fixed by turbine governor setting and V_k fixed by automatic voltage regulators acting on the machine excitation, or a bus where the voltage is fixed by supplying reactive power from static shunt capacitors or rotating synchronous compensators, e.g. at substations.

(ii) Nonvoltage-controlled bus. The total injected power $P_k + jQ_k$ is specified at this bus. In the physical power system this corresponds to a load centre such as a city or an industry, where the consumer demands his power requirements. Both P_k and Q_k are assumed to be unaffected by small variations in bus voltage.

(iii) Slack (swing) bus. This bus arises because the system losses are not known precisely in advance of the load-flow calculation. Therefore the total injected power cannot be specified at every single bus. It is usual to choose one of the available voltage-controlled buses as slack, and to regard its active power as unknown. The slack bus voltage is usually assigned as the system phase reference, and its complex voltage

$$E_s = V_s \underline{/\theta_s}$$

is therefore specified. The analogy in a practical power system is the generating station which has the responsibility of system frequency control.

Load-flow solves a set of simultaneous nonlinear algebraic power equations for the two unknown variables at each node in a system. A second set of variable equations, which are linear, are derived from the first set, and an iterative method is applied to this second set.

The basic algorithm which load-flow programs use is depicted in Fig. 2.6. System data, such as busbar power conditions, network connections and impedance, are read in and the admittance matrix formed. Initial voltages are specified to all buses; for base case load flows P, Q buses are set to $1 + j0$ while P, V busbars are set to $V + j0$.

The iteration cycle is terminated when the busbar voltages and angles are such that the specified conditions of load and generation are satisfied. This condition is accepted when power mismatches for all buses are less than a small tolerance, η_1, or voltage increments less than η_2. Typical figures for η_1 and η_2 are 0.01 p.u. and 0.001 p.u. respectively. The sum of the square of the absolute values of power mismatches is a further criterion sometimes used.

18

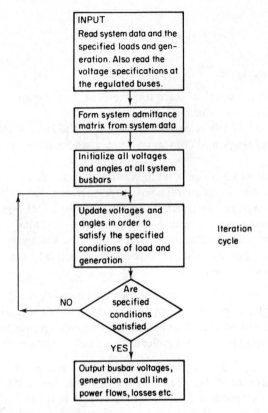

Figure 2.6
Flow diagram of basic load-flow algorithm

When a solution has been reached, complete terminal conditions for all buses are computed. Line power flows and losses and system totals can then be calculated.

2.7 NEWTON–RAPHSON METHOD OF SOLVING LOAD FLOWS

The generalised Newton–Raphson method is an iterative algorithm for solving a set of simultaneous nonlinear equations in an equal number of unknowns.

$$f_k(x_m) = 0 \quad \text{for} \quad k = 1 \to N \qquad \text{and } m = 1 \to N. \tag{2.7.1}$$

At each iteration of the $N - R$ method, the nonlinear problem is approximated by the linear matrix equation. The linearising approximation can best be visualised in the case of a single-variable problem.

In Fig. 2.7, x^p is an approximation to the solution, with error Δx^p at iteraction p. Then

$$f(x^p + \Delta x^p) = 0. \tag{2.7.2}$$

This equation can be expanded by Taylor's theorem:

$$f(x^p + \Delta x^p) = 0$$

$$= f(x^p) + \Delta x^p f'(x^p) + \frac{(\Delta x^p)^2}{2!} f''(x^p) + \dots . \tag{2.7.3}$$

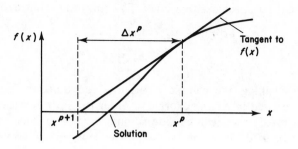

Figure 2.7
Single-variable linear approximation

If the initial estimate of the variable x^p is near the solution value, Δx^p will be relatively small and all terms of higher powers can be neglected. Hence

$$f(x^p) + \Delta x^p f'(x^p) = 0 \tag{2.7.4}$$

or

$$\Delta x^p = \frac{-f(x^p)}{f'(x^p)}. \tag{2.7.5}$$

The new value of the variable is then obtained from

$$x^{p+1} = x^p + \Delta x^p. \tag{2.7.6}$$

Equation (2.7.4) may be rewritten as

$$f(x^p) = -J\Delta x^p. \tag{2.7.7}$$

The method is readily extended to the set of N equations in N unknowns. J becomes the square Jacobian matrix of first-order partial differentials of the functions $f_k(x_m)$. Elements of $[J]$ are defined by

$$J_{km} = \frac{\partial f_k}{\partial x_m} \tag{2.7.8}$$

and represent the slopes of the tangent hyperplanes which approximate the functions $f_k(x_m)$ at each iteration point.

The Newton–Raphson algorithm will converge quadratically if the functions have continuous first derivatives in the neighbourhood of the solution, the Jacobian matrix is nonsingular, and the initial approximations of x are close to the actual solutions. However the method is sensitive to the behaviours of the functions $f_k(x_m)$ and hence to their formulation. The more linear they are, the more rapidly and reliably Newton's method converges. Nonsmoothness, i.e. humps, in any one of the functions in the region of interest, can cause convergence delays, total failure or misdirection to a nonuseful solution.

2.7.1 Equations Relating to Power System Load Flow

The network governing equations are

$$I_k = \sum_{m \in k} y_{km} R_m \qquad \text{for all } k \tag{2.7.9}$$

where I_k is the current injected into a bus k. The power at a bus is then given by

$$S_k = P_k + jQ_k = E_k I_k^*$$
$$= E_k \sum_{m\in k} y_{km}^* E_m^*. \qquad (2.7.10)$$

Mathematically speaking, the complex load-flow equations are nonanalytic, and cannot be differentiated in complex form. In order to apply Newton's method, the problem is separated into real equations and variables. Polar or rectangular coordinates may be used for the bus voltages. Hence we obtain two equations

$$P_k = P(V, \theta) \qquad \text{or} \quad P(e, f)$$

and

$$Q_k = Q(V, \theta) \qquad \text{or} \quad Q(e, f).$$

In polar coordinates the real and imaginary parts of equation (2.7.10) are

$$P_k = \sum_{m\in k} V_k V_m (G_{km} \cos \theta_{km} + B_{km} \sin \theta_{km}) \qquad (2.7.11)$$

$$Q_k = \sum_{m\in k} V_k V_m (G_{km} \sin \theta_{km} - B_{km} \cos \theta_{km}) \qquad (2.7.12)$$

where

$$\theta_{km} = \theta_k - \theta_m.$$

Linear relationships are obtained for small variations in the variables θ and V by forming the total differentials, the resulting equations being as follows:

- For a PQ busbar

$$\Delta P_k = \sum_{m\in k} \frac{\partial P_k}{\partial \theta_m} \Delta \theta_m + \sum_{m\in k} \frac{\partial P_k}{\partial V_m} \Delta V_m \qquad (2.7.13)$$

and

$$\Delta Q_k = \sum_{m\in k} \frac{\partial Q_k}{\partial \theta_m} \Delta \theta_m + \sum_{m\in k} \frac{\partial Q_k}{\partial V_m} \Delta V_m. \qquad (2.7.14)$$

- For a PV busbar, only equation (2.7.13) is used, since Q_k is not specified.
- For a slack busbar, no equations.

The voltage magnitudes appearing in equations (2.7.13) and (2.7.14) for PV and slack busbars are not variables, but are fixed at their specified values. Similarly θ at the slack busbar is fixed.

The complete set of defining equations is made up of two for each PQ busbar and one for each PV busbar. The problem variables are V and θ for each PQ busbar and θ for each PV busbar. The number of variables is therefore equal to the number of equations. Algorithm (2.7.7) then becomes:

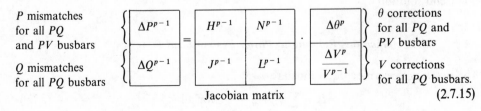

$$\left.\begin{array}{l} P \text{ mismatches} \\ \text{for all } PQ \\ \text{and } PV \text{ busbars} \end{array}\right\}\!\begin{array}{|c|} \hline \Delta P^{p-1} \\ \hline \Delta Q^{p-1} \\ \hline \end{array} = \begin{array}{|c|c|} \hline H^{p-1} & N^{p-1} \\ \hline J^{p-1} & L^{p-1} \\ \hline \end{array} \cdot \begin{array}{|c|} \hline \Delta\theta^p \\ \hline \Delta V^p \\ V^{p-1} \\ \hline \end{array}\!\left\{\begin{array}{l} \theta \text{ corrections} \\ \text{for all } PQ \text{ and} \\ PV \text{ busbars} \\ V \text{ corrections} \\ \text{for all } PQ \text{ busbars.} \end{array}\right.$$

Q mismatches for all PQ busbars

Jacobian matrix

$$(2.7.15)$$

The division of each ΔV_i^p by V_i^{p-1} does not numerically affect the algorithm, but simplifies some of the Jacobian matrix terms. For busbars k and m (not row k and column m in the matrix)

$$H_{km} = \frac{\partial P_k}{\partial \theta_m} = V_k V_m (G_{km} \sin \theta_{km} - B_{km} \cos \theta_{km})$$

$$N_{km} = V_m \frac{\partial P_k}{\partial V_m} = V_k V_m (G_{km} \cos \theta_{km} + B_{km} \sin \theta_{km})$$

$$J_{km} = \frac{\partial Q_k}{\partial \theta_m} = - V_k V_m (G_{km} \cos \theta_{km} + B_{km} \sin \theta_{km})$$

$$L_{km} = V_m \frac{\partial Q_k}{\partial V_m} = V_k V_m (G_{km} \sin \theta_{km} - B_{km} \cos \theta_{km})$$

and for $m = k$

$$H_{kk} = \frac{\partial P_k}{\partial \theta_k} = - Q_k - B_{kk} V_k^2$$

$$N_{kk} = V_k \frac{\partial P_k}{\partial V_k} = P_k + G_{kk} V_k^2$$

$$J_{kk} = \frac{\partial Q_k}{\partial \theta_k} = P_k - G_{kk} V_k^2$$

$$L_{kk} = V_k \frac{\partial Q_k}{\partial V_k} = Q_k - B_{kk} V_k^2.$$

In practice, some programs express these coefficients using voltages in rectangular form, i.e. $e_i + jf_i$. This only affects the speed of calculation of the mismatches and the matrix elements by eliminating the time-consuming trigonometrical functions.

In rectangular coordinates the complex power equations are given as

$$P_k + jQ_k = E_k \sum_{m \in k} Y_{km}^* E_m^* = (e_k + jf_k) \sum_{m \in k} (G_{km} - jB_{km})(e_m - jf_m)$$

and these are divided into real and imaginary parts

$$P_k = e_k \sum_{m \in k} (G_{km} e_m - B_{km} f_m) + f_k \sum_{m \in k} (G_{km} f_m + B_{km} e_m)$$

$$Q_k = f_k \sum_{m \in k} (G_{km} e_m - B_{km} f_m) - e_k \sum_{m \in k} (G_{km} f_m + B_{km} e_m).$$

At a voltage-controlled bus the voltage magnitude is fixed but not the phase angle. Hence both e_k and f_k vary at each iteration. It is necessary to provide another equation

$$V_k^2 = e_k^2 + f_k^2$$

to be solved with the real power equation for these buses.

Linear relationships are obtained for small variations in e and f by forming the total differentials

$$\Delta P_k = \sum_{m \in k} \frac{\partial P_k}{\partial e_m} \Delta e_m + \sum_{m \in k} \frac{\partial P_k}{\partial f_m} \Delta f_m$$

$$= \sum_{m \in k} S_{km} \Delta e_m + \sum_{m \in k} T_{km} \Delta f_m$$

for all buses except the slack bus;

$$\Delta Q_k = \sum_{m \in k} \frac{\partial Q_k}{\partial e_m} \Delta e_m + \sum_{m \in k} \frac{\partial Q_k}{\partial f_m} \Delta f_m$$

$$= \sum_{m \in k} U_{km} \Delta e_m + \sum_{m \in k} W_{km} \Delta f_m$$

for all nonvoltage-controlled buses; and

$$\Delta V_k^2 = \frac{\partial V_k^2}{\partial e_k} \Delta e_k + \frac{\partial V_k^2}{\partial f_k} \Delta f_k$$

$$= EE_k \Delta e_k + FF_k \Delta f_k$$

for voltage-controlled buses.

The Jacobian matrix has the form

$$
\begin{array}{|c|}
\hline
\Delta P \\
\hline
\Delta Q \\
\hline
\Delta V^2 \\
\hline
\end{array}
=
\begin{array}{|c|c|}
\hline
S & T \\
\hline
U & W \\
\hline
EE & FF \\
\hline
\end{array}
\cdot
\begin{array}{|c|}
\hline
\Delta e \\
\hline
\Delta f \\
\hline
\end{array}
\qquad (2.7.16)
$$

and the values of the partial differentials, which are the Jacobian elements, are given by

$$S_{km} = -W_{km} = G_{km} e_k + B_{km} f_k \qquad \text{for } m \neq k$$

$$T_{km} = U_{km} = G_{km} f_k - B_{km} e_k \qquad \text{for } m \neq k$$

$$S_{kk} = a_k + G_{kk} e_k + B_{kk} f_k$$

$$W_{kk} = a_k - G_{kk} e_k - B_{kk} f_k$$

$$T_{kk} = b_k - B_{kk} e_k + G_{kk} f_k$$

$$U_{kk} = -b_k - B_{kk} e_k + G_{kk} f_k$$

$$EE_k = 2e_k$$

$$FF_k = 2f_k.$$

For voltage-controlled buses, V is specified, but not the real and imaginary components of voltage, e and f. Approximations can be made, for example, by ignoring the off-diagonal elements in the Jacobian matrix, as the diagonal elements are the largest. Alternatively for the calculation of the elements the voltages can be considered as $E = 1 + j0$. The off-diagonal elements then become constant.

The polar coordinate representation appears to have computational advantages over rectangular coordinates. Real power mismatch equations are present for all buses except the slack bus, while reactive power mismatch equations are needed for nonvoltage-controlled buses only.

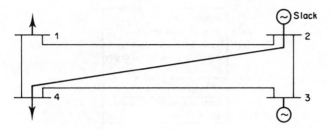

Figure 2.8
Sample system

The Jacobian matrix has the sparsity of the admittance matrix $[Y]$ and has positional but not numerical symmetry. To gain in computation, the form of $[\Delta\theta, \Delta V/V]^T$ is normally used for the variable voltage vector. Both increments are dimensionless and the Jacobian coefficients are now symmetric in structure though not in value. The values of $[J]$ are all functions of the voltage variables V and θ and must be recalculated for each iteration.

As an example, the Jacobian matrix equation for the four-busbar system of Fig. 2.8 is given as equation (2.7.17):

$$
\begin{vmatrix} \Delta P_1 \\ \Delta P_3 \\ \Delta P_4 \\ \Delta Q_1 \\ \Delta Q_4 \end{vmatrix}
=
\begin{vmatrix}
H_{11} & 0 & H_{14} & N_{11} & N_{14} \\
0 & H_{33} & H_{34} & 0 & N_{34} \\
H_{41} & H_{43} & H_{44} & N_{41} & N_{44} \\
J_{11} & 0 & J_{14} & L_{11} & L_{14} \\
J_{41} & J_{43} & J_{44} & L_{41} & L_{44}
\end{vmatrix}
\cdot
\begin{vmatrix} \Delta\theta_1 \\ \Delta\theta_3 \\ \Delta\theta_4 \\ \Delta V_1/V_1 \\ \Delta V_4/V_4 \end{vmatrix}
\qquad (2.7.17)
$$

The differences in bus powers are obtained from

$$\Delta P_k = P_k^{SP} - P_k \qquad (2.7.18)$$

$$\Delta Q_k = Q_k^{SP} - Q_k. \qquad (2.7.19)$$

A further improvement is to replace the reactive power residual ΔQ in the Jacobian matrix equations by $\Delta Q/V$. The performance of the Newton–Raphson method is closely associated with the degree of problem nonlinearity; the best left-hand defining functions are the most linear ones. If the system power equation (2.7.19) is divided throughout by V_k, only one term Q_k^{sp}/V_k on the right-hand side of this equation is nonlinear in V_k. For practical values of Q_k^{sp} and V_k, this nonlinear term is numerically relatively small. Hence it is preferable to use $\Delta Q/V$ instead of ΔQ in the Jacobian matrix equation.

Dividing ΔP by V is also helpful, but is less effective since the real power component of the problem is not strongly coupled with voltage magnitudes. A further alternative is to formulate current residuals at a bus. While computationally simple, this method shows poor convergence in the same way at Y matrix iterative methods.

A flow diagram of the basic Newton–Raphson algorithm is given in Fig. 2.9

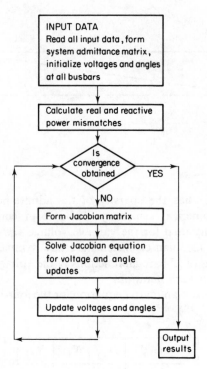

Figure 2.9
Flow diagram of the basic Newton–Raphson load-flow algorithm

2.8 TECHNIQUES WHICH MAKE THE NEWTON–RAPHSON METHOD COMPETITIVE IN LOAD FLOW

The efficient solution of equation (2.7.15) at each iteration is crucial to the success of the N–R method. If conventional matrix techniques were to be used, the storage ($\propto n^2$) and computing time ($\propto n^3$) would be prohibitive for large systems.

For most power system networks the admittance matrix is relatively sparse, and in the Newton–Raphson method of load flow the Jacobian matrix has this same sparsity.

The techniques which have been used to make the Newton–Raphson competitive with other load-flow methods involve the solution of the Jacobian matrix equation and the preservation of the sparsity of the matrix by ordered triangular factorisation.

2.8.1 Sparsity Programming

In conventional matrix programming, double subscript arrays are used for the location of elements. With sparsity programming [6] only the nonzero elements are stored, in one or more vectors, plus integer vectors for identification.

For the admittance matrix of order n the conventional storage requirements are n^2 words, but by sparsity programming $6b + 3n$ words are required, where b is the number of branches in the system. Typically $b = 1.5n$, and the total storage is $12n$ words. For a large system (say 500 buses) the ratio of storage requirements of conventional and sparse techniques is about 40:1.

2.8.2 Triangular Factorisation

To solve the Jacobian matrix equation (2.7.15), represented here as

$$[\Delta S] = [J][\Delta E]$$

for increments in voltage, the direct method is to find the inverse of $[J]$ and solve for $[\Delta E]$ from

$$[\Delta E] = [J]^{-1}[\Delta S]. \qquad (2.8.1)$$

In power systems $[J]$ is usually sparse but $[J]^{-1}$ is a full matrix.

The method of triangular factorisation solves for the vector $[\Delta E]$ by eliminating $[J]$ to an upper triangular matrix with a leading diagonal, and then back-substituting for $[\Delta E]$, i.e. eliminate to

$$[\Delta S'] = [U][\Delta E]$$

and back-substitute

$$[U]^{-1}[\Delta S'] = [\Delta E].$$

The triangulation of the Jacobian is best done by rows. Those rows below the one being operated on need not be entered until required. This means that the maximum storage is that of the resultant upper triangle and diagonal. The lower triangle can then be used to record operations.

The number of multiplications and additions to triangulate a full matrix is $\frac{1}{3}N^3$, compared to N^3 to find the inverse. With sparsity programming the number of operations varies as a factor of N. If rows are normalised N further operations are saved.

2.8.3 Optimal Ordering

In power system load flow, the Jacobian matrix is usually diagonally dominant which implies small round-off errors in computation. When a sparse matrix is triangulated, nonzero terms are added in the upper triangle. The number added is affected by the order of the row eliminations, and total computation time increases with more terms.

The pivot element is selected to minimise the accumulation of nonzero terms, and hence conserve sparsity, rather than minimising round-off error. The diagonals are used as pivots.

Optimal ordering of row eliminations to conserve sparsity is a practical impossibility due to the complexity of programming and time involved. However, semioptimal schemes are used and these can be divided into two sections.

(a) *Preordering* [7]. Nodes are renumbered before triangulation. No complicated programming or storage is required to keep track of row and column interchanges.

(i) Nodes are numbered in sequence of increasing number of connected lines.

(ii) Diagonal banding—nonzero elements are arranged about either the major or minor diagonals of the matrix.

(b) *Dynamic ordering* [8]. Ordering is effected at each row during the elimination.

(i) At each step in the elimination, the next row to be operated on is that with the fewest nonzero terms.

(ii) At each step in the elimination, the next row to be operated on is that which introduces the fewest new nonzero terms, one step ahead.

(iii) At each step in the elimination, the next row to be operated on is that which introduces the fewest new nonzero terms, two steps ahead. This may be extended to the fully optimal case of looking at the effect in the final step.

(iv) With cluster ordering, the network is subdivided into groups which are then optimally ordered. This is most efficient if the groups have a minimum of physical intertie. The matrix is then anchor banded.

The best method arises from a trade-off between a processing sequence which requires the least number of operations, and time and memory requirements.

The dynamic ordering scheme of choosing the next row to be eliminated as that with the fewest nonzero terms, appears to be better than all other schemes in sparsity conservation, number of arithmetic operations required, ordering times and total solution time.

However, there are conditions under which other ordering would be preferable, e.g. with system changes affecting only a few rows these rows should be numbered last; when the subnetworks have relatively few interconnections it is better to use cluster ordering.

2.8.4 Aids to Convergence

The N–R method can diverge very rapidly or converge to the wrong solution if the equations are not well behaved or if the starting voltages are badly chosen. Such problems can often be overcome by a variety of techniques. The simplest device is to impose a limit on the size of each $\Delta\theta$ and ΔV correction at each iteration. Figure 2.10 illustrates a case which would diverge without this device.

Another more complicated method is to calculate good starting values for the θs and Vs, which also reduces the number of iterations required.

In power system load flow, setting voltage-controlled buses to $V + j0$ and nonvoltage-controlled buses to $1 + j0$ may give a poor starting point for the N–R method.

If previously stored solutions for a network are available these should be used. One or two iterations of a Y matrix iterative method [2] can be applied before commencing the Newton method. This shows fast initial convergence unless the problem is ill-conditioned, in which case divergence occurs.

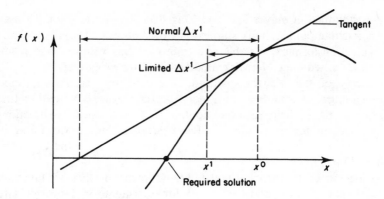

Figure 2.10
Example of diverging solution

A more reliable method is the use of one iteration of a d.c. load flow (i.e. neglecting losses and reactive power conditions) to provide estimates of voltage angles, followed by one iteration of a similar type of direct solution to obtain voltage magnitudes. The total computing time for both sets of equations is about 50% of one N–R iteration and the extra storage required is only in the programming statements. The resulting combined algorithm is faster and more reliable than the formal Newton method and can be used to monitor diverging or difficult cases, before commencing the N–R algorithm.

2.9 CHARACTERISTICS OF THE NEWTON–RAPHSON LOAD FLOW

With sparse programming techniques and optimally ordered triangular factorisation, the Newton method for solving load flow has become faster than other methods for large systems. The number of iterations is virtually independent of system size (from a flat voltage start and with no automatic adjustments) due to the quadratic characteristic of convergence. Most systems are solved in 2–5 iterations with no acceleration factors being necessary.

With good programming the time per iteration rises nearly linearly with the number of system buses N, so that the overall solution time varies as N. One Newton iteration is equivalent to about seven Gauss–Seidel iterations. For a 500-bus system, the conventional Gauss–Seidel method takes about 500 iterations and the speed advantage of the Newton method is then 15:1. Storage requirements of the Newton method are greater, however, but increase linearly with system size. It is therefore attractive for large systems.

The Newton method is very reliable in system solving, given good starting approximations. Heavily loaded systems with phase shifts up to 90° can be solved. The method is not troubled by ill-conditioned systems and the location of slack bus is not critical.

Due to the quadratic convergence of bus voltages, high accuracy (near exact solution) is obtained in only a few iterations. This is important for the use of load flow in short-circuit and stability studies. The method is readily extended to include tap-changing transformers, variable constraints on bus voltages, and reactive and optimal power scheduling. Network modifications are easily made.

The success of the Newton method is critical on the formulation of the problem-defining equations. Power mismatch representation is better than the current mismatch versions. To help negotiate nonlinearities in the defining functions, limits can be imposed on the permissible size of voltage corrections at each iteration. These should not be too small, however, as they may slow down the convergence for well-behaved systems.

The coefficients of the Jacobian matrix are not constant, they are functions of the voltage variables V and θ, and hence vary for each iteration. However, after a few iterations, as V and θ tend to their final values the coefficients will tend to constant values.

One modification to the Newton algorithm is to calculate the Jacobian for the first two or three iterations only and then use the final one for all the following iterations. Alternatively the Jacobian can be updated every two or more iterations. Neither of these modifications greatly affects the convergence of the algorithm, though much time is saved (but not storage).

2.10 DECOUPLED NEWTON LOAD FLOW

An inherent characteristic of any practical electric power transmission system operating in the steady-state condition is the strong interdependence between active powers and bus voltage angles, and between reactive powers and voltage magnitudes. Correspondingly, the coupling between these $P-\theta$ and $Q-V$ components of the problem is relatively weak. Many algorithms have been proposed which adopt this decoupling principle [9–11].

The voltage vectors method uses a series approximation for the sine terms which appear in the system-defining equations to calculate the Jacobian elements and arrive at two decoupled equations

$$[\mathscr{P}] = [T][\theta] \tag{2.10.1}$$

$$[\mathscr{L}] = [U][V - V_0] \tag{2.10.2}$$

where for the reference node $\theta_0 = 0$ and $V_k = V_0$. The values of $\mathscr{P}_k$ and $\mathscr{L}_k$ represent real and reactive power quantities respectively and $[T]$ and $[U]$ are given by

$$T_{km} = -\frac{V_k V_m}{Z_{km}^2 / X_{km}} \tag{2.10.3}$$

$$T_{kk} = -\sum_{m \in k} T_{km} \tag{2.10.4}$$

$$U_{km} = -\frac{1}{Z_{km}^2 / X_{km}}. \tag{2.10.5}$$

$$U_{kk} = -\sum_{m \in k} U_{km} \tag{2.10.6}$$

where Z_{km} and X_{km} are the branch impedance and reactance respectively. $[U]$ is constant valued and needs be triangulated once only for a solution. $[T]$ is recalculated and triangulated each iteration.

The two equations (2.10.1) and (2.10.2) are solved alternately until a solution is obtained. These equations can be solved using Newton's method, by expressing the Jacobian equations as

$$
\begin{array}{|c|}
\hline \Delta P \\ \hline \Delta Q/V \\ \hline
\end{array}
=
\begin{array}{|c|c|}
\hline T & \\ \hline & U \\ \hline
\end{array}
\cdot
\begin{array}{|c|}
\hline \Delta\theta \\ \hline \Delta V \\ \hline
\end{array}
\qquad (2.10.7)
$$

or

$$[\Delta P] = [T][\Delta\theta] \qquad (2.10.8)$$

$$[\Delta Q/V] = [U][\Delta V] \qquad (2.10.9)$$

where

$$[\Delta P] = [\Delta\mathscr{P}]$$

$$[\Delta Q/V] = [\Delta\mathscr{L}]$$

and T and U are therefore defined in equations (2.10.3) to (2.10.6).

The most successful decoupled load flow is that based on the Jacobian matrix equation for the formal Newton method, i.e.

$$
\begin{array}{|c|}
\hline \Delta P \\ \hline \Delta Q \\ \hline
\end{array}
=
\begin{array}{|c|c|}
\hline H & N \\ \hline J & L \\ \hline
\end{array}
\cdot
\begin{array}{|c|}
\hline \Delta\theta \\ \hline \Delta V \\ \hline
\end{array}
\qquad (2.10.10)
$$

If the submatrices N and J are neglected, since they represented the weak coupling between $P-\theta$ and $Q-V$, the following decoupled equations result:

$$[\Delta P] = [H][\Delta\theta] \qquad (2.10.11)$$

$$[\Delta Q] = [L][\Delta V]. \qquad (2.10.12)$$

It has been found that equation (2.10.12) is relatively unstable at some distance from the exact solution due to the nonlinear defining functions. An improvement in convergence is obtained by replacing this with the polar current-mismatch formulation [7]

$$[\Delta I] = [D][\Delta V]. \qquad (2.10.13)$$

Alternatively the right-hand side of both equations (2.10.11) and (2.10.12) is divided by voltage magnitude V:

$$[\Delta P/V] = [A][\Delta\theta] \qquad (2.10.14)$$

$$[\Delta Q/V] = [C][\Delta V]. \qquad (2.10.15)$$

The equations are solved successively using the most up-to-date values of V and θ available. $[A]$ and $[C]$ are sparse, nonsymmetric in value and are both functions of V and θ. They must be calculated and triangulated each iteration.

Further approximations that can be made are to assume that $E_k = 1.0$ p.u., for all buses, and $G_{km} \ll B_{km}$ in calculating the Jacobian elements. The off-diagonal terms then become symmetric about the leading diagonal.

The decoupled Newton method compares very favourably with the formal Newton method. While reliability is just as high for ill-conditioned problems, the decoupled method is simple and computationally efficient. Storage of the Jacobian and matrix triangulation is saved by a factor of four, or an overall saving of 30–40% on the formal Newton load flow. Computation time per iteration is also less than the Newton method.

However, the convergence characteristics of the decoupled method are linear, the quadratic characteristics of the formal Newton being sacrified. Thus, for high accuracies, more iterations are required. This is offset for practical accuracies by the fast initial convergence of the method. Typically, voltage magnitudes converge to within 0.3% of the final solution on the first iteration and may be used as a check for instability. Phase angles converge more slowly than voltage magnitudes but the overall solution is reached in 2–5 iterations. Adjusted solutions (the inclusion of transformer taps, phase shifters, interarea power transfers, Q and V limits) take many more iterations.

The Newton methods can be expressed as follows [12]:

$$
\left[\begin{array}{c} \Delta P/V \\ \hline \Delta Q/V \end{array} \right] = \left[\begin{array}{c|c} A_{11} & \varepsilon A_{12} \\ \hline \varepsilon A_{21} & A_{22} \end{array} \right] \cdot \left[\begin{array}{c} \Delta \theta \\ \hline \Delta V \end{array} \right]
\tag{2.10.16}
$$

where

$$\varepsilon = 1 \text{ for the full Newton–Raphson method}$$
$$\varepsilon = 0 \text{ for the decoupled Newton algorithm.}$$

A Taylor series expansion of the Jacobian about $\varepsilon = 0$ results in a first-order approximation of the Newton–Raphson method whereas the decoupled method is a zero-order approximation.

2.11   FAST-DECOUPLED LOAD FLOW

By further simplifications and assumptions, based on the physical properties of a practical system, the Jacobians of the decoupled Newton load flow can be made constant in value. This means that they need be triangulated only once per solution or for a particular network.

For ease of reference, the real and reactive power equations at a node k are reproduced here:

$$
P_k = V_k \sum_{m \in k} V_m (G_{km} \cos \theta_{km} + B_{km} \sin \theta_{km})
\tag{2.11.1}
$$

$$
Q_k = V_k \sum_{m \in k} V_m (G_{km} \sin \theta_{km} - B_{km} \cos \theta_{km})
\tag{2.11.2}
$$

where $\theta_{km} = \theta_k - \theta_m$.

A decoupled method which directly relates powers and voltages is derived using the series approximations for the trigonometric terms in equations (2.11.1) and (2.11.2):

$$\sin \theta = \theta - \frac{\theta^3}{6}$$

$$\cos \theta = 1 - \frac{\theta^2}{2}.$$

The equations, over all buses, can be expressed in their simplified matrix form

$$[A][\theta] = [P] \tag{2.11.3}$$

$$[C][V] = [Q] \tag{2.11.4}$$

where P and Q are terms of real and reactive power respectively and

$$A_{kk} = V_k \sum_{m \in k} V_m B_{km}$$

$$A_{km} = - V_k V_m B_{km} \qquad m \neq k$$

$$C_{kk} = \sum_{m \in k} t_{km} B_{km}$$

$$C_{km} = - B_{km} \qquad m \neq k$$

$t_{km} = $ tap ratio if a transformer is in the line.

A modification suggested is to replace equation (2.11.3) by

$$[\hat{A}][\hat{\theta}] = [\hat{P}]$$

where

$$\hat{A}_{km} = - B_{km} \qquad m \neq k$$

$$\hat{A}_{kk} = \sum_{m \in k} B_{km}$$

$$\hat{\theta}_k = \theta_k \cdot V_k$$

$$\hat{P}_k = P_k / V_k.$$

Hence $[\hat{A}]$ becomes constant valued.

A similar direct method is obtained from the decoupled voltage vectors method (equations (2.10.1) and (2.10.2)). If V_m, V_k are put as 1.0 p.u. for the calculation of matrix $[T]$, then $[T]$ becomes constant and need be triangulated once only. This same simplification can be used in the decoupled voltage vectors and Newton's method of equations (2.10.8) and (2.10.9).

Fast-decoupled load-flow algorithms [8] are also derived from the Jacobian matrix equations of Newton's method (equations (2.10.10)) and the decoupled version (equations (2.10.11) and (2.10.12)).

Let us make the following assumptions.

(i) $E_k, E_m = 1.0$ p.u.

(ii) $G_{km} \ll B_{km}$, and hence can be ignored (for most transmission line reactance/resistance ratios, $X/R \gg 1$).

(iii) $\cos(\theta_k - \theta_m) \doteq 1.0$
     $\sin(\theta_k - \theta_m) \doteq 0.0$
     since angle differences across transmission lines are small under normal loading conditions.

This leads to the decoupled equations

$$[\Delta P] = [\bar{B}][\Delta \theta] \quad \text{of order } N - 1) \tag{2.11.5}$$

$$[\Delta Q] = [\bar{B}][\Delta V] \quad \text{of order } (N - M) \tag{2.11.6}$$

where N is the number of busbars and M is the number of PV busbars. The elements of $[\bar{B}]$ are

$$\bar{B}_{km} = -B_{km} \qquad \text{for } m \neq k$$
$$\bar{B}_{kk} = \sum_{m \in k} B_{km}$$

and B_{km} are the imaginary parts of the admittance matrix. To simplify still further, line resistances may be neglected in the calculation of elements of $[\bar{B}]$.

An improvement over equations (2.11.5) and (2.11.6) is based on the decoupled equations (2.10.14) and (2.10.15) which have fewer nonlinear defining functions. Applying the same assumptions listed previously, we obtain the equations

$$[\Delta P/V] = [B^*][\Delta \theta] \tag{2.11.7}$$

$$[\Delta Q/V] = [B^*][\Delta V]. \tag{2.11.8}$$

A number of refinements make this method very successful.

(a) Omit from the Jacobian in equation (2.11.7) the representation of those network elements that predominantly affect MVAR or reactive power flow, e.g. shunt reactances and off-nominal in-phase transformer taps. Neglect also the series resistances of lines.

(b) Omit from the Jacobian of equation (2.11.8) the angle-shifting effects of phase shifters.

The resulting fast-decoupled load-flow equations are then

$$[\Delta P/V] = [B'][\Delta \theta] \tag{2.11.9}$$

$$[\Delta Q/V] = [B''][\Delta V] \tag{2.11.10}$$

where

$$B'_{km} = -\frac{1}{X_{km}} \qquad \text{for } m \neq k$$

$$B'_{kk} = \sum_{m \in k} \frac{1}{X_{km}}$$

$$B''_{km} = -B_{km} \qquad \text{for } m \neq k$$
$$B''_{kk} = \sum_{m \in k} B_{km}.$$

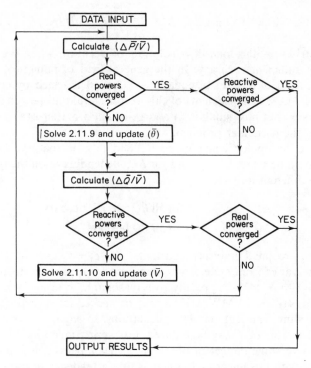

Figure 2.11
Flow diagram of the fast-decoupled load flow

The equations are solved alternatively using the most recent values of V and θ available as shown in Fig. 2.11 [8].

The matrices B' and B'' are real and are of order $(N-1)$ and $(N-M)$ respectively. B'' is symmetric in value and so is B' if phase shifters are ignored; it is found that the performance of the algorithm is not adversely affected. The elements of the matrices are constant and need to be evaluated and triangulated only once for a network.

Convergence is geometric, 2–5 iterations are required for practical accuracies, and more reliable than the formal Newton's method. This is because the elements of B' and B'' are fixed approximations to the tangents of the defining functions $\Delta P/V$ and $\Delta Q/V$, and are not susceptible to any 'humps' in the defining functions.

If $\Delta P/V$ and $\Delta Q/V$ are computed efficiently, then the speed for iterations of the fast-decoupled method is about five times that of the formal Newton–Raphson or about two-thirds that of the Gauss–Seidel method. Storage requirements are about 60% of the formal Newton, but slightly more than the decoupled Newton method.

Changes in system configurations are easily effected, and while adjusted solutions take many more iterations these are short in time and the overall solution time is still low.

The fast-decoupled Newton load flow can be used in optimisation studies for a network and is particularly useful for accurate information of both real and reactive power for multiple load-flow studies, as in contingency evaluation for system security assessment.

2.12   CONVERGENCE CRITERIA AND TESTS [13]

The problem arises in the load-flow solution of deciding when the process has converged with sufficient accuracy. In the general field of numerical analysis, the accuracy of solution of any set of equations $F(X) = 0$ is tested by computing the 'residual' vector $F(X^p)$. The elements of this vector should all be suitably small for adequate accuracy, but how small is to a large extent a matter of experience of the requirements of the particular problem.

The normal criterion for convergence in load flow is that the busbar power mismatches should be small, i.e. ΔQ_i and /or ΔP_i, depending upon the type of busbar i, and can take different forms, e.g.

$$
\begin{aligned}
|\Delta P_i| \leqslant c_1 \qquad & \text{for all } PQ \text{ and } PV \text{ busbars} \\
|\Delta Q_i| \leqslant c_2 \qquad & \text{for all } PQ \text{ busbars}
\end{aligned}
\tag{2.12.1}
$$

where c_1 and c_2 are small empirical constants, and $c_1 = c_2$ usually. The value of c used in practice varies from system to system and from problem to problem. In a large system, $c = 1$ MW/MVAR typically gives reasonable accuracy for most purposes. Higher accuracy, say $c = 0.1$ MW/MVAR may be needed for special studies, such as load flows preceding transient stability calculations. In smaller systems, or systems at light load, the value of c may be reduced. For approximate load flow, c may be increased, but with some danger of obtaining a meaningless solution if it becomes too large. Faced with this uncertainty, there is thus a tendency to use smaller values of c than are strictly necessary. The criterion (equation (2.12.1)) is probably the most common in use. A popular variant on it is

$$
\sum_i \Delta P_i^2 + \sum_k \Delta Q_k^2 \leqslant c_3
\tag{2.12.2}
$$

and other similar expressions are also being used.

In the Newton–Raphson algorithms the calculation and testing of the mismatches at each iteration are part of the algorithm.

The set of equations defining the load-flow problem has multiple solutions, only one of which corresponds to the physical mode of operation of the system. It is extremely rare for there to be more than one solution in the neighbourhood of the initial estimates for the busbar voltages ($(1 + j0)$ p.u., in the absence of anything better), and apart from the possibility of data errors, a sensible-looking mathematically converged solution is normally accepted as being the correct one. However, infrequent cases of very ill-conditioned networks and systems operating close to their stability limits arise where two or more mathematically converged solutions of feasible appearance can be obtained by different choices of starting voltages, or by different load-flow algorithms.

A load-flow problem whose data corresponds to a physically unstable system operating condition (often due to data errors, or in the investigation of unusual operating modes, or in system planning studies) usually diverges. However, the more powerful solution methods, and in particular Newton–Raphson, will sometimes produce a converged solution, and it is not always easy in such cases to recognise that the solution is a physically unstable operating condition. Certain simple checks,

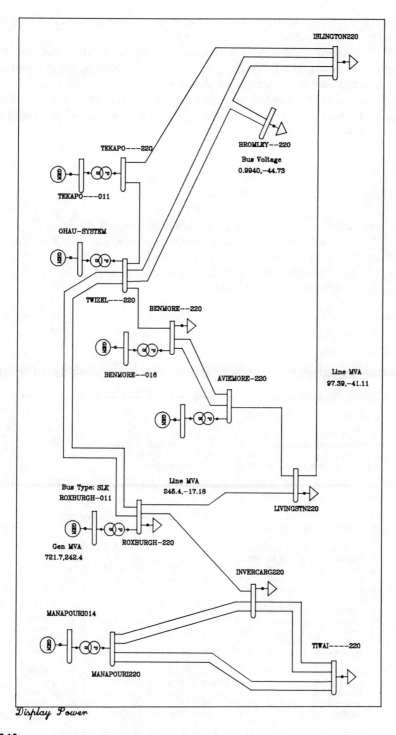

Figure 2.12
Reduced primary a.c. system for the South Island of New Zealand

e.g. on the transmission angle and the voltage drop across each line, can be included in the program to automatically monitor the solution.

Finally, a practical load-flow program should include some automatic test to discontinue the solution if it is diverging, to avoid unnecessary waste of computation, and to avoid overflow in the computer. A suitable test is to check at each iteration whether any voltage magnitude is outside the arbitrary range 0.5–1.5 p.u., since it is highly unlikely in any practical power system that a meaningful voltage solution lies outside this range.

2.13 NUMERICAL EXAMPLE

The test network illustrated in Fig. 2.12, drawn by *Display Power* as described in Chapter 11, involves the main generating and loading points of New Zealand's South Island, with the h.v.d.c. convertor represented as a load, i.e. by specified P and Q.

The following computer print out illustrates the numerical input and output information for the specified conditions.

LOAD FLOW PROGRAM

DEPARTMENT OF ELECTRICAL & ELECTRONIC ENGINEERING, UNIVERSITY OF CANTERBURY, NEW ZEALAND

SYSTEM NO.   2   23 MAR 90

THE SLACK BUS IS	6		
MAXIMUM NUMBER OF ITERATIONS	10	NUMBER OF BUSES	17
POWER TOLERANCE	.00100	NUMBER OF LINES	20
PRINTOUT INDICATOR	000000000	NO OF TRANSFORMERS	6

B U S   D A T A

BUS	NAME	TYPE	VOLTS	LOAD MW	LOAD MVAR	GENERATION MW	GENERATION MVAR	MINIMUM MVAR	MAXIMUM MVAR	SHUNT SUSCEPTANCE
1	INVERCARG220	0	0.0000	200.00	51.00	0.00	0.00	0.00	0.00	0.000
2	ROXBURGH–220	0	0.0000	150.00	60.00	0.00	0.00	0.00	0.00	0.000
3	MANAPOURI220	0	0.0000	0.00	0.00	0.00	0.00	0.00	0.00	0.000
4	MANAPOURI014	1	1.0450	0.00	0.00	690.00	0.00	0.00	0.00	0.000
5	TIWAI——220	0	0.0000	420.00	185.00	0.00	0.00	0.00	0.00	0.000
6	ROXBURGH–011	1	1.0500	0.00	0.00	0.00	0.00	0.00	0.00	0.000
7	BENMORE—220	0	0.0000	500.00	200.00	0.00	0.00	0.00	0.00	0.000
8	BENMORE—016	1	1.0600	0.00	0.00	0.00	0.00	0.00	0.00	0.000
9	AVIEMORE–220	0	0.0000	0.00	0.00	0.00	0.00	0.00	0.00	0.000
10	AVIEMORE–011	1	1.0450	0.00	0.00	200.00	0.00	0.00	0.00	0.000
11	OHAU–SYSTEM	1	1.0500	0.00	0.00	350.00	0.00	0.00	0.00	0.000
12	LIVINGSTN220	0	0.0000	150.00	60.00	0.00	0.00	0.00	0.00	0.000
13	ISLINGTON220	1	1.0000	500.00	300.00	0.00	0.00	0.00	0.00	0.000
14	BROMLEY—220	0	0.0000	100.00	60.00	0.00	0.00	0.00	0.00	0.000
15	TEKAPO——011	0	1.0500	0.00	0.00	150.00	0.00	0.00	0.00	0.000
16	TEKAPO——220	0	0.0000	0.00	0.00	0.00	0.00	0.00	0.00	0.000
17	TWIZEL——220	0	0.0000	0.00	0.00	0.00	0.00	0.00	0.00	0.000

L I N E   D A T A

BUS	NAME	BUS	NAME	RESISTANCE	REACTANCE	SUSCEPTANCE
1	INVERCARG220	3	MANAPOURI220	0.01300	0.09000	0.25000
1	INVERCARG220	3	MANAPOURI220	0.01300	0.09000	0.25000
3	MANAPOURI220	5	TIWAI——220	0.01000	0.10000	0.29000
3	MANAPOURI220	5	TIWAI——220	0.01000	0.10000	0.29000
1	INVERCARG220	5	TIWAI——220	0.00200	0.01000	0.04000
1	INVERCARG220	5	TIWAI——220	0.00200	0.01000	0.04000
1	INVERCARG220	2	ROXBURGH-220	0.01000	0.11000	0.17000
2	ROXBURGH-220	17	TWIZEL——220	0.01600	0.14000	0.24000
2	ROXBURGH-220	17	TWIZEL——220	0.01600	0.14000	0.24000
2	ROXBURGH-220	12	LIVINGSTN220	0.03000	0.12000	0.18000
7	BENMORE—220	17	TWIZEL——220	0.00400	0.03000	0.07000
12	LIVINGSTN220	9	AVIEMORE-220	0.00700	0.03000	0.05000
9	AVIEMORE-220	7	BENMORE—220	0.00400	0.05000	0.02000
9	AVIEMORE-220	7	BENMORE—220	0.00400	0.05000	0.02000
12	LIVINGSTN220	13	ISLINGTON220	0.03000	0.18000	0.35000
17	TWIZEL——220	16	TEKAPO—220	0.00200	0.01000	0.02000
16	TEKAPO—220	13	ISLINGTON220	0.02000	0.13000	0.35000
17	TWIZEL——220	14	BROMLEY—220	0.02000	0.14000	0.45000
14	BROMLEY—220	13	ISLINGTON220	0.00200	0.01000	0.05000
17	TWIZEL——220	13	ISLINGTON220	0.02000	0.14000	0.45000

T R A N S F O R M E R   D A T A

BUS	NAME	BUS	NAME	RESISTANCE	REACTANCE	TAP	CODE
3	MANAPOURI220	4	MANAPOURI014	0.00060	0.01600	1.000	0
2	ROXBURGH-220	6	ROXBURGH-011	0.00200	0.04000	1.000	0
17	TWIZEL——220	11	OHAU-SYSTEM	0.00400	0.03200	1.000	0
9	AVIEMORE-220	10	AVIEMORE-011	0.00150	0.04500	1.000	0
7	BENMORE—220	8	BENMORE-016	0.00120	0.03200	1.000	0
16	TEKAPO——220	15	TEKAPO—011	0.00300	0.05600	1.000	0

SOLUTION CONVERGED IN 5 P–D AND 5 Q–V ITERATIONS

LOAD		GENERATION		AC LOSSES		MISMATCH		SHUNTS
MW	MVAR	MW	MVAR	MW	MVAR	MW	MVAR	MVAR
2020.00	916.00	2113.71	1420.67	93.92	504.69	–0.21	–0.02	0.00

POWER TRANSFERS

BUS DATA

BUS	NAME	VOLTS	ANGLE	GENERATION		LOAD		SHUNT		BUS	NAME	MW	MVAR
				MW	MVAR	MW	MVAR	MVAR					
1	INVERCARG220	0.936	–12.26	0.00	0.00	200.00	51.00	0.00		3	MANAPOURI220	–174.88	–40.45
										3	MANAPOURI220	–174.88	–40.45
										5	TIWAI——220	49.34	39.65
										5	TIWAI——220	49.34	39.65
										2	ROXBURGH–220	51.09	–49.40
										MISMATCH		–0.014	–0.004
2	ROXBURGH–220	0.982	–16.02	0.00	0.00	150.00	60.00	0.00		1	INVERCARG220	–50.59	39.24
										17	TWIZEL——220	184.16	–25.51
										17	TWIZEL——220	184.16	–25.51
										12	LIVINGSTN220	245.35	–17.18
										6	ROXBURGH–011	–713.14	–31.03
										MISMATCH		0.076	–0.009
3	MANAPOURI220	1.002	–2.84	0.00	0.00	0.00	0.00	0.00		1	INVERCARG220	179.54	49.24
										1	INVERCARG220	179.54	49.24
										5	TIWAI——220	163.87	54.14
										5	TIWAI——220	163.87	54.14
										4	MANAPOURI014	–686.91	–206.77
										MISMATCH		0.088	–0.003
4	MANAPOURI014	1.045	3.12	690.00	288.73	0.00	0.00	0.00		3	MANAPOURI220	689.98	288.73
										MISMATCH		0.020	0.000
5	TIWAI——220	0.931	–12.53	0.00	0.00	420.00	185.00	0.00		3	MANAPOURI220	–160.72	–49.83
										3	MANAPOURI220	–160.72	–49.83
										1	INVERCARG220	–49.24	–42.66
										1	INVERCARG220	–49.24	–42.66
										MISMATCH		–0.067	–0.016
6	ROXBURGH–011	1.050	0.000	723.71	242.37	0.00	0.00	0.00		2	ROXBURGH–220	723.71	242.37
										MISMATCH		0.000	0.000
7	BENMORE——220	0.993	–36.85	0.00	0.00	500.00	200.00	0.00		17	TWIZEL——220	–323.19	6.63
										9	AVIEMORE–220	–88.64	1.28
										9	AVIEMORE–220	–88.64	1.28
										8	BENMORE——016	0.53	–209.19
										MISMATCH		–0.061	–0.005
8	BENMORE——016	1.060	–37.00	0.00	223.40	0.00	0.00	0.00		7	BENMORE——220	0.01	223.40
										MISMATCH		–0.006	0.000
9	AVIEMORE–220	0.996	–34.28	0.00	0.00	0.00	0.00	0.00		12	LIVINGSTN220	21.37	92.02
										7	BENMORE——220	88.96	0.73
										7	BENMORE——220	88.96	0.73
										10	AVIEMORE–011	–199.26	–93.49
										MISMATCH		–0.023	–0.000
10	AVIEMORE–011	1.045	–29.41	200.00	115.46	0.00	0.00	0.00		9	AVIEMORE–220	199.99	115.46
										MISMATCH		0.007	0.000
11	OHAU–SYSTEM	1.050	–25.43	350.00	113.38	0.00	0.00	0.00		17	TWIZEL——220	350.00	113.38
										MISMATCH		–0.004	0.000

12	LIVINGSTN220	0.966	-34.27	0.00	0.00	150.00	60.00	0.00

2	ROXBURGH-220	-226.60	75.10
9	AVIEMORE-220	-20.71	-94.00
13	ISLINGTON220	97.39	-41.11
MISMATCH		-0.082	0.006

13	ISLINGTON220	1.000	-45.17	0.00	437.32	500.00	300.00	0.00

12	LIVINGSTN220	-94.14	26.75
16	TEKAPO——220	-176.86	26.13
14	BROMLEY——220	-61.68	67.19
17	TWIZEL——220	-167.23	17.24
MISMATCH		-0.085	0.000

14	BROMLEY——220	0.994	-44.73	0.00	0.00	100.00	60.00	0.00

17	TWIZEL——220	-161.84	11.30
13	ISLINGTON220	61.85	-71.30
MISMATCH		-0.009	0.002

15	TEKAPO——011	1.008	-26.72	150.00	0.00	0.00	0.00	0.00

16	TEKAPO——220	149.98	0.00
MISMATCH		0.017	-0.000

16	TEKAPO——220	1.007	-31.47	0.00	0.00	0.00	0.00	0.00

17	TWIZEL——220	-34.17	5.86
13	ISLINGTON220	183.50	-18.25
15	TEKAPO——011	-149.32	12.39
MISMATCH		-0.007	0.001

17	TWIZEL——220	1.007	-31.27	0.00	0.00	0.00	0.00	0.00

2	ROXBURGH-220	-178.50	51.27
2	ROXBURGH-220	-178.50	51.27
7	BENMORE——220	327.44	18.21
16	TEKAPO——220	34.19	-7.77
14	BROMLEY——220	167.37	-17.69
13	ISLINGTON220	173.14	-21.21
11	OHAU-SYSTEM	-345.09	-74.09
MISMATCH		-0.052	0.010

THE MAXIMUM MISMATCH IS   0.0881 MVA ON BUS   3 (MANAPOURI220)
THE SLACK BUS GENERATION IS   723.709 MW   242.372 MVAR

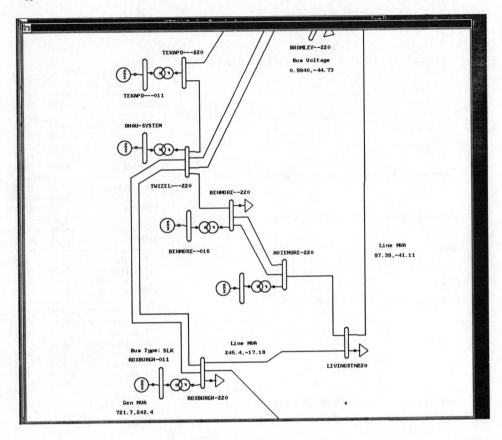

Figure 2.13
Screen display of part of the system shown in Fig. 2.12.

An example of the screen display while running *Display Power* is shown in Fig. 2.13.

2.14   REFERENCES

[1] B. Stott, 1974. Review of load-flow calculation methods, *Proc. IEEE* **62** 916–929.
[2] J. B. Ward and H. W. Hale, 1956. Digital computer solution of power-flow problems, *Trans. AIEE* **PAS-75** 398–404.
[3] H. E. Brown, G. K. Carter, H. H. Happ and C. E. Person, 1963. Power-flow solution by impedance matrix iterative method, *IEEE Trans.* **PAS-82** 1–10.
[4] J. E. Van Ness and J. H. Griffin, 1961. Elimination methods for load-flow studies, *Trans. AIEE.* **PAS-80** 299–304.
[5] W. F. Tinney, C. E. Hart, Power flow solution by Newton's method, *IEEE Trans.* **PAS-86** 1449–1460.
[6] E. C. Ogbuobiri, W. F. Tinney and J. W. Walker, 1970. Sparsity-directed decomposition for Gaussian elimination on matrices, *IEEE Trans.* **PAS-89** 141–150.

[7] B. Stott and E. Hobson, 1971. Solution of large power-system networks by ordered elimination: a comparison of ordering schemes, *Proc. IEE* **118** 125–134.

[8] W. F. Tinney and J. W. Walker, 1967. Direct solutions of sparse network equations by optimally ordered triangular factorization, *Proc. IEEE* **55** 1801–1809.

[9] S. T. Despotovic, 1974. A new decoupled load-flow method, *IEEE Trans.* **PAS-93** 884–891.

[10] B. Stott, 1972. Decoupled Newton load flow, *IEEE Trans.* **PAS-91** 1955–1959.

[11] B. Stott and O. Alsac, 1974. Fast decoupled load flow, *IEEE Trans.* **PAS-93** 859–869.

[12] J. Medanic and B. Avramovic, 1975. Solution of load-flow problems in power systems by ε-coupling method, *Proc. IEE* **122** 801–805.

[13] B. Stott, 1972. Power-system load flow (MSc lecture notes) University of Manchester Institute of Science and Technology.

3. THREE-PHASE LOAD FLOW

3.1 INTRODUCTION

For most purposes in the steady-state analysis of power systems, the system unbalance can be ignored and the single-phase analysis described in Chapter 2 is adequate. However, in practice it is uneconomical to balance the load completely or to achieve perfectly balanced transmission system impedances, as a result of untransposed high-voltage lines and lines sharing the same right of way for considerable lengths.

Among the effects of power system unbalance are: negative sequence currents causing machine rotor overheating, zero sequence currents causing relay maloperations and increased losses due to parallel untransposed lines.

The use of long-distance transmission motivated the development of analytical techniques for the assessment of power system unbalance. Early techniques [1, 2] were restricted to the case of isolated unbalanced lines operating from known terminal conditions. However, a realistic assessment of the unbalanced operation of an interconnected system, including the influence of any significant load unbalance, requires the use of three-phase load-flow algorithms, [3–5]. The object of the three-phase load flow is to find the state of the three-phase power system under the specified conditions of load, generation and system configuration. Three-phase load flow studies are also required to provide initial conditions for electromagnetic transients and harmonic studies.

The rules for the combination of three-phase models of system components into overall network admittance matrices, discussed in Appendix I, are used as the framework for the three-phase load flow described in this chapter.

The storage and computational requirements of a three-phase load-flow program are much greater than those of the corresponding single-phase case. The need for efficient algorithms is therefore significant even though, in contrast to single-phase analysis, the three-phase load flow is likely to remain a planning, rather than an operational exercise.

The basic characteristics of the fast-decoupled Newton–Raphson algorithms described in Chapter 2, have been shown [6] to apply equally to the three-phase load-flow problem. Consequently, this algorithm is now used as a basis for the development of an efficient three-phase load-flow program. When the program is used for post-operational studies of important unbalanced situations on the power system, additional practical features such as automatic transformer tapping and generator VAR limiting are necessary.

3.2 THREE-PHASE MODELS OF SYNCHRONOUS MACHINES

Synchronous machines are designed for maximum symmetry of the phase winding and are therefore adequately modelled by their sequence impedances. Such impedances contain all the information that is required to analyse the steady-state unbalanced behaviour of the synchronous machine.

The representation of the generator in phase components may be derived from the sequence impedance matrix $(Z_g)_{012}$ as follows:

$$[Z_g]_{abc} = [T_s][Z_g]_{012}[T_s]^{-1} \tag{3.2.1}$$
$$= [T_s][Z_g]_{012}[T_s]^* \tag{3.2.2}$$

where

$$[T_s] = \begin{bmatrix} 1 & 1 & 1 \\ 1 & a^2 & a \\ 1 & a & a^2 \end{bmatrix} \tag{3.2.3}$$

and a is the complex operator $e^{j2\pi/3}$. The phase component impedance matrix is thus

$$[Z_g]_{abc} = \begin{bmatrix} Z_0 + Z_1 + Z_2 & Z_0 + aZ_1 + a^2Z_2 & Z_0 + a^2Z_1 + aZ_2 \\ Z_0 + a^2Z_1 + aZ_2 & Z_0 + Z_1 + Z_2 & Z_0 + aZ_1 + a^2Z_1 \\ Z_0 + aZ_1 + a^2Z_2 & Z_0 + a^2Z_1 + aZ_2 & Z_0 + Z_1 + Z_2 \end{bmatrix} \tag{3.2.4}$$

The phase component model of the generator is illustrated in Fig. 3.1(a) The machine excitation acts symmetrically on the three phases and the voltages at the internal or excitation busbar form a balanced three-phase set, i.e.

$$E_k^a = E_k^b = E_k^c \tag{3.2.5}$$

and

$$\theta_k^a = \theta_k^b + \frac{2\pi}{3} = \theta_k^c - \frac{2\pi}{3}. \tag{3.2.6}$$

For three-phase load flow the voltage regulator must be accurately modelled as it influences the machine operation under unbalanced conditions. The voltage regulator monitors the terminal voltages of the machine and controls the excitation voltage according to some predetermined function of the terminal voltages. Often the positive sequence is extracted from the three-phase voltage measurement using a sequence filter.

Before proceeding further it is instructive to consider the generator modelling from a symmetrical component frame of reference. The sequence network model of the generator is illustrated in Fig. 3.1(b). As the machine excitation acts symmetrically on the three phases, positive sequence voltages only are present at the internal busbar.

The influence of the generator upon the unbalanced system is known if the voltages at the terminal busbar are known. In terms of sequence voltages, the positive sequence voltage may be obtained from the excitation and the positive sequence voltage drop caused by the flow of positive sequence currents through the positive sequence

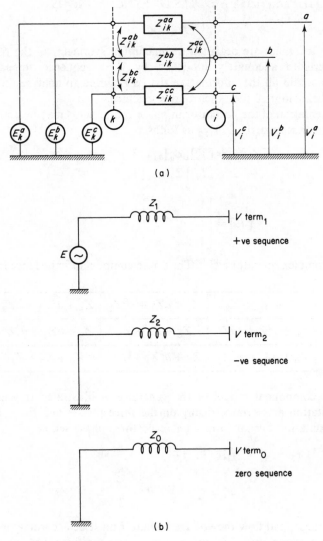

Figure 3.1
Synchronous machine models. (a) Phase component representation. (b) Symmetrical component
representation

reactance. The negative and zero sequence voltages are derived from the flow of their
respective currents through their respective impedances. It is important to note that
the negative and zero sequence voltages are not influenced by the excitation or
positive sequence impedance.

  There are infinite combinations of machine excitation and machine positive
sequence reactance which will satisfy the conditions at the machine terminals and
give the correct positive sequence voltage. Whenever the machine excitation must be
known (as in fault studies) the actual positive sequence impedance must be used. For
load flow however, the excitation is not of any particular interest and the positive

sequence impedance may be arbitrarily assigned to any value [3]. The positive sequence impedance is usually set to zero for these studies.

Thus the practice with regard to three-phase load flow in phase coordinates, is to set the positive sequence reactance to a small value in order to reduce the excitation voltage to the same order as the usual system voltages with a corresponding reduction in the angle between the internal busbar and the terminal busbar. Both these features are important when a fast decoupled algorithm is used.

Therefore, in forming the phase component generator model using equation (3.2.4), an arbitrary value may be used for Z_1 but the actual values are used for Z_0 and Z_2. There is no loss of relevant information as the influence of the generator upon the unbalanced system is accurately modelled.

The nodal admittance matrix, relating the injected currents at the generator busbars to their nodal voltages, is given by the inverse of the series impedance matrix derived from equation (3.2.4).

3.3 THREE-PHASE MODELS OF TRANSMISSION LINES

Transmission line parameters are calculated from the line geometrical characteristics. The calculated paramters are expressed as a series impedance and shunt admittance per unit length of line. The effects of ground currents and earth wires are included in the calculation of these parameters.

Series impedance. A three-phase transmission line with a ground wire is illustrated in Fig. 3.2(a). The following equations can be written for phase a:

$$V_a - V'_a = I_a(R_a + j\omega L_a) + I_b(j\omega L_{ab}) + I_c(j\omega L_{ac})$$
$$+ j\omega L_{ag}I_g - j\omega L_{an}I_n + V_n$$
$$V_n = I_n(R_n + j\omega L_n) - I_a j\omega L_{na} - I_b j\omega L_{nb} - I_c j\omega L_{nc} - I_g j\omega L_{ng}$$

and substituting

$$I_n = I_a + I_b + I_c + I_g$$
$$V_a - V'_a = I_a(R_a + j\omega L_a) + I_b j\omega L_{ab} + I_c j\omega L_{ac}$$
$$+ j\omega L_{ag}I_g - j\omega L_{an}(I_a + I_b + I_c + I_g) + V_n.$$

Regrouping and substituting for V_n, i.e.

$$\Delta V_a = V_a - V'_a$$
$$= I_a(R_a + j\omega L_a - j\omega L_{an} + R_n + j\omega L_n - j\omega L_{na})$$
$$+ I_b(j\omega L_{ab} - j\omega L_{an} + R_n + j\omega L_n - j\omega L_{nb})$$
$$+ I_c(j\omega L_{ac} - j\omega L_{an} + R_n + j\omega L_n - j\omega L_{nc})$$
$$+ I_g(j\omega L_{ag} - j\omega L_{an} + R_n + j\omega L_n - j\omega L_{ng})$$
$$\Delta V_a = I_a(R_a + j\omega L_a - 2j\omega L_{an} + R_n + j\omega L_n)$$
$$+ I_b(j\omega L_{ab} - j\omega L_{bn} - j\omega L_{an} + R_n + j\omega L_n)$$
$$+ I_c(j\omega L_{ac} - j\omega L_{cn} - j\omega L_{an} + R_n + j\omega L_n)$$
$$+ I_g(j\omega L_{ag} - j\omega L_{gn} - j\omega L_{an} + R_n + j\omega L_n)$$

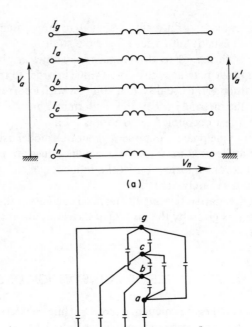

(a)

(b)

Figure 3.2
(a) Three-phase transmission series impedance equivalent. (b) Three-phase transmission shunt
impedance equivalent

or

$$\Delta V_a = Z_{aa-n} I_a + Z_{ab-n} I_b + Z_{ac-n} I_c + Z_{ag-n} I_g \qquad (3.3.1)$$

and writing similar equations for the other phases the following matrix equation
results:

$$
\begin{bmatrix}
\Delta V_a \\
\Delta V_b \\
\Delta V_c \\
\hline
\Delta V_g
\end{bmatrix}
=
\left[
\begin{array}{ccc|c}
Z_{aa-n} & Z_{ab-n} & Z_{ac-n} & Z_{ag-n} \\
Z_{ba-n} & Z_{bb-n} & Z_{bc-n} & Z_{bg-n} \\
Z_{ca-n} & Z_{cb-n} & Z_{cc-n} & Z_{cg-n} \\
\hline
Z_{ga} & Z_{gb-n} & Z_{gc-n} & Z_{gg-n}
\end{array}
\right]
\begin{bmatrix}
I_a \\
I_b \\
I_c \\
\hline
I_g
\end{bmatrix}
\qquad (3.3.2)
$$

Since we are interested only in the performance of the phase conductors it is more
convenient to use a three-conductor equivalent for the transmission line. This is
achieved by writing matrix equation (3.3.2) in partitioned form as follows:

$$
\begin{bmatrix}
\Delta V_{abc} \\
\hline
\Delta V_g
\end{bmatrix}
=
\left[
\begin{array}{c|c}
Z_A & Z_B \\
\hline
Z_C & Z_D
\end{array}
\right]
\begin{bmatrix}
I_{abc} \\
\hline
I_g
\end{bmatrix}
\qquad (3.3.3)
$$

From (3.3.3)

$$\Delta V_{abc} = Z_A I_{abc} + Z_B I_g \tag{3.3.4}$$

$$\Delta V_g = Z_C I_{abc} + Z_D I_g. \tag{3.3.5}$$

From equations (3.3.4) and (3.3.5) and assuming that the ground wire is at zero potential

$$\Delta V_{abc} = Z_{abc} I_{abc} \tag{3.3.6}$$

where

$$Z_{abc} = Z_A - Z_B Z_D^{-1} Z_C = \begin{array}{|c|c|c|} \hline Z'_{aa-n} & Z'_{ab-n} & Z'_{ac-n} \\ \hline Z'_{ba-n} & Z'_{bb-n} & Z'_{bc-n} \\ \hline Z'_{ca-n} & Z'_{cb-n} & Z'_{cc-n} \\ \hline \end{array}$$

Shunt admittance. With reference to Fig. 3.2(b) the potentials of the line conductors are related to the conductor charges by the matrix equation [7]

$$\begin{array}{|c|} \hline V_a \\ \hline V_b \\ \hline V_c \\ \hline V_g \\ \hline \end{array} = \begin{array}{|c|c|c|c|} \hline P_{aa} & P_{ab} & P_{ac} & P_{ag} \\ \hline P_{ba} & P_{bb} & P_{bc} & P_{bg} \\ \hline P_{ca} & P_{cb} & P_{cc} & P_{cg} \\ \hline P_{ga} & P_{gb} & P_{gc} & P_{gg} \\ \hline \end{array} \cdot \begin{array}{|c|} \hline Q_a \\ \hline Q_b \\ \hline Q_c \\ \hline Q_g \\ \hline \end{array} \tag{3.3.7}$$

Similar considerations as for the series impedance matrix lead to

$$V_{abc} = P'_{abc} Q_{abc} \tag{3.3.8}$$

where P'_{abc} is a 3×3 matrix which includes the effects of the ground wire. The capacitance matrix of the transmission line of Fig. 3.2 is given by

$$C'_{abc} = P'^{-1}_{abc} = \begin{array}{|c|c|c|} \hline C_{aa} & -C_{ab} & -C_{ac} \\ \hline -C_{ba} & C_{bb} & -C_{bc} \\ \hline -C_{ca} & -C_{cb} & C_{cc} \\ \hline \end{array}$$

The series impedance and shunt admittance lumped-π model representation of the three-phase line is shown in Fig. 3.3(a) and its matrix equivalent is illustrated in Fig. 3.3(b). These two matrices can be represented by compound admittances (Fig. 3.3(c)) as described in Appendix I.

Following the rules developed for the formation of the admittance matrix using the compound concept, the nodal injected currents of Fig. 3.3(c) can be related to

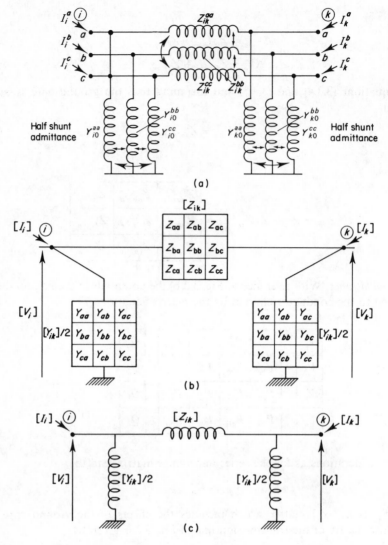

Figure 3.3
Lumped-π model of a short three-phase line series impedance. (a) Full circuit representation. (b) Matrix equivalent. (c) Using three-phase compound admittances

the nodal voltages by the equation

$$
\begin{array}{|c|}
\hline
[I_i] \\
\hline
[I_k] \\
\hline
\end{array}
=
\begin{array}{|c|c|}
\hline
[Z]^{-1}+[Y]/2 & -[Z]^{-1} \\
\hline
-[Z]^{-1} & [Z]^{-1}+[Y]/2 \\
\hline
\end{array}
\cdot
\begin{array}{|c|}
\hline
[V_i] \\
\hline
[V_k] \\
\hline
\end{array}
\qquad (3.3.9)
$$

$$6 \times 1 \qquad\qquad 6 \times 6 \qquad\qquad 6 \times 1$$

This forms the element admittance matrix representation for the short line between busbars i and k in terms of 3×3 matrix quantities.

This representation may not be accurate enough for electrically long lines. The physical length at which a line is no longer electrically short depends on the wavelength, therefore if harmonic frequencies are being considered, this physical length may be quite small. Using transmission line and wave propagation theory more exact models may be derived. However, for normal system frequency analysis, it is considered sufficient to model a long line as a series of two or three nominal-π sections.

3.3.1 Mutually Coupled Three-phase Lines

When two or more transmission lines occupy the same right of way for a considerable length, the electrostatic and electromagnetic coupling between those lines must be taken into account.

Consider the simplest case of two mutually coupled three-phase lines. The two coupled lines are considered to form one subsystem composed of four system busbars. The coupled lines are illustrated in Fig. 3.4, where each element is a 3×3 compound admittance and all voltages and currents are 3×1 vectors.

The coupled series elements represent the electromagnetic coupling while the coupled shunt elements represent the capacitive or electrostatic coupling. These coupling parameters are lumped in a similar way to the standard line parameters.

With the admittances labelled as in Fig. 3.4 and applying the rules of linear transformation for compound networks the admittance matrix for the subsytem is defined as follows:

$$
\begin{array}{c}
\begin{bmatrix} I_A \\ I_B \\ I_C \\ I_D \end{bmatrix} =
\begin{bmatrix}
Y_{11} + Y_{33} & Y_{12} + Y_{34} & -Y_{11} & -Y_{12} \\
Y_{12}^{T} + Y_{34}^{T} & Y_{22} + Y_{44} & -Y_{12}^{T} & -Y_{22} \\
-Y_{11} & -Y_{12} & Y_{11} + Y_{55} & Y_{12} + Y_{56} \\
-Y_{12}^{T} & -Y_{22} & Y_{12}^{T} + Y_{56}^{T} & Y_{22} + Y_{66}
\end{bmatrix}
\cdot
\begin{bmatrix} V_A \\ V_B \\ V_C \\ V_D \end{bmatrix}
\end{array}
\qquad (3.3.10)
$$

$$12 \times 1 \qquad\qquad 12 \times 12 \qquad\qquad 12 \times 1$$

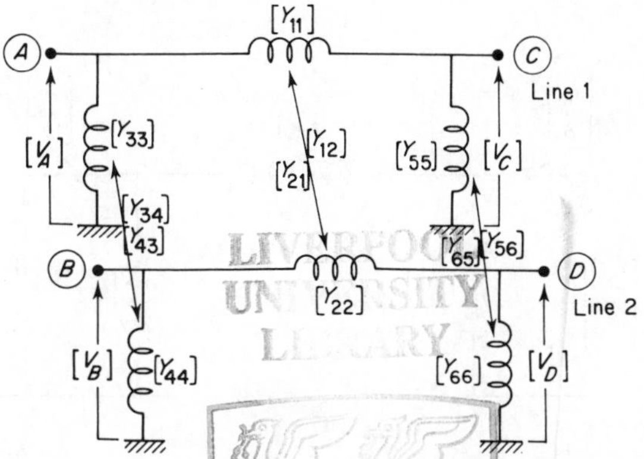

Figure 3.4
Two coupled three-phase lines

It is assumed here that the mutual coupling is bilateral. Therefore, $Y_{21} = Y_{12}^T$ and so on.

The subsystem may be redrawn as Fig. 3.5. The pairs of coupled 3×3 compound admittances are now represented as a 6×6 compound admittance. The matrix representation is also shown. Following this representation and the labelling of the admittance blocks in the figure, the admittance matrix may be written in terms of the 6×6 compound coils as

$$\underbrace{\begin{bmatrix} \begin{bmatrix} I_A \\ I_B \end{bmatrix} \\ \begin{bmatrix} I_C \\ I_D \end{bmatrix} \end{bmatrix}}_{12 \times 1} = \underbrace{\begin{bmatrix} [Z_s]^{-1} + [Y_{s1}] & -[Z_s]^{-1} \\ -[Z_s]^{-1} & [Z_s]^{-1} + [Y_{s2}] \end{bmatrix}}_{12 \times 12} \underbrace{\begin{bmatrix} \begin{bmatrix} V_A \\ V_B \end{bmatrix} \\ \begin{bmatrix} V_C \\ V_D \end{bmatrix} \end{bmatrix}}_{12 \times 1}. \tag{3.3.11}$$

This is clearly identical to equation (3.3.10) with the appropriate matrix partitioning.

The representation of Fig. 3.5 is more concise and the formation of equation (3.3.11)

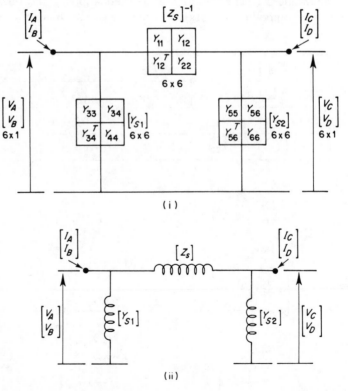

Figure 3.5
6×6 compound admittance representation of two coupled three-phase lines. (i) 6×6 Matrix representation; (ii) 6×6 Compound admittance representation

from this representation is straightforward, being exactly similar to that which results from the use of 3×3 compound admittances for the normal single three-phase line.

The data which must be available, to enable coupled lines to be treated in a similar manner to single lines, the series impedance and shunt admittance matrices. These matrices are of order 3×3 for a single line, 6×6 for two coupled lines, 9×9 for three and 12×12 for four coupled lines.

Once the matrices $[Z_s]$ and $[Y_s]$ are available, the admittance matrix for the subsystem is formed by application of equation (3.3.11).

When all the busbars of the coupled lines are distinct, the subsystem may be combined directly into the system admittance matrix. However, if the busbars are not distinct then the admittance matrix as derived from equation (3.3.11) must be modified. This is considered in the following section.

3.3.2 Consideration of Terminal Connections

The admittance matrix as derived above must be reduced if there are different elements in the subsystem connected to the same busbar. As an example consider two parallel transmission lines as illustrated in Fig. 3.6.

The admittance matrix derived previously related the currents and voltages at the four busbar $A1$, $A2$, $B1$ and $B2$. This relationship is given by

$$\begin{bmatrix} I_{A1} \\ I_{A2} \\ I_{B1} \\ I_{B2} \end{bmatrix} = [Y_{A1A2B1B2}] \cdot \begin{bmatrix} V_{A1} \\ V_{A2} \\ B_{B1} \\ V_{B2} \end{bmatrix} \tag{3.3.12}$$

The nodal injected current at busbar A, (I_A), is given by

$$I_A = I_{A1} + I_{A2}$$

similarly

$$I_B = I_{B1} + I_{B2}.$$

Also from inspection of Fig. 3.6,

$$V_A = V_{A1} = V_{A2}$$
$$V_B = V_{B1} = V_{B2}.$$

The required matrix equation relates the nodal injected currents, I_A and I_B, to the

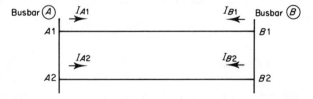

Figure 3.6
Mutually coupled parallel transmission lines

voltages at these busbar. This is readily derived from equation (3.3.12) and the conditions specified above. This is simply a matter of adding appropriate rows and columns and yields

$$
\left| \begin{array}{c} I_A \\ I_B \end{array} \right| = [Y_{AB}] \left| \begin{array}{c} V_A \\ V_B \end{array} \right| \tag{3.3.13}
$$

This matrix $[Y_{AB}]$ is the required nodal admittance matrix for the subsystem.

It should be noted that the matrix in equation (3.3.12) must be retained as it is required in the calculation of the individual line power flows.

3.3.3 Shunt Elements

Shunt reactors and capacitors are used in a power system for reactive power control. The data for these elements are usually given in terms of their rated MVA and rated kV; the equivalent phase admittance in p.u. is calculated from these data.

Consider, as an example, a three-phase capacitor bank shown in Fig. 3.7. A similar triple representation as that for a line section is illustrated. The final two forms are the most compact and will be used exclusively from this point on.

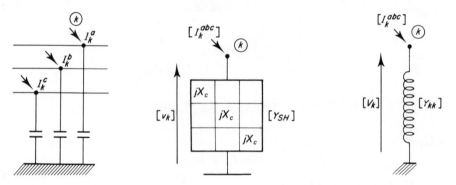

Figure 3.7
Representation of a shunt capacitor bank

The admittance matrix for shunt elements is usually diagonal as there is normally no coupling between the components of each phase. This matrix is then incorporated directly into the system admittance matrix, contributing only to the self-admittance of the particular bus.

3.3.4 Series Elements

Any element connected directly between two buses may be considered a series element. Series elements are often taken as being a section in a line sectionalisation which is described later in the chapter.

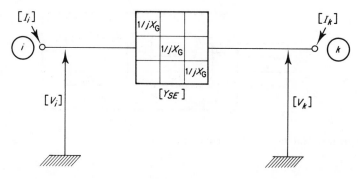

Figure 3.8
Graphic representation of series capacitor bank between nodes i and k

A typical example is the series capacitor bank which is usually taken as uncoupled, i.e. the admittance matrix is diagonal.

This can be represented graphically as in Fig. 3.8.

The admittance matrix for the subsystem can be written by inspection as

$$[Y] = \begin{array}{|c|c|} \hline [Y_{SE}] & -[Y_{SE}] \\ \hline -[Y_{SE}] & [Y_{SE}] \\ \hline \end{array} \tag{3.3.14}$$

3.4 THREE-PHASE MODELS OF TRANSFORMERS

The inherent assumption that the transformer is a balanced three-phase device is justified in the majority of practical situations, and traditionally, three-phase transformers are represented by their equivalent sequence networks.

More recently, however, methods have been developed [7, 8] to enable all three-phase transformer connections to be accurately modelled in phase coordinates. In phase coordinates no assumptions are necessary although physically justifiable assumptions are still used in order to simplify the model. The primitive admittance matrix, used as a basis for the phase coordinate transformer model is derived from the primitive or unconnected network for the transformer windings and the method of linear transformation enables the admittance matrix of the actual connected network to be found.

3.4.1 Primitive Admittance Model of Three-phase Transformers

Many three-phase transformers are wound on a common core and all windings are therefore coupled to all other windings. Therefore, in general, a basic two-winding three-phase transformer has a primitive or unconnected network consisting of six coupled coils. If a tertiary winding is also present the primitive network consists of nine coupled coils. The basic two-winding transformer shown in Fig. 3.9 is now

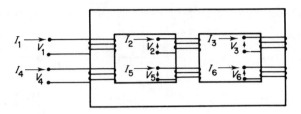

Figure 3.9
Diagrammatic representation of two-winding transformer

considered, the addition of further windings being a simple but cumbersome extension of the method.

The primitive network, Fig. 3.10, can be represented by the primitive admittance matrix which has the following general form:

$$
\begin{bmatrix} I_1 \\ I_2 \\ I_3 \\ I_4 \\ I_5 \\ I_6 \end{bmatrix}
=
\begin{bmatrix}
y_{11} & y_{12} & y_{13} & y_{14} & y_{15} & y_{16} \\
y_{21} & y_{22} & y_{23} & y_{24} & y_{25} & y_{26} \\
y_{31} & y_{32} & y_{33} & y_{34} & y_{35} & y_{36} \\
y_{41} & y_{42} & y_{43} & y_{44} & y_{45} & y_{46} \\
y_{51} & y_{52} & y_{53} & y_{54} & y_{55} & y_{56} \\
y_{61} & y_{62} & y_{63} & y_{64} & y_{65} & y_{66}
\end{bmatrix}
\cdot
\begin{bmatrix} V_1 \\ V_2 \\ V_3 \\ V_4 \\ V_5 \\ V_6 \end{bmatrix}
\tag{3.4.1}
$$

The elements of matrix $[Y]$ can be measured directly, i.e. by energising coil i and short-circuiting all other coils, column i of $[Y]$ can be calculated from $y_{ki} = I_k/V_i$.

Considering the reciprocal nature of the mutual couplings in equation (3.4.1) 21

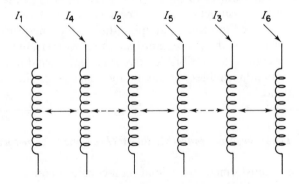

Figure 3.10
Primitive network of two-winding transformer. Six coupled coil primitive network. (Note the dotted coupling represents parasitic coupling between phases.)

short-circuit measurements would be necessary to complete the admittance matrix. Such a detailed representation is seldom required.

By assuming that the flux paths are symmetrically distributed between all windings equation (3.4.1) may be simplified to equation (3.4.2):

$$
\begin{bmatrix} I_1 \\ I_2 \\ I_3 \\ I_4 \\ I_5 \\ I_6 \end{bmatrix}
=
\begin{bmatrix}
y_p & y'_m & y'_m & -y_m & y''_m & y''_m \\
y'_m & y_p & y'_m & y''_m & -y_m & y''_m \\
y'_m & y'_m & y_p & y''_m & y''_m & -y_m \\
-y_m & y''_m & y''_m & y_s & y'''_m & y'''_m \\
y''_m & -y_m & y''_m & y'''_m & y_s & y'''_m \\
y''_m & y''_m & -y_m & y'''_m & y'''_m & y_s
\end{bmatrix}
\cdot
\begin{bmatrix} V_1 \\ V_2 \\ V_3 \\ V_4 \\ V_5 \\ V_6 \end{bmatrix}
\qquad (3.4.2)
$$

where

y'_m is the mutual admittance between primary coils;
y''_m is the mutual admittance between primary and secondary coils on different cores;
y'''_m is the mutual admittance between secondary coils.

For three separate single-phase units all the primed values are effectively zero. In three-phase units the primed values, representing parasitic interphase coupling, do have a noticeable effect. This effect can be interpretd through the symmetrical component equivalent circuits.

If the values in equation (3.4.2) are available then this representation of the primitive network should be used. If interphase coupling can be ignored, the coupling between a primary and a secondary coil is modelled as for the single-phase unit, giving rise to the primitive network of Fig. 3.11.

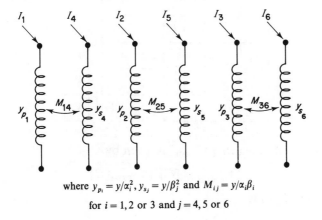

where $y_{p_i} = y/\alpha_i^2$, $y_{s_j} = y/\beta_j^2$ and $M_{ij} = y/\alpha_i\beta_i$

for $i = 1, 2$ or 3 and $j = 4, 5$ or 6

Figure 3.11
Primitive network

The new admittance matrix equation is

$$
\begin{bmatrix} I_1 \\ I_2 \\ I_3 \\ I_4 \\ I_5 \\ I_6 \end{bmatrix}
=
\begin{bmatrix}
y_{p1} & & & M_{14} & & \\
& y_{p2} & & & M_{25} & \\
& & y_{p3} & & & M_{36} \\
M_{41} & & & y_{s4} & & \\
& M_{52} & & & y_{s5} & \\
& & M_{63} & & & y_{s6}
\end{bmatrix}
\cdot
\begin{bmatrix} V_1 \\ V_2 \\ V_3 \\ V_4 \\ V_5 \\ V_6 \end{bmatrix}
\tag{3.4.3}
$$

3.4.2 Models for Common Transformer Connections

The network admittance matrix for any two-winding three-phase transformer can now be formed by the method of linear transformation.

As a simple example, consider the formation of the admittance matrix for a star–star connection with both neutrals solidly earthed in the absence of interphase mutuals. This example is chosen as it is the simplest computationally.

The connection matrix is derived from consideration of the actual connected network. For the star–star (or wye–wye) transformer illustrated in Fig. 3.12, the connection matrix $[C]$ relating the branch voltages (i.e. voltages of the primitive network) to the node voltages (i.e. voltages of the actual network) is a 6×6 identity matrix, i.e.

$$
\begin{bmatrix} V_1 \\ V_2 \\ V_3 \\ V_4 \\ V_5 \\ V_6 \end{bmatrix}
=
\begin{bmatrix}
1 & & & & & \\
& 1 & & & & \\
& & 1 & & & \\
& & & 1 & & \\
& & & & 1 & \\
& & & & & 1
\end{bmatrix}
\cdot
\begin{bmatrix} v_a \\ v_b \\ v_c \\ V_a \\ V_b \\ V_c \end{bmatrix}
$$

The nodal admittance matrix $[Y]_{\text{NODE}}$ is given by

$$
[Y]_{\text{NODE}} = [C]^{\text{T}}[Y]_{\text{PRIM}}[C].
\tag{3.4.4}
$$

Substituting for $[C]$ yields

$$
[Y]_{\text{NODE}} = [Y]_{\text{PRIM}}.
\tag{3.4.5}
$$

Let us now consider the wye G–delta connection illustrated in Fig. 3.13.

The following connection can be written by inspection between the primitive branch

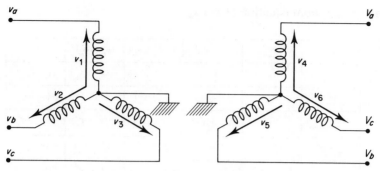

Figure 3.12
Network connection diagram for three-phase star–star transformer

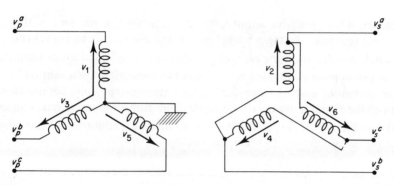

Figure 3.13
Network connection diagram for wye G–delta transformer

voltages and the node voltages:

$$
\begin{array}{|c|}
V_1 \\
V_2 \\
V_3 \\
V_4 \\
V_5 \\
V_6
\end{array}
=
\begin{array}{|cccccc|}
1 & 0 & 0 & 0 & 0 & 0 \\
0 & 0 & 0 & 1 & -1 & 0 \\
0 & 1 & 0 & 0 & 0 & 0 \\
0 & 0 & 0 & 0 & 1 & -1 \\
0 & 0 & 1 & 0 & 0 & 0 \\
0 & 0 & 0 & -1 & 0 & 1
\end{array}
\cdot
\begin{array}{|c|}
V_p^a \\
V_p^b \\
V_p^c \\
V_s^a \\
V_s^b \\
V_s^c
\end{array}
\tag{3.4.6}
$$

or

$$[V]_{\text{branch}} = [C][V]_{\text{node}}. \tag{3.4.7}$$

We can also write

$$[Y]_{\text{NODE}} = [C]^{\text{T}}[Y]_{\text{PRIM}}[C] \tag{3.4.8}$$

and using $[Y]_{PRIM}$ from equation (3.4.2):

$$[Y]_{NODE} =
\begin{array}{|cccccc|l}
\hline
y_p & y'_m & y'_m & -(y_m+y''_m) & (y_m+y''_m) & 0 & a \\
y'_m & y_p & y'_m & 0 & -(y_m+y''_m) & (y_m+y''_m) & b \\
y'_m & y'_m & y_p & (y_m+y''_m) & 0 & -(y_m+y''_m) & c \\
-(y_m+y''_m) & 0 & (y_m+y''_m) & 2(y_s-y''_m) & -(y_s-y''_m) & -(y_s-y''_m) & A \\
(y_m+y''_m) & -(y_m+y''_m) & 0 & -(y_s-y''_m) & 2(y_s-y''_m) & -(y_s-y''_m) & B \\
0 & (y_m+y''_m) & -(y_m+y''_m) & -(y_s-y''_m) & -(y_s-y''_m) & 2(y_s-y''_m) & C \\
\hline
\end{array}$$

$$(3.4.9)$$

Moreover, if the primitive admittances are expressed in per unit, with both the primary and secondary voltages being one per unit, the wye–delta transformer model must include an effective turns ratio of $\sqrt{3}$. The upper right and lower left quadrants of matrix (3.4.9) must be divided by $\sqrt{3}$ and the lower right quadrant by 3.

In the particular case of three-single phase transformer units connected in wye G–delta all the y' and y'' terms will disappear. Ignoring off-nominal taps (but keeping in mind the effective $\sqrt{3}$ turns ratio in per unit) the nodal admittance matrix equation relating the nodal currents to the nodal voltages is

$$
\begin{bmatrix} I^a_p \\ I^b_p \\ I^c_p \\ I^A_s \\ I^B_s \\ I^C_s \end{bmatrix}
=
\begin{bmatrix}
y & & & -y/\sqrt{3} & y/\sqrt{3} & \\
& y & & & -y/\sqrt{3} & y/\sqrt{3} \\
& & y & y/\sqrt{3} & & -y/\sqrt{3} \\
-y/\sqrt{3} & & y/\sqrt{3} & \tfrac{2}{3}y & -\tfrac{1}{3}y & -\tfrac{1}{3}y \\
y/\sqrt{3} & -y/\sqrt{3} & & -\tfrac{1}{3}y & \tfrac{2}{3}y & -\tfrac{1}{3}y \\
& y/\sqrt{3} & -y/\sqrt{3} & -\tfrac{1}{3}y & -\tfrac{1}{3}y & \tfrac{2}{3}y
\end{bmatrix}
\cdot
\begin{bmatrix} V^a_p \\ V^b_p \\ V^c_p \\ V^A_s \\ V^B_s \\ V^C_s \end{bmatrix}
$$

$$(3.4.10)$$

where Y is the transformer leakage admittance in p.u.

An equivalent circuit can be drawn, corresponding to this admittance model of the transformer, as illustrated in Fig. 3.14.

The large shunt admittances to earth from the nodes of the star connection are apparent in the equivalent circuit. These shunts are typically around 10 p.u. (for a 10% leakage reactance transformer).

The models for the other common connections can be derived following a similar procedure.

In general, any two-winding three-phase transformer may be represented using two coupled compound coils. The network and admittance matrix for this

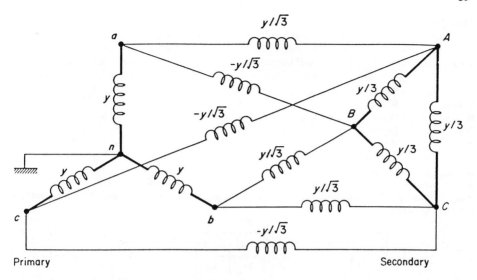

Figure 3.14
Equivalent circuit for star–delta transformer

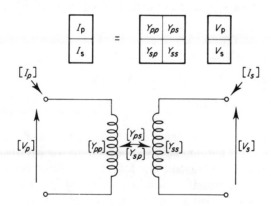

Figure 3.15
Two-winding three-phase transformer as two coupled compound coils

representation is illustrated in Fig. 3.15. It should be noted that

$$[Y_{sp}] = [Y_{ps}]^{\mathrm{T}}$$

as the coupling between the two compound coils is bilateral.

Often, because more detailed information is not required, the parameters of all three phases are assumed balanced. In this case the common three-phase connections are found to be modelled by three basic submatrices.

The submatrices, $[Y_{pp}]$, $[Y_{ps}]$ etc., are given in Table 3.1 for the common connections.

Table 3.1
Characteristic submatrices used in forming the transformer admittance matrices

Trans. connection		Self admittance		Mutual admittance
Bus P	Bus S	Y_{PP}	Y_{SS}	Y_{PS}, Y_{SP}
Wye G	Wye G	Y_I	Y_I	$-Y_I$
Wye G	Wye	$Y_{II/3}$	$Y_{II/3}$	$-Y_{II/3}$
Wye G	Delta	Y_I	Y_{II}	$+Y_{III}$
Wye	Wye	$Y_{II/3}$	$Y_{II/3}$	$-Y_{II/3}$
Wye	Wye	$Y_{II/3}$	Y_{II}	Y_{III}
Delta	Delta	Y_{II}	Y_{II}	$-Y_{II}$

Basic submatrices used in node admittance formulation of common three-phase transformer connections, where:

$$Y_I = \begin{bmatrix} y_t & & \\ & y_t & \\ & & y_t \end{bmatrix} \qquad Y_{II} = \begin{bmatrix} 2y_t & -y_t & -y_t \\ -y_t & 2y_t & -y_t \\ -y_t & -y_t & 2y_t \end{bmatrix} \qquad Y_{III} = \begin{bmatrix} -y_t & y_t & \\ & -y_t & y_t \\ y_t & & -y_t \end{bmatrix}$$

Finally, these submatrices must be modified to accounts for off-nominal tap ratio as follows.

(i) Divide the self-admittance of the primary by α^2.

(ii) Divide the self-admittance of the secondary by β^2.

(iii) Divide the mutual admittance matrices by $(\alpha\beta)$.

It should be noted that in the p.u. system a delta winding has an off-nominal tap of $\sqrt{3}$.

For transformers with ungrounded wye connections, or with neutrals connected through an impedance, an extra coil is added to the primitive network for each unearthed neutral and the primitive admittance matrix increases in dimension. By noting that the injected current in the neutral is zero, these extra terms can be eliminated from the connected network admittance matrix.

Once the admittance matrix has been formed for a particular connection it represents a simple subsystem composed of the two busbars interconnected by the transformer.

3.4.3 Sequence Components Modelling of Three-phase Transformers

In most cases lack of data will prevent the use of the general model based on the primitive admittance matrix and will justify the conventional approach in terms of symmetrical components. Let us now derive the general sequence components equivalent circuits and the assumptions introduced in order to arrive at the conventional models.

With reference to the wye G–delta common-core transformer of Fig. 3.13 represented by equation (3.4.9), and partitioning this matrix to separate self and mutual elements the following transformations apply.

Primary side:

$$y^P_{120} = T_s^{-1} \begin{array}{|c|c|c|} \hline y_p & y'_m & y'_m \\ \hline y'_m & y_p & y'_m \\ \hline y'_m & y'_m & y_p \\ \hline \end{array} T_s$$

where

$$[T_s] = \begin{array}{|c|c|c|} \hline 1 & 1 & 1 \\ \hline 1 & a^2 & a \\ \hline 1 & a & a^2 \\ \hline \end{array} \quad \text{and} \quad a = e^{j2\pi/3}.$$

Therefore

$$y^P_{120} = \begin{array}{|c|c|c|} \hline y_p - y'_m & 0 & 0 \\ \hline 0 & y_p - y'_m & 0 \\ \hline 0 & 0 & y_p + 2y'_m \\ \hline \end{array} \tag{3.4.11}$$

Secondary side:

The delta connection on the secondary side introduces an effective $\sqrt{3}$ turns ratio and the sequence components admittance matrix is

$$y^s_{120} = \tfrac{1}{3} T_s^{-1} \begin{array}{|c|c|c|} \hline 2(y_s - y'''_m) & -(y_s - y'''_m) & -(y_s - y'''_m) \\ \hline -(y_s - y'''_m) & 2(y_s - y'''_m) & -(y_s - y'''_m) \\ \hline -(y_s - y'''_m) & -(y_s - y'''_m) & 2(y_s - y'''_m) \\ \hline \end{array} T_s$$

$$= \begin{array}{|c|c|c|} \hline y_s - y'''_m & 0 & 0 \\ \hline 0 & y_s - y'''_m & 0 \\ \hline 0 & 0 & 0 \\ \hline \end{array} \tag{3.4.12}$$

Mutual terms:

The mutual admittance submatrix of equation (3.4.9), modified for effective turns ratio, is transformed as follows:

$$y^M_{120} = \frac{\sqrt{3}}{3} T_s^{-1} \begin{array}{|c|c|c|} \hline -(y_m + y''_m) & (y_m + y''_m) & 0 \\ \hline 0 & -(y_m + y''_m) & (y_m + y''_m) \\ \hline (y_m + y''_m) & 0 & -(y_m + y''_m) \\ \hline \end{array} T_s$$

$$= \begin{array}{|c|c|c|} \hline -(y_m + y''_m)\angle 30° & 0 & 0 \\ \hline 0 & -(y_m + y''_m)\angle{-30°} & 0 \\ \hline 0 & 0 & 0 \\ \hline \end{array} \quad (3.4.13)$$

Recombining the sequence components submatrices yields

	V^p_1	V^p_2	V^p_0	V^s_1	V^s_2	V^s_0
I^p_1	$y_p - y'_m$			$-(y_m + y''_m)\angle 30$		0
I^p_2		$y_p - y'_m$			$-(y_m + y''_m)\angle{-30}$	0
I^p_0			$y_p + 2y'_m$			0
I^s_1	$-(y_m + y''_m)\angle 30$			$y_s - y''_m$		0
I^s_2		$-(y_m + y''_m)\angle{-30}$			$y_s - y''_m$	0
I^s_0						0

$$(3.4.14)$$

Equations (3.4.14) can be represented by the three sequence network of Figs. 3.16, 3.17 and 3.18 respectively.

In general, therefore, the three sequence impedances are different on a common-core transformer.

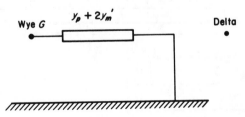

Figure 3.16
Zero-sequence node admittance model for a common-core grounded wye–delta transformer [7] (©1982 IEEE)

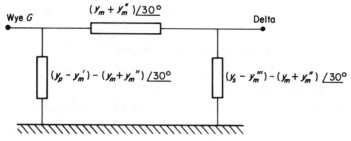

Figure 3.17
Positive-sequence node admittance model for a common-core grounded wye–delta transformer [7]
(© 1982 IEEE)

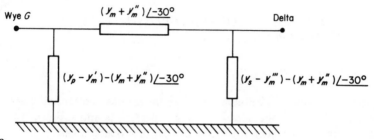

Figure 3.18
Negative-sequence node admittance model for a common-core grounded wye–delta transformer [7]
(©1982 IEEE)

Table 3.2
Typical symmetrical-component models for the six most common connections of three-phase transformers (4). (© 1982 IEEE)

Bus P	Bus Q	Pos Seq	Neg Seq	Zero Seq
Wye G	Wye G	P —z_{sc}— Q	P —z_{sc}— Q	P —z_{sc}— Q
Wye G	Wye	P —z_{sc}— Q	P —z_{sc}— Q	P —z_{sc}— Q
Wye G	Delta	P —z_{sc}— Q	P —z_{sc}— Q	P —z_{sc}— Q
Wye	Wye	P —z_{sc}— Q	P —z_{sc}— Q	P —z_{sc}— Q
Wye	Delta	P —z_{sc}— Q	P —z_{sc}— Q	P —z_{sc}— Q
Delta	Delta	P —z_{sc}— Q	P —z_{sc}— Q	P —z_{sc}— Q

The complexity of these equivalent models is normally eliminated by the following simplifications.

- The 30° phase shifts of wye–delta connections are ignored.
- The interphase mutulas admittances are assumed equal, i.e. $y'_m = y''_m = y'''_m$. These are all zero with uncoupled single-phase units.
- The differences $(y_p - y_m)$ and $(y_s - y_m)$ are very small and are therefore ignored.

With these simplifications, Table 3.2, illustrates the sequence impedance models of three-phase transformes in conventional steady-state balanced transmission system studies.

3.5 FORMULATION OF THE THREE-PHASE LOAD-FLOW PROBLEM

3.5.1 Notation

A clear and unambiguous identification of the three-phase vector and matrix elements requires a suitable symbolic notation using superscripts and subscripts.

The a.c. system is considered to have a total of n busbars where

- $n = nb + ng$
- nb is the number of actual system busbars
- ng is the number of synchronous machines.

Subscripts i, j, etc refer to system busbars as shown in the following examples.

- $i = 1, nb$ identifies all actual system busbars, i.e. all load busbars plus all generator terminal busbars.
- $i = nb + 1, nb + ng - 1$ identifies all generator internal busbars with the exception of the slack machine.
- $i = nb + ng$ identifies the internal busbar at the slack machine.

The following subscripts are also used for clarity.

- reg—refers to a voltage regulator
- int—refers to an internal busbar at a generator
- gen—refers to a generator.

Superscripts p, m identify the three phases at a particular busbar.

3.5.2 Specified Variables

The following variables form a minimum and sufficient set to define the three-phase system under steady-state operation.

- The slack generator internal busbar voltage magnitude $V_{\text{int}\,j}$ where $j = nb + ng$. (The angle $\theta_{\text{int}\,j}$ is taken as a reference.)

- The internal busbar voltage magnitude $V_{\text{int}\,j}$ and angles $\theta_{\text{int}\,j}$ at all other generators, i.e. $j = nb + 1, nb + ng - 1$.
- The three voltage magnitudes (V_i^p) and angles (θ_i^p) at every generator terminal busbar and every load busbar in the system, i.e. $i = 1, nb$ and $p = 1, 3$.

Only two variables are associated with each generator internal busbar as the three-phase voltages are balanced and there is no need for retaining the redundant voltages and angles as variables. However, these variables are retained to facilitate the calculation of the real and reactive power mismatches. The equations necessary to solve for the above set of variables are derived from the specified operating conditions, i.e.

- The individual phase real and reactive power loading at every system busbar.
- The voltage regulator specification for every synchronous machine.
- The total real power generation of each synchronous machine, with the exception of the slack machine.

The usual load-flow specification of a slack machine, i.e. fixed voltage in phase and magnitude, is applicable to the three-phase load flow.

3.5.3 Derivation of Equations

The three-phase system behaviour is described by the equation

$$[I] - [Y][V] = 0 \tag{3.5.1}$$

where the system admittance matrix $[Y]$ represents each phase independently and models all inductive and capacitive mutual couplings between phases and between circuits. The mathematical statement of the specified conditions is derived in terms of the system admittance matrix

$$[Y] = [G] + j[B]$$

as follows.

(i) For each of the three phases (p) at every load and generator terminal busbar (i),

$$\Delta P_i^p = (P_i^p)^{sp} - P_i^p$$

$$= (P_i^p)^{sp} - V_i^p \sum_{k=1}^{n} \sum_{m=1}^{3} V_k^m [G_{ik}^{pm} \cos \theta_{ik}^{pm} + B_{ik}^{pm} \sin \theta_{ik}^{pm}] \tag{3.5.2}$$

and

$$\Delta Q_i^p = (Q_i^p)^{sp} - Q_i^p$$

$$= (Q_i^p)^{sp} - V_i^p \sum_{k=1}^{n} \sum_{m=1}^{3} V_k^m [G_{ik}^{pm} \sin \theta_{ik}^{pm} - B_{ik}^{pm} \cos \theta_{ik}^{pm}]. \tag{3.5.3}$$

(ii) For every generator j,

$$(\Delta V_{\text{reg}})_j = f(V_k^1, V_k^2, V_k^3) \tag{3.5.4}$$

where k is the bus number of the jth generator's terminal busbar.

(iii) For every generator j, with the exception of the slack machine, i.e. $j \neq nb + ng$,

$$(\Delta P_{\text{gen}})_j = (P^{sp}_{\text{gen}})_j - (P_{\text{gen}})_j$$

$$= (P^{sp}_{\text{gen}})_j - \sum_{p=1}^{3} V_{\text{int}\,j} \sum_{k=1}^{n} \sum_{m=1}^{3} V^m_k [G^{pm}_{jk} \cos \theta^{pm}_{jk} + B^{pm}_{jk} \sin \theta^{pm}_{jk}] \quad (3.5.5)$$

where, although the summation for k is over all system busbars, the mutual terms G_{jk} and B_{jk} are nonzero only when k is the terminal busbar of the jth generator.

It should be noted that the real power specified for the generator is the total real power at the internal or excitation busbar whereas in actual practice the specified quantity is the power leaving the terminal busbar. This in effect means that the generator's real power loss is ignored.

The generator losses have no significant influence on the system operation and may be calculated from the sequence impedances at the end of the load-flow solution, when all generator sequence currents have been found. Any other method would require the real power mismatch to be written at busbars remote from the variable in question, that is, the angle at the internal busbar. In addition, inspection of equations (3.5.2) and (3.5.5) will show that the equations are identical except for the summation over the three phases at the generator internal busbar.

That is, the sum of the powers leaving the generator may be calculated in exactly the same way and by the same sabroutines as the power mismatches at other system busbars. This is possible because the generator internal busbar is not connected to any other element in the system. Inspection of the Jacobian submatrices derived later will show that this feature is retained throughout the study. In terms of programming the generators present no additional complexity.

Equations (3.5.2) to (3.5.5) form the mathematical formulation of the three-phase load flow as a set of independent algebraic equations in terms of the system variables.

The solution to the load-flow problem is the set of variables which, upon substitution, make the left-hand-side mismatches in equations (3.5.2) to (3.5.5) equal to zero.

3.6 FAST-DECOUPLED THREE-PHASE ALGORITHM

The standard Newton–Raphson algorithm may be used to solve equations (3.5.2) to (3.5.5). This involves an iterative solution of the matrix equation

$$\begin{bmatrix} \Delta P \\ \Delta P_{\text{gen}} \\ \Delta Q \\ \Delta V_{\text{reg}} \end{bmatrix} = \begin{bmatrix} A & E & I & M \\ B & F & J & N \\ C & G & K & P \\ D & H & L & R \end{bmatrix} \cdot \begin{bmatrix} \Delta \theta \\ \Delta \theta_{\text{int}} \\ \Delta V/V \\ \Delta V_{\text{int}}/V_{\text{int}} \end{bmatrix} \quad (3.6.1)$$

for the right-hand-side vector of variable updates. The right-hand-side matrix in equation (3.6.1) is the Jacobian matrix of first-order partial derivatives.

Following decoupled single-phase load-flow practice, the effects of $\Delta \theta$ on reactive power flows and ΔV on real power flows are ignored. Equation (3.6.1) may therefore be simplified by assigning

$$[I] = [M] = [J] = [N] = 0$$

and

$$[C] = [G] = 0.$$

In addition, the voltage regulator specification is assumed to be in terms of the terminal voltage magnitudes only and therefore

$$[D] = [H] = 0.$$

Equation (3.6.1) may then be written in decoupled form as

$$\begin{bmatrix} \Delta P_i^p \\ \Delta P_{\text{gen}\,j} \end{bmatrix} = \begin{bmatrix} A & E \\ B & F \end{bmatrix} \begin{bmatrix} \Delta \theta_k^m \\ \Delta \theta_{\text{int}\,l} \end{bmatrix} \tag{3.6.2}$$

for $i, k = 1, nb$ and $j, l = 1, ng - 1$ (i.e. excluding the slack generator), and as

$$\begin{bmatrix} \Delta Q_i^p \\ \Delta V_{\text{reg}\,j} \end{bmatrix} = \begin{bmatrix} K & P \\ L & R \end{bmatrix} \begin{bmatrix} \Delta V_k^m / V_k^m \\ \Delta V_{\text{int}\,l} / V_{\text{int}\,l} \end{bmatrix} \tag{3.6.3}$$

for $i, k = 1, nb$ and $j, l = 1, ng$ (i.e. including the slack generator).

To enable further development of the algorithm it is necessary to consider the Jacobian submatrices in more detail. In deriving these Jacobians from equations (3.5.2) to (3.5.5) it must be remembered that

$$V_l^1 = V_l^2 = V_l^3 = V_{\text{int}\,l}$$

$$\theta_l^1 = \theta_l^2 - \frac{2\pi}{3} = \theta_l^3 + \frac{2\pi}{3} = \theta_{\text{int}\,l}$$

when l refers to a generator internal busbar.

The coefficients of matrix equation (3.6.2) are

$$[A_{ik}^{pm}] = (\partial \Delta P_i^p / \partial \theta_k^m)$$

or

$$A_{ik}^{pm} = V_i^p V_k^m [G_{ik}^{pm} \sin \theta_{ik}^{pm} - B_{ik}^{pm} \cos \theta_{ik}^{pm}]$$

except for

$$A_{kk}^{mm} = -B_{kk}^{mm}(V_k^m)^2 - Q_k^m$$

$$[B_{jk}^m] = [\partial \Delta P_{\text{gen}\,j} / \partial \theta_k^m]$$

$$= \sum_{p=1}^{3} V_{\text{int}\,j} V_k^m [G_{jk}^{pm} \sin \theta_{jk}^{pm} - B_{jk}^{pm} \cos \theta_{jk}^{pm}]$$

$$[E_{il}^p] = [\partial P_i^p / \partial \theta_{\text{int}\,l}]$$

$$= \sum_{m=1}^{3} V_{\text{int}\,l} V_i^p [G_{il}^{pm} \sin \theta_{il}^{pm} - B_{il}^{pm} \cos \theta_{il}^{pm}]$$

$$[F_{jl}] = [\partial P_{\text{gen}\,j} / \partial \theta_{\text{int}\,l}]$$

where $[F_{jl}] = 0$ for all $j \neq l$ because the jth generator has no connection with the lth generator's internal busbar, and

$$[F_{ll}] = \sum_{p=1}^{3} (-B_{ll}^{pp}(V_{int\,l})^2 - Q_l^p)$$

$$+ \sum_{\substack{m=1 \\ m \neq p}}^{3} \sum_{p=1}^{3} (V_{int\,l})^2 [G_{ll}^{pm} \sin \theta_{ll}^{pm} - B_{ll}^{pm} \cos \theta_{ll}^{pm}].$$

The coefficients of matrix equation (3.6.3) are

$$-[K_{ik}^{pm}] = V_k^m [\partial \Delta Q_i^p / \partial V_k^m]$$

where

$$K_{ik}^{pm} = V_k^m V_i^p [G_{ik}^{pm} \sin \theta_{ik}^{pm} - B_{ik}^{pm} \cos \theta_{ik}^{pm}]$$

except

$$K_{kk}^{mm} = -B_{kk}^{mm}(V_k^m)^2 + Q_k^m$$
$$-[L_{jk}^{m}] = V_k^m [\partial \Delta V_{reg\,j} / \partial V_k^m].$$

Let $[L_{jk}^{m}] = V_k^m [L_{jk}^{m}]'$ where k is the terminal busbar of the jth generator and $L_{jk}^{m} = 0$ otherwise.

$$-[P_{ll}^{p}] = V_{int\,l} [\partial \Delta Q_i^p / V_{int\,l}]$$

$$= V_{int\,l} \sum_{m=1}^{3} V_i^p [G_{il}^{pm} \sin \theta_{il}^{pm} - B_{il}^{pm} \cos \theta_{il}^{pm}]$$

$$-[R_{jl}] = [\partial \Delta V_{reg\,j} / \partial V_{int\,l}]$$
$$= 0 \qquad \text{for all } j, l \text{ as the voltage regulator}$$
specification does not explicitly
include the variables V_{int}.

Although the above expressions appear complex, their meaning and derivation are similar to those of the usual single-phase Jacobian elements.

3.6.1 Jacobian Approximations

Approximations similar to those applied to the single-phase load flow are applicable to the Jacobian elements as follows.

(i) At all nodes (i.e. all phases of all busbars)

$$Q_k^m \ll B_{kk}^{mm}(V_k^m)^2.$$

(ii) Between connected nodes of the same phase

$$\cos \theta_{ik}^{mm} \approx 1 \quad \text{i.e.} \quad \theta_{ik}^{mm} \text{ is small}$$

and

$$G_{ik}^{mm} \sin \theta_{ik}^{mm} \ll B_{ik}^{mm}.$$

(iii) Moreover the phase-angle unbalance at any busbar will be small and hence an additional approximation applies to the three-phase system, i.e.

$$\theta_{kk}^{pm} \approx \pm 120° \quad \text{for} \quad p \neq m.$$

(iv) Finally, as a result of (ii) and (iii) the angle between different phases of connected busbars will be approximately 120°, i.e.

$$\theta_{ik}^{pm} \approx \pm 120° \quad \text{for} \quad p \neq m$$

or

$$\cos \theta_{ik}^{pm} \approx -0.5$$

and

$$\sin \theta_{ik}^{pm} \approx \pm 0.866.$$

These values are modified for the $\pm 30°$ phase shift inherent in the star–delta connection of three-phase transformers.

The final approximation (iv), necessary if the Jacobians are to be kept constant, is the least valid, as the cosine and sine values change rapidly with small angle variations around 120°. This accounts for the slower convergence of the phase unbalance at busbars as compared with that of the voltage magnitudes and angles.

It should be emphasised that these approximations apply to the Jacobian elements only, i.e. they do not prejudice the accuracy of the solution nor do they restrict the type of problem which may be attempted.

Applying approximations (i) to (iv) to the Jacobians and substituting into equations (3.6.2) and (3.6.3) yields

$$
\begin{bmatrix} \Delta P_i^p \\ \Delta P_{gen\,j} \end{bmatrix} =
\begin{bmatrix} [V_i^p M_{ik}^{pm} V_k^m] & \left[\displaystyle\sum_{m=1}^{3} V_i^p M_{il}^{pm} V_{int\,l} \right] \\ \left[\displaystyle\sum_{p=1}^{3} V_{int\,j} M_{jk}^{pm} V_k^m \right] & \left[\displaystyle\sum_{m=1}^{3} \sum_{p=1}^{3} V_{int\,j} M_{jl}^{pm} V_{int\,l} \right] \end{bmatrix} \cdot
\begin{bmatrix} \Delta\theta_k^m \\ \Delta\theta_{int\,l} \end{bmatrix}
\tag{3.6.4}
$$

and

$$
\begin{bmatrix} \Delta Q_i^p \\ \Delta V_{reg\,j} \end{bmatrix} =
\begin{bmatrix} [V_i^p M_{ik}^{pm} V_k^m] & \left[\displaystyle\sum_{m=1}^{3} V_i^p M_{il}^{pm} V_{int\,l} \right] \\ V_k^m [L'] & [0] \end{bmatrix} \cdot
\begin{bmatrix} \Delta V_k^m / V_k^m \\ \Delta V_{int\,l} / V_{int\,l} \end{bmatrix}
\tag{3.6.5}
$$

where

$$M_{ik}^{pm} = G_{ik}^{pm} \sin \theta_{ik}^{pm} - B_{ik}^{pm} \cos \theta_{ik}^{pm}$$

with

$$\theta_{kk}^{mm} = 0$$
$$\theta_{ik}^{mm} = 0$$
$$\theta_{ik}^{pm} = \pm 120° \quad \text{for} \quad p \neq m.$$

All terms in the matrix $[M]$ are constant, being derived solely from the system admittance matrix. Matrix $[M]$ is the same as matrix $[-B]$ except for the off-diagonal terms which connect nodes of different phases. These are modified by allowing for the nominal 120° angle and also including the $G_{ik}^{pm} \sin \theta_{ik}^{pm}$ terms.

The similarity in structure of all Jacobian submatrices reduces the programming complexity normally found in three-phase load flows. This uniformity has been achieved primarily by the method used to implement the three-phase generator constraints.

The above derivation closely parallels the single-phase fast-decoupled algorithm, but the added complexity of the notation obscures this feature. At the present stage the Jacobian elements in equations (3.6.4) and (3.6.5) are identical except for those terms which involve the additional features of the generator modelling.

These functions are more linear in terms of the voltage magnitude $[\bar{V}]$ than are the functions $[\Delta\bar{P}]$ and $[\Delta\bar{Q}]$. In the Newton–Raphson and related constant Jacobian methods the reliability and speed of convergence improve with the linearity of the defining functions. With this aim, equations (3.6.4) and (3.6.5) are modified as follows.

- The left-hand side defining functions are redefined as $[\Delta P_i^p/V_i^p]$, $[\Delta P_{\text{gen}\,j}/V_{\text{int}\,j}]$ and $[\Delta Q_i^p/V_i^p]$.

- In equation (3.6.4), the remaining right-hand-side V terms are set to 1 p.u.

- In equation (3.6.5), the remaining right-hand-side V terms are cancelled by the corresponding terms in the right-hand-side vector.

These modifications yield the following expressions.

$$
\begin{bmatrix} \Delta P_i^p/V_i^p \\ \Delta P_{\text{gen}\,j}/V_{\text{int}\,j} \end{bmatrix} = \underbrace{\begin{bmatrix} M_{ik}^{pm} & \sum_{m=1}^{3} M_{il}^{pm} \\ \sum_{p=1}^{3} M_{jk}^{pm} & \sum_{p=1}^{3}\sum_{m=1}^{3} M_{jl}^{pm} \end{bmatrix}}_{[B']} \cdot \begin{bmatrix} \Delta\theta_k^p \\ \Delta\theta_{\text{int}\,l} \end{bmatrix}
\tag{3.6.6}
$$

$$
\begin{bmatrix} \Delta Q_i^p/V_i^p \\ \Delta V_{\text{reg}\,j} \end{bmatrix} = \underbrace{\begin{bmatrix} M_{ik}^{pm} & \sum_{m=1}^{3} M_{il}^{pm} \\ [L_{jk}^{m}]' & 0 \end{bmatrix}}_{[B'']} \cdot \begin{bmatrix} \Delta V_k^m \\ \Delta V_{\text{int}\,l} \end{bmatrix}.
\tag{3.6.7}
$$

Recalling that $[L_{jk}^{m}]' = [\partial\Delta V_{\text{reg}\,j}/\partial V_k^m]$, as V_{reg} is normally a simple linear function of the terminal voltages, $[L']$ will be a constant matrix.

Therefore, the Jacobian matrices $[B']$ and $[B'']$ in equations (3.6.6) and (3.6.7) have been approximated to constants.

Zero diagonal elements in equation (3.6.7) may result from the ordering of the equations and variables. This feature causes no problems if these diagonals are not used as pivots until the rest of the matrix has been factorised (by which time, fill-in terms will have appeared on the diagonal). This causes a minor loss of efficiency as it inhibits optimal ordering for the complete matrix. Although this could be avoided by reordering the equations, the extra program complexity is not justified.

Based on the reasoning of Stott and Alsac [9], which proved successful in the single-phase load flow the $[B']$ matrix in equation (3.6.6) is further modified by omitting the representation of those elements that predominantly affect MVAR flows.

The capacitance matrix and its physical significance is illustrated in Fig. 3.19, for

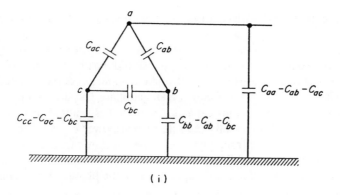

(i)

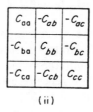

(ii)

Figure 3.19
Shunt capacitance matrices

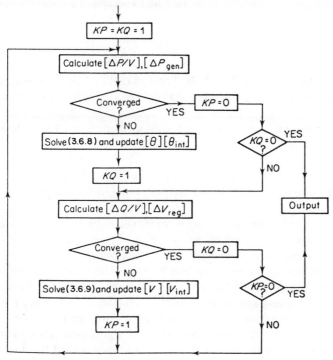

Figure 3.20
Iteration sequence for three-phase a.c. load flow

a single three-phase line. With n capacitively coupled parallel lines the matrix will be $3n \times 3n$.

In single-phase load flows the shunt capacitance is the positive sequence capacitance which is determined from both the phase-to-phase and the phase-to-earth capacitances of the line. It therefore appears that the entire shunt capacitance matrix predominantly affects MVAR flows only. Thus, following single-phase fast-decoupling practice the representation of the entire shunt capacitance matrix is omitted in the formulation of $[B']$. This increases dramatically the rate or real power convergence.

With capacitively coupled three-phase lines the interline capacitance influences the positive sequence shunt capacitance. However, as the values of interline capacitances are small in comparison with the self-capacitance of the phases, their inclusion makes no noticeable difference. The effective tap of $\sqrt{3}$ introduced by the star–delta transformer connection is interpreted as a nominal tap and is therefore included when forming the $[B']$ matrix.

A further difficulty arises from the modelling of the star-g/delta transformer connection. The equivalent circuit, illustrated in Section 3.4 shows that large shunt admittances are effectively introduced into the system. When these are excluded from $[B']$, as for a normal shunt element, divergence results. The entire transformer model must therefore be included in both $[B']$ and $[B'']$.

With the modifications described above the two final algorithmic equations may be concisely written, i.e.

$$\begin{bmatrix} \Delta P/V \\ \Delta P_{gen}/V_{int} \end{bmatrix} = [B'_m] \cdot \begin{bmatrix} \Delta\theta \\ \Delta\theta_{int} \end{bmatrix} \tag{3.6.8}$$

$$\begin{bmatrix} \Delta Q/V \\ \Delta V_{reg} \end{bmatrix} = [B''_m] \cdot \begin{bmatrix} \Delta V \\ \Delta V_{int} \end{bmatrix}. \tag{3.6.9}$$

The constant Jacobians $[B'_m]$ and $[B''_m]$ correspond to fixed approximated tangent slopes to the multidimensional surfaces defined by the left-hand-side defining functions.

Equations(3.6.8) and (3.6.9) are then solved according to the iteration sequence illustrated in Fig. 3.20.

3.6.2 Generator Models and the Fast-decoupled Algorithm

The derivation of the fast-decoupled algorithm involves the use of several assumptions to enable the Jacobian matrices to be approximated to constant. The same assumptions have been applied to the excitation busbars associated with the generator model as are applied to the usual system busbars. The validity of the assumptions regarding voltage magnitudes and the angles between connected busbars depends upon the machine loading and positive sequence reactance. As discussed in Section 3.5 this reactance may be set to any value without altering the load-flow solution and a value may therefore be selected to give the best algorithmic performance.

When the actual value of positive sequence reactance is used the angle across the generator and the magnitude of the excitation voltage both become comparatively large under full load operation. Angles in excess of $45°$ and excitation voltages greater than 2.0 p.u. are not uncommon. Despite this considerable divergence from assumed conditions, convergence is surprisingly good. Convergence difficulties may occur at the slack generator and then only when it is modelled with a high synchronous reactance (1.5 p.u. on machine rating) and with greater than 70% full load power.

All other system generators, where the real power is specified, converge reliably but somewhat slowly under similar conditions.

The deterioration in convergence rate and the limitation on the slack generator loading may be avoided by setting the generator positive sequence reactance to an artificially low value (say 0.01 p.u. on machine rating), a procedure which does not involve any loss of relevant system information.

3.7 STRUCTURE OF THE COMPUTER PROGRAM

The main components of the computer program are illustrated in Fig. 3.21. The approximate number of FORTRAN statements for each block is indicated in parenthesis. The main features of each block are described in the following sections.

3.7.1 Data Input

The input data routine implements the system modelling techniques described in Sections 3.2 to 3.4 and Appendix I to form the system admittance model from the

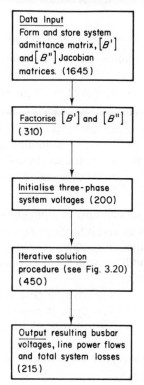

Figure 3.21
Program structure

raw data for each system component. Examples of the raw data are given in Section 3.9, with reference to a particular test system.

The structure and content of the constant Jacobians B' and B'' are based upon the system admittance matrix and are thus formed simultaneously with this matrix.

Both the system admittance matrix and the Jacobian matrices are stored and processed using sparsity techniques which are structured in 3×3 matrix blocks to take full advantage of the inherent block structure of the three-phase system matrices.

3.7.2 Factorisation of Constant Jacobians

The heart of the load-flow program is the repeat solutions of equations (3.6.8) and (3.6.9) as illustrated in Fig. 3.20. These equations are solved using sparsity techniques and near optimal ordering as discussed in Chapter 2 (Section 2.7) or like those embodied in Zollenkopf's bifactorisation [10]. The constant Jacobians are factorised before the iteration sequence is initiated. The solution of each equation within the iterative procedure is relatively fast, consisting only of the forward and back substitution processes.

3.7.3 Starting Values

Starting values are assigned as follows.

- The nonvoltage-controlled busbars are assigned 1 p.u. on all phases.
- At generator terminal busbars all voltages are assigned values according to the voltage regulator specifications.
- All system busbar angles are assigned 0, $-120°$, $+120°$ for the three phases respectively.
- The generator internal voltages and angles are calculated from the specified real power and, in the absence of better estimates, by assuming zero reactive power. For the slack machine the real power is estimated as the difference between total load and total generation plus a small percentage (say 8%) of the total load to allow for losses.

For cases where convergence is excessively slow or difficult it is advisable to use the results of a single-phase load flow to establish the starting values. The values will, under normal steady-state unbalance, provide excellent estimates for all voltages and angles including generator internal conditions which are calculated from the single-phase real and reactive power conditions.

Moreover, as a three-phase iteration is more costly than a single-phase iteration, this practice can be generally recommended to provide more efficient overall convergence and to enable the more obvious data errors to be detected at an early stage.

For the purpose of investigating the load-flow performance, flat voltage and angle values are used in the examples that follow.

3.7.4 Iterative Solution

The iterative solution process (Fig.3.20) yields the values of the system voltages which satisfy the specified system conditions of load, generation and system configuration.

3.7.5 Output Results

The three-phase busbar voltages, the line power flows and the total system losses are calculated and printed out. An example is given in Table 3.8 of Section 3.9. In addition the sequence components of busbar voltages are also calculated as these provide a more direct measure of the unbalance present in the system under study.

3.8 PERFORMANCE OF THE ALGORITHM

This section attempts to identify those features which influence the convergence with particular reference to several small- to medium-sized test systems.

The performance of the 'three-phase' algorithm is examined under both balanced and unbalanced conditions, and comparisons are made with the performance of the single-phase fast-decoupled algorithm.

3.8.1 Performance under Balanced Conditions

A symmetrical three-phase system, operating with balanced loading, is accurately modelled by the positive sequence system and either a three-phase or a single-phase load flow may be used to analyse the system. Under these conditions it is possible to compare the performance of the three-phase and single-phase fast-decoupled algorithms.

The three-phase system transmission lines are represented by balanced full 3×3 matrices. Transformers are modelled with balanced parameters on all phases and

Table 3.3
Convergence results

Case	Number of busbars	Single-phase load flow	Balanced three-phase load flow $\lambda\lambda$	$\lambda\Delta$	Typical three-phase unbalance
1	5	4.3	4.3	4.3	6.6
2	6	3.3	3.3	3.3	8.8
3	14	3.3	3.3	3.3	6.5
4	17	3.3	3.3	3.3	8.7
5	30	3.3	3.3	3.3	6.6

Convergence tolerance is 0.1 MW/MVAR. The numerical results, (i, j), should be interpreted as follows:
i—refers to the number of real power-angle update iterations.
j—refers to the number of reactive power-voltage update iterations.

generators are modelled by their phase parameter matrices as derived from their sequence impedances.

Typical numbers of iterations to convergence for both the single-phase and three-phase algorithms, given in Table 3.3, indicate that the algorithms behave identically. Features such as the transformer connection and the negative and zero sequence generator impedances have no effect on the convergence rate of the three-phase system under balanced conditions. This is not unexpected as, under balanced conditions, only the positive sequence network has any power flow and there is no coupling between sequence networks. The negative- and zero-sequence information inherent in the three-phase system model of the balanced systems has no influence on system operation and this is reflected into the performance of the algorithm.

3.8.2 Performance with Unbalanced Systems

The number of iterations to convergence for the same test systems, under realistic steady-state unbalanced operation, are also given in Table 3.3. The convergence rate deteriorates compared with the balanced case, requiring on average twice as many iterations.

The graphs of Fig. 3.22 show that initial convergence of the three-phase mismatches is very close to that of the single-phase load flow. However, as the solution is approached the three-phase convergence becomes slower. It appears that although the voltage and angle unbalance are introduced from the first iteration, they have only a secondary effect on the convergence until the positive sequence power flows are approaching convergence.

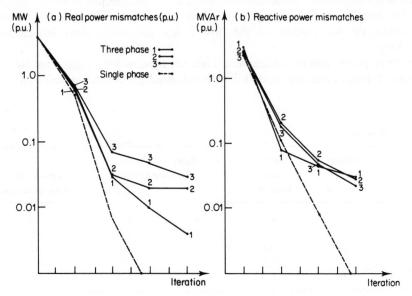

Figure 3.22
Power convergence patterns for three-phase and single-phase load flow

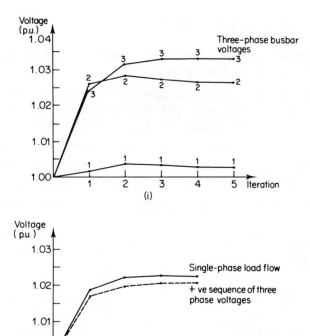

Figure 3.23
Voltage convergence patterns for three-phase and single-phase load flows: (i) three-phase voltages;
(ii) single-phase and three-phase positive sequence voltages

   This feature is further illustrated in Fig. 3.23(i) where the convergence pattern of the
three-phase voltages is shown. The convergence pattern of the positive sequence
component of the unbalanced voltages is shown in Fig. 3.23(ii) together with the
convergence pattern of the voltage at the same busbar for the corresponding
single-phase load flow. The latter figure illustrates that the positive sequence voltage
of the three-phase unbalanced load flow has an almost identical convergence pattern
to the corresponding single-phase fast-decoupled load flow. The final convergence of
the system unbalance is somewhat slow but is reliable.
   The following features are peculiar to a three-phase load flow and their influence
on convergence is of interest:

- asymmetry of the system parameters
- unbalance of the system loading
- influence of the transformer connection
- mutual coupling between parallel transmission lines.

These features have been examined with reference to a small six-bus test system.

3.9   TEST SYSTEM AND RESULTS

A single-line diagram of the test system under consideration is illustrated in Fig. 3.24. Some features of interest are listed below.

- An example of a line sectionalisation is included. One section contains four mutually coupled three-phase power lines. The other section contains two sets of two mutually coupled three-phase lines.
- All parallel lines are represented in their unbalanced mutually coupled state.
- Both transformers are star–delta connected with the star neutrals solidly earthed. Tap ratios are present on both primary and secondary sides.

The system is redrawn in Fig. 3.25 using 3×3 compound coil notation and

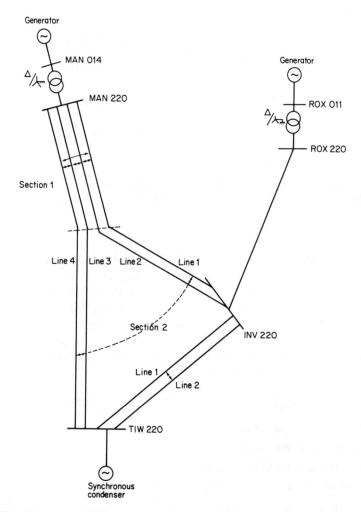

Figure 3.24
Test system single-line diagram

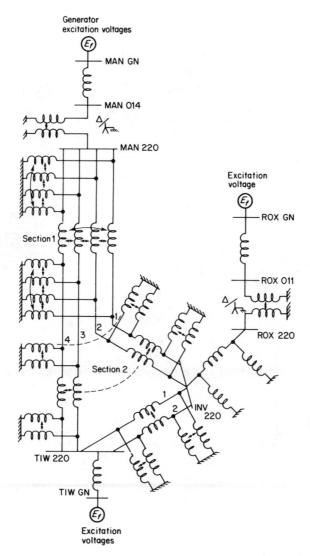

Figure 3.25
Test system 3 × 3 compound coil representation

substituting for the generator and line models. Following this, Fig. 3.26 illustrates the system graphically in terms of 3 × 3, 6 × 6 and 12 × 12 matrix blocks, representing the various system elements. The matrix quantities illustrated in Fig. 3.26 are given by, or derived from, the input data to the load-flow program.

For the purpose of input data organisation and the formation of the system admittance matrix, the system is divided into eight natural subsystems. These are illustrated in the exploded system diagram for Fig. 3.27.

Once the matrices defined in Figs. 3.26 and 3.27 are known, the admittance matrix for each subsystem can be formed following the procedures outlined in Appendix I. The subsystems are then combined to form the overall system admittance matrix.

80

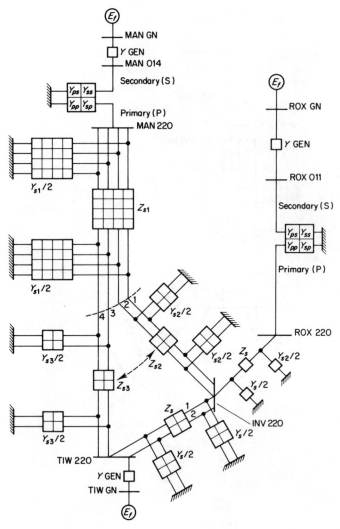

Figure 3.26
Test system 3×3 matrix representation

The input data, which enables all the matrices in Fig. 3.27 to be formed, is listed for each subsystem in the following sections. The data is all in p.u. to a base of 33.3 MVA.

3.9.1 Input Data

3.9.1.1 Generator Data—Subsystems 1 and 2

Subsystems 1 and 2 represent two synchronous generators. The input data to the computer program consists of the three-sequence impedances, the voltage regulator

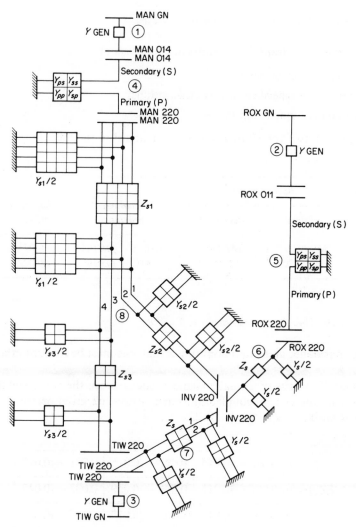

Figure 3.27
Test system exploded into eight systems

specification and the total real power generation at all generators except one which is the slack machine.

Table of generator data

Generator No. Name	Zero R0	Impedance X0	Pos. R1	Impedance X1	Neg. R2	Impedance X2	P p.u.	Voltage regulator V_{phase_2}
1 MAN014	0.0	0.080	0.0	0.010	0.0	0.021	15.000	1.045
2 ROX011	0.0	0.150	0.0	0.010	0.0	0.091	SLACK	1.050

The effect of subsystem 3 (the synchronous condenser) is not included in the numerical example.

3.9.1.2 Transformer Data—Subsystems 4 and 5

The input data for a transformer subsystem consists of:

- leakage impedance in p.u. $(r + jx)$
- transformer type (specified in case descriptions)
- primary and secondary tap ratios.

The data for the two transformations is summarised in the following table:

Busbar names		Leakage reactance	Tap ratio primary
primary	secondary		
MAN220	MAN014	$0.0006 + j0.0164$	0.045
ROX220	ROX011	$0.0020 + j0.038$	0.022

3.9.1.3 Line Data—Subsystems 6, 7 and 8

The series impedance and shunt admittance matrices must be read into the computer program.

Subsystem 6 consists of a single-balanced line between the two terminal busbars. The phases are taken as uncoupled and the matrices are given below.

Terminal busbars INV 220–ROX 220

$$[Z_S] = \begin{array}{c} a \\ b \\ c \end{array} \begin{vmatrix} 0.006 + j0.045 & 0.002 + j0.015 & 0.001 + j0.017 \\ 0.002 + j0.015 & 0.006 + j0.050 & 0.002 + j0.017 \\ 0.001 + j0.017 & 0.002 + j0.017 & 0.007 + j0.047 \end{vmatrix}$$

$$[Y_S] = \begin{array}{c} a \\ b \\ c \end{array} \begin{vmatrix} 0.0 + j0.35 & 0.0 - j0.6 & 0.0 - j0.04 \\ 0.0 - j0.06 & 0.0 + j0.352 & 0.0 - j0.06 \\ 0.0 - j0.04 & 0.0 - j0.06 & 0.0 + j0.34 \end{vmatrix}$$

Both these matrices are in p.u. for the total length.

Subsystem 7 consists of a pair of parallel, mutually coupled three-phase lines. These lines are represented in their natural coupled unbalanced state.

Terminal busbars:

- line 1 INV220–TIW220
- line 2 INV220–TIW220.

The series impedance matrix for the length (Z_S) is:

		Line 1			Line 2		
		a	b	c	a	b	c
Line 1	a	0.0023 +j0.0147					
	b	0.0012 +j0.008	0.0021 +j0.015				
	c	0.0011 +j0.007	0.001 +j0.008	0.0024 +j0.0148			
Line 2	a	0.0009 +j0.0062	0.0008 +j0.0061	0.0008 +j0.0058	0.0023 +j0.0147		
	b	0.0008 +j0.0061	0.0007 +j0.0059	0.0007 +j0.0056	0.0014 +j0.009	0.0026 +j0.015	
	c	0.0008 +j0.0058	0.0007 +j0.0056	0.0006 +j0.0054	0.0012 +j0.009	0.001 +j0.009	0.0021 +j0.013

The shunt admittance matrix for the total length is:

		Line 1			Line 2		
		a	b	c	a	b	c
Line 1	a	+j0.045					
	b	−j0.008	+j0.040				
	c	−j0.009	−j0.011	+j0.035			
Line 2	a	−j0.007	−j0.003	−j0.003	+j0.044		
	b	−j0.003	−j0.005	−j0.002	−j0.01	+j0.040	
	c	−j0.002	−j0.002	−j0.004	−j0.01	−j0.011	+j0.036

The lower diagonal half only is shown as all line matrices are symmetrical.

Subsystem 8 consists of sectionalised mutually coupled lines. Section 1 consists of four mutually coupled three-phase lines and has 12 × 12 characteristic matrices, $[Z_{S1}]$ and $[Y_{S1}]$, as indicated in the system diagrams. These are given in Figs. 3.28 and 3.29 in per unit length of line and section 1 is taken as having a length of 0.75 units.

Section 2 consists of two sets of two mutually coupled three-phase lines. To ensure consistent dimensionality with section 1, the second section is considered as being composed of four mutually coupled three-phase lines, the elements representing the coupling between the two separate double-circuit lines being set to zero. The

		Line 1 a	Line 1 b	Line 1 c	Line 2 a	Line 2 b	Line 2 c	Line 3 a	Line 3 b	Line 3 c	Line 4 a	Line 4 b	Line 4 c
Line 1	a	0.0156 + j0.1088											
	b	0.008 + j0.032	0.015 + j0.1080										
	c	0.007 + j0.022	0.008 + j0.032	0.0160 + j0.1095									
Line 2	a	0.003 + j0.025	0.004 + j0.025	0.002 + j0.025	0.0156 + j0.1088								
	b	0.0025 + j0.022	0.0042 + j0.028	0.004 + j0.028	0.008 + j0.032	0.0150 + j0.1080							
	c	0.002 + j0.025	0.004 + j0.028	0.0042 + j0.028	0.007 + j0.032	0.008 + j0.032	0.0160 + j0.1095						
Line 3	a	0.0015 + j0.012	0.0012 + j0.01	0.001 + j0.011	0.001 + j0.01	0.0008 + j0.007	0.0009 + j0.007	0.0133 + j0.0904					
	b	0.0012 + j0.01	0.0015 + j0.012	0.0012 + j0.01	0.0008 + j0.007	0.001 + j0.01	0.0008 + j0.007	0.006 + j0.04	0.140 + j0.08				
	c	0.001 + j0.011	0.0012 + j0.015	0.0015 + j0.012	0.0009 + j0.007	0.0008 + j0.007	0.001 + j0.01	0.005 + j0.02	0.006 + j0.04	0.0130 + j0.085			
Line 4	a	0.0009 + j0.009	0.0008 + j0.01	0.0008 + j0.009	0.0008 + j0.006	0.0006 + j0.004	0.0005 + j0.003	0.003 + j0.025	0.002 + j0.02	0.002 + j0.01	0.0133 + j0.0904		
	b	0.0008 + j0.01	0.0006 + j0.008	0.0006 + j0.008	0.0006 + j0.004	0.0008 + j0.006	0.0006 + j0.004	0.002 + j0.02	0.003 + j0.025	0.002 + j0.02	0.006 + j0.04	0.0140 + j0.08	
	c	0.0008 + j0.009	0.0006 + j0.008	0.0006 + j0.008	0.0005 + j0.003	0.0006 + j0.004	0.0008 + j0.006	0.002 + j0.01	0.002 + j0.02	0.003 + j0.025	0.005 + j0.03	0.006 + j0.04	0.0130 + j0.085

Figure 3.28 Series impedance matrix $[Z_{s1}]$ for section 1

Note: Lower diagonal only given as matrix is symmetrical.

	Line 1			Line 2			Line 3			Line 4		
	a	b	c	a	b	c	a	b	c	a	b	c
a	$+j0.2967$											
Line 1 b	$-j0.06$	$+j0.299$										
c	$-j0.05$	$-j0.06$	$+j0.31$									
a	$-j0.04$	$-j0.03$	$-j0.035$	$+j0.2967$								
Line 2 b	$-j0.045$	$-j0.035$	$-j0.032$	$-j0.06$	$+j0.299$							
c	$-j0.04$	$-j0.032$	$-j0.028$	$-j0.05$	$-j0.06$	$+j0.3$						
a	$-j0.02$	$-j0.22$	$-j0.018$	$-j0.018$	$-j0.012$	$-j00.009$	$+j0.2569$					
Line 3 b	$-j0.022$	$-j0.018$	$-j0.15$	$-j0.012$	$-j0.012$	$-j0.01$	$-j0.50$	$+j0.26$				
c	$-j0.018$	$-j0.015$	$-j0.018$	$-j0.009$	$-j0.01$	$-j0.014$	$-j0.045$	$-j0.042$	$+j0.251$			
a	$-j0.15$	$-j0.009$	$-j0.009$	$-j0.01$	$-j0.009$	$-j0.008$	$-j0.043$	$-j0.04$	$-j0.032$	$+j0.2569$		
Line 4 b	$-j0.009$	$-j0.008$	$-j0.009$	$-j0.009$	$-j0.008$	$-j0.007$	$-j0.032$	$-j0.038$	$-j0.028$	$-j0.050$	$+j0.026$	
c	$-j0.009$	$-j0.008$	$-j0.008$	$-j0.008$	$-j0.007$	$-j0.006$	$-j0.028$	$-j0.032$	$-j0.025$	$-j0.045$	$-j0.041$	$+j0.251$

Figure 3.29
Shunt admittance matrix $[Y_{s1}]$ for section 1

Note: Lower diagonal only shown as matrix is symmetrical.

	Line 1 a	Line 1 b	Line 1 c	Line 2 a	Line 2 b	Line 2 c	Line 3 a	Line 3 b	Line 3 c	Line 4 a	Line 4 b	Line 4 c
Line 1 a	0.0156 $+j0.1088$											
Line 1 b	0.008 $+j0.032$	0.015 $+j0.1080$										
Line 1 c	0.007 $+j0.022$	0.008 $+j0.032$	0.0160 $+j0.1095$									
Line 2 a	0.003 $+j0.025$	0.004 $+j0.025$	0.002 $+j0.025$	0.0156 $+j0.1088$								
Line 2 b	0.0025 $+j0.022$	0.0042 $+j0.028$	0.004 $+j0.028$	0.008 $+j0.032$	0.0150 $+j0.1080$							
Line 2 c	0.002 $+j0.025$	0.004 $+j0.028$	0.0042 $+j0.028$	0.007 $+j0.032$	0.008 $+j0.032$	0.0160 $+j0.1095$						
Line 3 a							0.0133 $+j0.0904$					
Line 3 b							0.006 $+j0.04$	0.0140 $+j0.08$				
Line 3 c							0.005 $+j0.03$	0.006 $+j0.04$	0.0130 $+j0.085$			
Line 4 a							0.003 $+j0.025$	0.002 $+j0.02$	0.002 $+j0.01$	0.0133 $+j0.0904$		
Line 4 b							0.002 $+j0.02$	0.003 $+j0.025$	0.002 $+j0.02$	0.006 $+j0.04$	0.0140 $+j0.08$	
Line 4 c							0.002 $+j0.01$	0.002 $+j0.02$	0.003 $+j0.025$	0.005 $+j0.03$	0.006 $+j0.04$	0.0130 $+j0.085$

Figure 3.30 Series impedance matrix $[Z_S]$ for section 2

Figure 3.31
Shunt capacitance matrix $[Y_S]$ for section 2

Note: Lower diagonal only shown as matrix is symmetrical.

	Line 1 a	Line 1 b	Line 1 c	Line 2 a	Line 2 b	Line 2 c	Line 3 a	Line 3 b	Line 3 c	Line 4 a	Line 4 b	Line 4 c
Line 1 a	$-j0.2967$											
Line 1 b	$-j0.06$	$+j0.299$										
Line 1 c	$-j0.05$	$-j0.06$	$+j0.31$									
Line 2 a	$-j0.04$	$-j0.03$	$-j0.035$	$+j0.2967$								
Line 2 b	$-j0.045$	$-j0.035$	$-j0.032$	$-j0.06$	$+j0.299$							
Line 2 c	$-j0.04$	$-j0.032$	$-j0.028$	$-j0.05$	$-j0.06$	$+j0.3$						
Line 3 a							$+j0.2569$					
Line 3 b							$-j0.50$	$+j0.26$				
Line 3 c							$-j0.045$	$-j0.042$	$+j0.251$			
Line 4 a							$-j0.043$	$-j0.04$	$-j0.032$	$+j0.2569$		
Line 4 b							$-j0.032$	$-j0.038$	$-j0.028$	$-j0.050$	$+j0.026$	
Line 4 c							$-j0.028$	$-j0.032$	$-j0.025$	$-j0.045$	$-j0.041$	$+j0.251$

characteristic matrices for Section 2 become

$$
\begin{array}{c}
[Z_S] \\
12 \times 12 \\
\text{Section 2}
\end{array}
=
\begin{array}{c}
1 \\ 2 \\ 3 \\ 4
\end{array}
\begin{array}{|c|c|}
\hline
[Z_{S2}] & [0] \\
\hline
[0] & [Z_{S3}] \\
\hline
\end{array}
\qquad
\begin{array}{c}
[Y_S] \\
12 \times 12 \\
\text{Section 2}
\end{array}
=
\begin{array}{c}
1 \\ 2 \\ 3 \\ 4
\end{array}
\begin{array}{|c|c|}
\hline
[Y_{S2}] & [0] \\
\hline
[0] & [Y_{S3}] \\
\hline
\end{array}
$$

where [0] is a matrix of zeros. The submatrices are labelled as those in Fig. 3.24.

These 12×12 matrices are given in Fig. 3.30 and Fig. 3.31 in per unit length of line. Section 2 is taken as having a length of 0.25 units.

Once the overall admittance matrix for the combined sections has been found it must be stored in full. This is to enable calculation of the power flows in the four individual lines. The matrix is modified, as described in Section 3.3.2. This modified matrix is the subsystem admittance matrix to be combined into the overall system admittance matrix.

3.9.2 Test Cases and Typical Results

The following cases have been examined.

(i) Balanced system with balanced loading and no mutual coupling between parallel three-phase lines. Generator transformers are star-g/star-g.

(ii) As for case (i) but with balanced mutual coupling introduced for all parallel three-phase lines as indicated in Fig. 3.24.

(iii) As for case (ii) but with balanced loading.

(iv) As for case (ii) but with system unbalance introduced by line capacitance unbalance only.

Table 3.4
Number of iterations to convergence for six-bus test system

Case	Convergence tolerance (MW/MVAR)		
	10.0	1.0	0.1
i	2.1	2.2	3.3
ii	2.1	2.2	3.3
iii	2.1	6.5	10,10
iv	2.1	5.4	8.8
v	2.1	5.4	9.9
vi	2.1	5.4	9.9
vii	2.1	4.3	10.9
viii	2.1	3.3	8.7
ix	4.3	11.9	17.16
x	4.3	10.10	16.16

Table 3.5
Sequence components of busbar voltages

Busbar	+ ve sequence		− ve sequence		Zero sequence	
	V_1	θ^1	V_2	θ_2	V_0	θ_0
INV220	1.020	−0.16	0.028	2.42	0.021	−0.85
ROX220	1.037	−0.13	0.028	2.37	0.025	−1.13
MAN220	1.058	−0.09	0.015	1.84	0.014	−0.77
MAN014	1.039	−0.01	0.008	1.85	0.012	−0.76
TIW220	1.015	−0.17	0.028	2.40	0.021	−0.74
ROX011	1.055	−0.03	0.019	2.39	0.019	−1.12
MAN.GN	1.056	0.03	0.0	—	0.0	—
ROX.GN	1.066	0.0	0.0	—	0.0	—

Case (vii)

Busbar	+ ve sequence		− ve sequence		Zero sequence	
	V_1	θ_1	V_2	θ_2	V_0	θ_0
INV220	1.034	0.36	0.023	−3.12	0.004	0.23
ROX220	1.049	0.40	0.023	3.04	0.005	−0.80
MAN220	1.071	0.43	0.015	2.39	0.001	0.20
MAN014	1.050	−0.01	0.006	2.93	0.0	—
TIW220	1.029	0.36	0.023	3.11	0.005	0.69
ROX011	1.064	−0.02	0.016	−2.70	0.0	—
MAN.GN	1.067	0.03	0.0	—	0.0	—
ROX.GN	1.074	0.0	0.0	—	0.0	—

Case (viii)

Busbar	+ ve sequence		− ve sequence		Zero sequence	
	V_1	θ_1	V_2	θ_2	V_0	θ_0
INV220	1.011	0.37	0.100	−2.69	0.083	−2.62
ROX220	1.043	0.40	0.086	−2.70	0.031	−2.36
MAN220	1.065	0.44	0.058	−2.65	0.017	−2.50
MAN014	1.061	−0.01	0.032	−2.11	0.0	—
TIW220	1.007	0.36	0.098	−2.68	0.080	−2.59
ROX011	1.081	−0.02	0.060	−2.16	0.0	—
MAN.GN	1.086	0.03	0.0	—	0.0	—
ROX.GN	1.096	0.0	0.0	—	0.0	—

Case (x)

(v) As for case (ii) but with system unbalance introduced by line series impedance unbalance only.

(vi) Combined system capacitance and series impedance unbalance with balanced loading. Generator transformers star-g/star-g.

(vii) As for case (vi) but with unbalanced loading.

(viii) As for case (vii) but with delta/star-g for the generator transformers.

(ix) As for case (viii) but with large unbalanced real power loading at INV220.

(x) As for case (viii) but with large unbalanced reactive power loading at INV220.

The number of iterations to convergence, given in Table 3.4, clearly indicates that system unbalance causes a deterioration in convergence. Such deterioration is largely independent of the cause of the unbalance, but is very dependent on the severity or degree of the unbalance.

In all these cases the degree of system unbalance is significant as may be assessed from the sequence components of the busbar voltages, which are given in Table 3.5 for cases (vii), (viii), and (x). The latter case is only included to demonstrate the convergence properties of the algorithm.

Table 3.6
Table of busbar data

No.	Busbar name	Phase A P-load	Phase A Q-load	Phase B P-load	Phase B Q-load	Phase C P-load	Phase C Q-load
1	INV220	50.000	15.000	45.000	14.000	48.300	16.600
2	ROX220	48.000	20.000	47.000	12.000	51.300	28.300
3	MAN220	0.0	0.0	0.0	0.0	0.0	0.0
4	MAN014	0.0	0.0	0.0	0.0	0.0	0.0
5	TIW220	150.000	80.000	157.000	78.000	173.000	72.000
6	ROX011	0.0	0.0	0.0	0.0	0.0	0.0

Table 3.7
Busbar results

No.	Busbar name	Phase A Volt	Phase A Ang	Phase B Volt	Phase B Ang	Phase C Volt	Phase C Ang	Generation Total	
1	INV220	1.0173	21.36	1.0509	−98.16	1.0351	139.44	0.0	0.0
2	ROX220	1.0319	23.30	1.0730	−96.18	1.0449	141.76	0.0	0.0
3	MAN220	1.0693	25.34	1.0816	−95.21	1.0641	144.34	0.0	0.0
4	MAN014	1.0450	−0.79	1.0545	−120.64	1.0522	118.84	0.0	0.0
5	TIW220	1.0137	21.08	1.0434	−98.61	1.0316	138.98	0.0	0.0
6	ROX011	1.0500	−1.79	1.0653	−120.57	1.0771	118.12	0.0	0.0
7	MAN.GN	1.0669	1.69	1.0669	−118.31	1.0669	121.69	500.000	185.804
8	ROX.GN	1.0738	0.0	1.0738	−120.00	1.0738	120.00	281.277	108.106

Table 3.8
Computed power flows

| Sending end busbar | | Receiving end busbar | | Sending end | | Receiving end | |
No.	Name	No.	Name	MW	MVAR	MW	MVAR
4	MAN014	7	MAN.GN	−163.583	−62.676	164.077	71.179
				−160.184	−47.925	159.968	55.047
				−176.232	−50.050	175.955	59.577
6	ROX011	8	ROX.GN	−95.416	−37.762	96.329	41.642
				−87.270	−34.303	87.620	35.449
				−98.590	−27.893	97.327	31.011
3	MAN220	5	TIW220	34.710	10.919	−34.135	−19.871
				33.997	8.911	−32.640	−19.255
				38.172	6.260	−37.730	−15.979
3	MAN220	5	TIW220	36.209	15.598	−36.075	−24.989
				29.544	4.985	−28.602	−16.504
				40.282	4.235	−39.851	−13.410
3	MAN220	1	INV220	41.950	7.870	−41.154	−14.703
				50.720	8.167	−49.293	−16.798
				47.746	18.296	−48.539	−24.434
3	MAN220	1	INV220	44.368	6.728	−43.347	−13.739
				52.547	9.863	−51.290	−18.097
				48.269	16.689	−48.704	−23.079
1	INV220	5	TIW220	35.058	10.315	−34.987	−11.785
				43.883	22.383	−43.806	−23.593
				34.740	18.915	−34.720	−20.106
1	INV220	5	TIW220	44.852	22.010	−44.801	−23.424
				52.175	17.444	−51.939	−18.659
				60.745	21.385	−60.691	−22.413
1	INV220	2	ROX220	−22.706	−9.412	22.491	3.271
				−20.242	−9.462	20.467	2.548
				−23.275	−4.725	23.737	−1.362
1	INV220	2	ROX220	−22.706	−9.412	22.491	3.271
				−20.242	−9.462	20.467	2.548
				−23.275	−4.725	23.737	−1.362
3	MAN220	4	MAN014	−157.242	−42.113	163.587	62.660
				−166.786	−31.943	160.186	47.958
				−174.468	−45.462	176.229	50.033
2	ROX220	6	ROX011	−92.984	−26.544	95.461	37.757
				−87.935	−17.105	87.271	34.312
				−98.772	−25.566	98.588	27.888

Total generation	781.27 MW	293.91 MVAR	
Total load	768.60 MW	335.90 MVAR	
System losses	11.67 MW	−41.98 MVAR	
Mismatch	0.0013 MW	−0.0096 MVAR	

Note that the initial convergence of the algorithm is fast even in cases of extreme steady-state unbalance. The reliability of the algorithm is not prejudiced by significant unbalance although convergence to small tolerances becomes slow.

The influence of the three-phase transformer connection may be seen in the sequence voltages of cases (vii) and (viii). The star-g/delta connection provides no through path for zero-sequence currents and the zero-sequence machine current is zero. This is reflected in the zero-sequence voltages at the machine terminal voltages.

The sequence voltages also illustrate the position of angle reference at the slack generator internal busbar. In addition, it may be seen that at all generator internal busbars the negative- and zero-sequence voltages are zero reflecting the balanced and symmetrical nature of the machine excitations.

As an example of the numerical results, the busbar loadings for case (viii) are given in Table 3.6 and the resulting busbar voltages and line power flows are presented in Tables 3.7 and 3.8.

Besides the significant unbalance other features to be noticed are:

- the approximate 30° phase shift due to the star–delta connected transformers
- balanced voltages at the generator-internal busbars
- balanced angles at the generator-internal busbars
- an apparent gain in active power flow in any one phase. (This power flows through the mutual coupling terms between phases. The overall active power shows a net loss as expected for a realistic system.)

3.10   REFERENCES

[1] A. Holley, C. Coleman and R. B. Shipley, 1964. Untransposed e.h.v. line computations, *IEEE Trans.* **PAS-83** 291.

[2] M. H. Hesse, 1966. Circulating currents in parallel untransposed multicircuit lines: I—Numerical evaluations, *IEEE Trans.* **PAS-85** 802. II—Methods of eliminating current unbalance, *IEEE Trans.* **PAS-95** 812.

[3] A. H. El-Abiad and D. C. Tarisi, 1967. Load-flow solution of untransposed EHV network *PICA, Pittsburgh, Pa* 377–384.

[4] R. G. Wasley and M. A. Slash, 1974. Newton–Raphson algorithm for three-phase load flow, *Proc. IEE* **121** 630.

[5] K. A. Birt, J. J. Graffy, J. D. McDonald and A. H. El-Abiad, 1976. Three-phase load-flow program, *IEEE Trans.* **PAS-95** 59.

[6] J. Arrillaga and B. J. Harker, 1978. Fast decoupled three-phase load flow, *Proc. IEE* **125** 734–740.

[7] M. S. Chen and W. E. Dillon, 1974. Power system modelling, *Proc. IEEE* **62** 901.

[8] M. A. Laughton, 1968. Analysis of unbalanced polyphase networks by the method of phase coordinates. Part I, system representation in phase frame of references, *Proc. IEE* **115** 1163–1172.

[9] B. Stott and O. Alsac 1978. Fast decoupled load flow, *IEEE Trans* **PAS-93** 859.

[10] K. Zollenkopf, 1970. Bifactorization—basic computational algorithm and programming techniques, *Conference on Large Sets of Sparse Linear Equations, Oxford* pp 75–96.

4. A. C.–D. C. LOAD FLOW
Single-Phase Algorithm

4.1 INTRODUCTION

The first half of this chapter (Sections 4.1–4.8) will deal with the single-phase algorithm, while the remainder (Sections 4.9 onwards) will cover the three-phase algorithm.

High-voltage d.c. (h.v.d.c) transmission is now an acceptable alternative to a.c. and is proving an economical solution not only for very long distance but also for underground and submarine transmission as well as a means of interconnecting systems of different frequency or with problems of stability or fault level.

The growing number of schemes in existence and under consideration demands corrresponding modelling facilities for planning and operational purposes.

The basic load flow has to be substantially modified to be capable of modelling the operating state of the combined a.c. and d.c. systems under the specified conditions of load, generation and d.c. system control strategies.

Having established the superiority of the fast-decoupled a.c. load flow [1] the integration of h.v.d.c. transmission is now described with reference to such an algorithm.

A sequential approach [2,3] is used, where the a.c. and d.c. equations are solved separately and thus the integration into existing load-flow programs is carried out without significant modification or restructuring of the a.c. solution technique. For the a.c. iterations each converter is modelled simply by the equivalent real or reactive power injection at the terminal busbar. The terminal busbar voltages obtained from the a.c. iteration are then used to solve the d.c. equations and consequently new power injections are obtained. This process continues iteratively to convergence.

4.2 FORMULATION OF THE PROBLEM

The operating state of the combined power system is defined by the vector $[\bar{V}, \bar{\theta}, \bar{x}]^{\mathrm{T}}$ where $\bar{V}$ is a vector of the voltage magnitudes at all a.c. system busbars, $\bar{\theta}$ is a vector of the angles at all a.c. system busbars (except the reference bus which is assigned $\theta = 0$), and $\bar{x}$ is a vector of d.c. variables. The use of $\bar{V}$ and $\bar{\theta}$ as a.c. system variables was described in Chapter 2 and the selection of d.c. variables $\bar{x}$ is discussed in Section 4.3.

The development of a Newton–Raphson-based algorithm requires the formulation of n independent equations in terms of the n variables.

The equations which relate the a.c. system variables are derived from the specified a.c. system operating conditions. The only modification required to the usual real and reactive power mismatches occurs for those equations which relate to the converter terminal busbars. These equations become

$$P_{term}^{sp} - P_{term}(ac) - P_{term}(dc) = 0 \qquad (4.2.1)$$

$$Q_{term}^{sp} - Q_{term}(ac) - Q_{term}(dc) = 0 \qquad (4.2.2)$$

where

$P_{term}(ac)$ is the injected power at the terminal busbar as a function of the a.c. system variables

$P_{term}(dc)$ is the injected power at the terminal busbar as function of the d.c. system variables

P_{term}^{sp} is the usual a.c. system load at the busbar

and similarly for $Q_{term}(dc)$, $Q_{term}(ac)$ and Q_{term}^{sp}.

The injected powers $Q_{term}(dc)$, and $P_{term}(dc)$ are functions of the converter a.c. terminal busbar voltage and of the d.c. system variables, i.e.

$$P_{term}(dc) = f(V_{term}, \bar{x}) \qquad (4.2.3)$$

$$Q_{term}(dc) = f(V_{term}, \bar{x}). \qquad (4.2.4)$$

The equations derived from the specified a.c. system conditions may therefore be summarised as

$$\begin{bmatrix} \Delta\bar{P}(\bar{V}, \bar{\theta}) \\ \Delta\bar{P}_{term}(\bar{V}, \bar{\theta}, \bar{x}) \\ \Delta\bar{Q}(\bar{V}, \bar{\theta}) \\ \Delta\bar{Q}_{term}(\bar{V}, \bar{\theta}, \bar{x}) \end{bmatrix} = 0 \qquad (4.2.5)$$

where the mismatches at the converter terminal busbars are indicated separately.

A further set of independent equations are derived from the d.c. system conditions. These are designated

$$\bar{R}(V_{term}, \bar{x})_k = 0 \qquad (4.2.6)$$

for $k = 1$, number of converters present.

The d.c. system equations (4.2.3), (4.2.4) and (4.2.6) are made independent of the a.c. system angles $\bar{\theta}$ by selecting a separate angle reference for the d.c. system variables as defined in Fig. 4.2 later. This improves the algorithmic performance by effectively decoupling the angle dependence of a.c. and d.c. systems.

The general a.c.–d.c. load-flow problem may therefore be summarised as the solution of

$$\begin{bmatrix} \Delta\bar{P}(\bar{V}, \theta) \\ \Delta\bar{P}_{term}(\bar{V}, \bar{\theta}, \bar{x}) \\ \Delta\bar{Q}(\bar{V}, \bar{\theta}) \\ \Delta\bar{Q}_{term}(\bar{V}, \bar{\theta}, \bar{x}) \\ \bar{R}(V_{term}, \bar{x}) \end{bmatrix} = 0 \qquad (4.2.7)$$

where the subscript 'term' refers to the converter a.c. terminal busbar.

4.3 D.C. SYSTEM MODEL

The selection of variables $\bar{x}$ and formulation of the equations require several basic assumptions which are generally accepted [1] in the analysis of steady state d.c. converter operation.

(i) The three a.c. voltages at the terminal busbar are balanced and sinusoidal.

(ii) The converter operation is perfectly balanced.

(iii) The direct current and voltage are smooth.

(iv) The converter transformer is lossless and the magnetising admittance is ignored.

4.3.1 Converter Variables

Under balanced conditions similar converter bridges attached to the same a.c. terminal busbar will operate identically regardless of the transformer connection. They may therefore be replaced by an equivalent single bridge for the purpose of single-phase load-flow analysis. With reference to Fig. 4.1 the set of variables illustrated, representing fundamental frequency or d.c. quantities permits a full description of the converter system operation.

An equivalent circuit for the converter is shown in Fig. 4.2 which includes the modification explained in Section 4.2 as regards the position of angle reference.

The variables, defined with reference to Fig. 4.2, are as follows:

$V_{term}\underline{/\phi}$ converter terminal busbar nodal voltage (phase angle referred to converter reference)

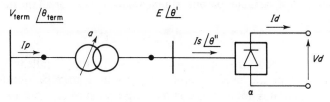

Figure 4.1
Basic d.c. converter (angles refer to a.c. system reference)

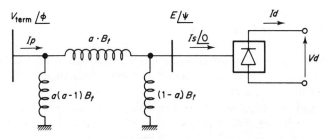

Figure 4.2
Single-phase equivalent circuit for basic converter (angles referred to d.c. reference)

E/ψ fundamental frequency component of the voltage waveform at the converter transfomer secondary

I_p, I_s fundamental frequency component of the current waveshape on the primary and secondary of the converter transformer respectively

α firing delay angle

a transformer off-nominal tap ratio

V_d average d.c. voltage

I_d converter direct current.

These ten variables—nine associated with the converter, plus the a.c. terminal voltage magnitude V_{term}—form a possible choice of $\bar{x}$ for the formulation of equations (4.2.3), (4.2.4) and (4.2.6).

The minimum number of variables required to define the operation of the system is the number of independent variables. Any other system variable or parameter (e.g. P_{dc} and Q_{dc}) may be written in terms of these variables.

Two independent variables are sufficient to model a d.c. converter, operating under balanced conditions, from a known terminal voltage source. However, the control requirements of h.v.d.c. converters are such that a range of variables, or functions of them (e.g. constant power), are the specified conditions. If the minimum number of variables are used, then the control specifications must be translated into equations in terms of these two variables. These equations will often contain complex nonlinearities, and present difficulties in their derivation and program implementation. In addition, the expressions used for P_{dc} and Q_{dc} in equations (4.2.1.) and (4.2.2.) may be rather complex and this will make the programming more difficult.

For these reasons, a nonminimal set of variables is recommended, i.e. all variables which are influenced by control action are retained in the model. This is in contrast to a.c. load flows where, due to the restricted nature of control specifications, the minimum set is normally used.

The following set of variables permits simple relationships for all the normal control strategies:

$$[\bar{x}] = [V_d, I_d, a, \cos \alpha, \phi]^T.$$

Variable ϕ is included to ensure a simple expression for Q_{dc}. While this is important in the formulation of the unified solution, variable ϕ may be omitted with the sequential solution as it is not involved in the formulation of any control specification; $\cos \alpha$ is used as a variable rather than α to linearise the equations and thus improve convergence.

4.3.2 D.C. per Unit System

To avoid translating from per unit to actual value and to enable the use of comparable convergence tolerances for both a.c. and d.c. system mismatches, a per unit system is also used for the d.c. quantities.

Computational simplicity is achieved by using common power and voltage base parameters on both sides of the converter, i.e. the a.c. and d.c. sides. Consequently, in order to preserve consistency of power in per unit, the direct current base, obtained from $(MVA_B)/V_B$, has to be $\sqrt{3}$ times larger than the a.c. current base.

This has the effect of changing the coefficients involved in the a.c.–d.c. current relationships. For a perfectly smooth direct current and neglecting the commutation overlap, the r.m.s. fundamental components of the phase current is related to I_d by the approximation (Appendix II)

$$I_s = \frac{\sqrt{6}}{\pi} I_d. \qquad (4.3.1)$$

Translating equation (4.3.1) to per unit yields

$$I_s(\text{p.u.}) = \frac{\sqrt{6}}{\pi} \sqrt{3} \cdot I_d(\text{p.u.})$$

and if commutation overlap is taken into account, this equation becomes

$$I_s(\text{p.u.}) = k \frac{3\sqrt{2}}{\pi} I_d(\text{p.u.}) \qquad (4.3.2)$$

where k is very close to unity. In load-flow studies, equation (4.3.2) can be made sufficiently accurate in most cases by letting $k = 0.995$.

4.3.3   Derivation of Equations

The following relationships are derived for the variables defined in Fig. 4.2. The equations are in per unit.

(i) The fundamental current magnitude on the converter side is related to the direct current by the eqation

$$I_s = k \frac{3\sqrt{2}}{\pi} I_d. \qquad (4.3.3)$$

(ii) The fundamental current magnitudes on both sides of the lossless transformer are related by the off-nominal tap, i.e.

$$I_p = aI_s. \qquad (4.3.4)$$

(iii) The d.c. voltage may be expressed in terms of the a.c. source commutating voltage referred to the transformer secondary, i.e.

$$V_d = \frac{3\sqrt{2}}{\pi} aV_{\text{term}} \cos \alpha - \frac{3}{\pi} I_d X_c. \qquad (4.3.5)$$

The converter a.c. source commutating voltage is the busbar voltage on the system side of the converter transformer, V_{term}.

(iv) The d.c. current and voltage are related by the d.c. system configuration

$$f(V_d, I_d) = 0. \qquad (4.3.6)$$

For example, for a simple rectifier supplying a passive load,

$$V_d - I_d R_d = 0.$$

(v) The assumptions listed at the beginning of this section prevent any real power of harmonic frequencies at the primary and secondary busbars. Therefore, the real power equation relates the d.c. power to the transformer secondary power in terms of fundamental components only, i.e.

$$V_d I_d = E I_s \cos \psi. \tag{4.3.7}$$

(vi) As the transformer is lossless, the primary real power may also be equated to the d.c. power, i.e.

$$V_d I_d = V_{\text{term}} I_p \cos \phi. \tag{4.3.8}$$

(vii) The fundamental component of current flow across the converter transformer can be expressed as

$$I_s = B_t \sin \psi - B_t a V_{\text{term}} \sin \phi \tag{4.3.9}$$

where $j B_t$ is the transformer leakage susceptance.

So far, a total of seven equations have been derived and no other independent equation may be written relating the total set of nine converter variables.

Variables I_p, I_s, E and ψ can be eliminated as they play no part in defining control specifications. Thus equations (4.3.3), (4.3.4), (4.3.7) and (4.3.8) can be combined into

$$V_d - k_1 a V_{\text{term}} \cos \phi = 0 \tag{4.3.10}$$

where $k_1 = k(3\sqrt{2}/\pi)$.

The final two independent equations required are derived from the specified control mode.

The d.c. model may thus be summarised as follows:

$$\bar{R}(\bar{x}, V_{\text{term}})_k = 0 \tag{4.3.11}$$

where

$$R(1) = V_d - k_1 a V_{\text{term}} \cos \phi$$

$$R(2) = V_d - k_1 a V_{\text{term}} \cos \alpha + \frac{3}{\pi} I_d X_c$$

$$R(3) = f(V_d, I_d)$$

$$R(4) = \text{control equation}$$

$$R(5) = \text{control equation}$$

$$\bar{x} = [V_d, I_d, a, \cos \alpha, \phi]^{\mathrm{T}}$$

V_{term} can either be a specified quantity or an a.c. system variable. The equations for P_{dc} and Q_{dc} may now be written as

$$Q_{\text{term}}(dc) = V_{\text{term}} I_p \sin \phi \tag{4.3.12}$$
$$= V_{\text{term}} k_1 a I_d \sin \phi$$

and

$$P_{\text{term}}(dc) = V_{\text{term}} I_p \cos \phi \tag{4.3.13}$$
$$= V_{\text{term}} k_1 a I_d \cos \phi$$

or

$$P_{\text{term}}(dc) = V_d I_d. \tag{4.3.14}$$

4.3.4 Incorporation of Control Equations

Each additional converter in the d.c. system contributes two independent variables to the system and thus two further constraint equations must be derived from the control strategy of the system to define the operating state. For example, a classical two-terminal d.c. link has two converters and therefore requires four control equations. The four equations must be written in terms of ten d.c. variables (five for each converter).

Any function of the ten d.c. system variables is valid (mathematically) control equation so long as each equation is independent of all other equations. In practice, there are restrictions limiting the number of alternatives. Some control strategies refer to the characteristics of power transmission (e.g. constant power or constant current), others introduce constraints such as minimum delay or extinction angles.

Examples of valid control specifications are:

- Specified converter transformer tap $a - a^{sp} = 0$
- Specified d.c. voltage $V_d - V_d^{sp} = 0$
- Specified d.c. current $I_d - I_d^{sp} = 0$
- Specified minimum firing angle $\cos \alpha - \cos \alpha_{\min} = 0$
- Specified d.c. power transmission $V_d I_d - P_{dc}^{sp} = 0$.

These control equations are simple and are easily incorporated into the solution algorithm. In addition to the usual control modes, nonstandard modes such as specified a.c. terminal voltage may also be included as converter control equations.

During the iterative solution procedure the uncontrolled converter variables may go outside prespecified limits. When this occurs, the offending variable is usually held to its limit value and an appropriate control variable is freed [4].

4.3.5 Inverter Operation

All equations presented so far are equally applicable to inverter operation. However, during inversion it is the extinction advance angle (γ) which is the subject of control action and not the firing angle (α). For convenience therefore, equation $R(2)$ of (4.3.11) may be rewritten as

$$V_d - k_1 a V_{\text{term}} \cos(\pi - \gamma) - \frac{3}{\pi} X_c I_d = 0. \tag{4.3.15}$$

This equation is valid for rectification or inversion. Under inversion, V_d (as calculated by equation (4.3.15)) will be negative.

To specify operation with constant extinction angle the following equation is used:

$$\cos(\pi - \gamma) - \cos(\pi - \gamma^{sp}) = 0$$

where γ^{sp} is usually $\gamma_{\min}$ for minimum reactive power consumption of the inverter.

4.4 SEQUENTIAL SOLUTION TECHNIQUES

The following three equations are solved iteratively to convergence:

$$[\Delta \bar{P}/\bar{V}] = [B'][\Delta \bar{\theta}] \tag{4.4.1}$$

$$[\Delta \bar{Q}/\bar{V}] = [B''][\Delta \bar{V}] \tag{4.4.2}$$

$$[\bar{R}] = [A][\Delta \bar{x}]. \tag{4.4.3}$$

This iteration sequence, referred to as P, Q, DC, is illustrated in the flow chart of Fig. 4.3 and may be summarised as follows.

(i) Calculate $\Delta \bar{P}/\bar{V}$, solve equation (4.4.1) and update $\bar{\theta}$.

(ii) Calculate $\Delta \bar{Q}/\bar{V}$, solve equation (4.4.2) and update $\bar{V}$.

(iii) Calculate d.c. residuals, $\bar{R}$, solve equation (4.4.3) and update $\bar{x}$.

(iv) Return to (i).

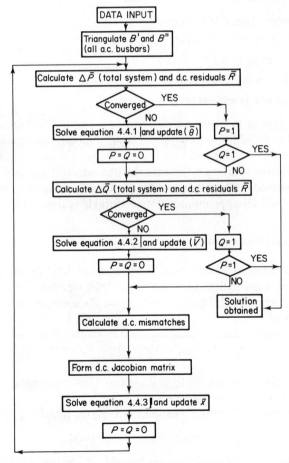

Figure 4.3
Flow chart for sequential single-phase a.c.–d.c. load flow

With the sequential method the d.c. equations need not be solved for the entire iterative process. Once the d.c. residuals have converged, the d.c. system may be modelled simply as fixed real and reactive power injections at the appropriate converter terminal busbar. The d.c. residuals must still be checked after each a.c. iteration to ensure that the d.c. system remains converged.

Alternatively, the d.c. equations can be solved after each real power as well as after each reactive power iteration and the resulting sequence is referred to as *P, DC, Q, DC*. As in the previous methods, the d.c. equations are solved until all mismatches are within tolerance.

4.5 EXTENSION TO MULTIPLE AND/OR MULTITERMINAL D.C. SYSTEM

The basic algorithm has been developed in previous sections for a single d.c. converter. Each additional converter adds a further five d.c. variables and a corresponding set of five equations. The number of a.c. system Jacobian elements which become modified in the unified solutions is equal to the number of converters.

As an example, consider the system shown in Fig. 4.4. The system represents the North and South Islands of the New Zealand 220 kV a.c. system. At present converters 1, 2 and 3 are in operation. Converters 1 and 2 form the 600 MW, 500 kV d.c. link between the two islands. Converter 3 represents a 420 MW aluminium smelter. A further three-terminal d.c. interconnection has been added (converters 2, 5 and 6) to illustrate the flexibility of the algorithm.

Normally, converter 4 will operate in the rectifier mode with converters 5 and 6 in the inversion mode.

The reactive power-d.c. Jacobian for the unified method has the following structure:

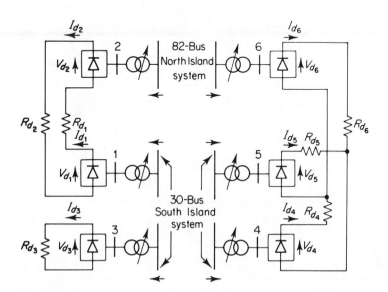

Figure 4.4
Multiterminal d.c. system

$$
\begin{array}{c}
\Delta \bar{Q}/\bar{V} \\
\dfrac{\Delta Q_{term\,1}}{V_{term\,2}} \\
\dfrac{\Delta Q_{term\,2}}{V_{term\,2}} \\
\dfrac{\Delta Q_{term\,3}}{V_{term\,3}} \\
\dfrac{\Delta Q_{term\,4}}{V_{term\,4}} \\
\dfrac{\Delta Q_{term\,5}}{V_{term\,5}} \\
\dfrac{\Delta Q_{term\,6}}{V_{term\,6}} \\
\Delta \bar{R}_1 \\
\Delta \bar{R}_2 \\
\Delta \bar{R}_3 \\
\Delta \bar{R}_4 \\
\Delta \bar{R}_5 \\
\Delta \bar{R}_6
\end{array}
\;=\;
\underbrace{
\left[
\begin{array}{c|c}
\begin{matrix}
AA''_{11} & & & & & \\
 & AA''_{22} & & & & \\
 & & AA''_{33} & & & \\
 & & & AA''_{44} & & \\
 & & & & AA''_{55} & \\
 & & & & & AA''_{66}
\end{matrix}
&
\begin{matrix}
 & & & \\
 & & & \\
 & A & & \\
 & (30 \times 30) & & \\
 & & & \\
 & & &
\end{matrix}
\\ \hline
\begin{matrix}
 & & & & & \\
 & & B''_{MOD} & & & \\
 & & & & & \\
 & & & & &
\end{matrix}
&
\begin{matrix}
BB''_{11} & & & & & \\
 & BB''_{22} & & & & \\
 & & BB''_{33} & & & \\
 & & & BB''_{44} & & \\
 & & & & BB''_{55} & \\
 & & & & & BB''_{66}
\end{matrix}
\end{array}
\right]
}_{B''}
\;
\begin{array}{c}
\Delta \bar{V} \\
\Delta V_{term\,1} \\
\Delta V_{term\,2} \\
\Delta V_{term\,3} \\
\Delta V_{term\,4} \\
\Delta V_{term\,5} \\
\Delta V_{term\,6} \\
\Delta \bar{X}_1 \\
\Delta \bar{X}_2 \\
\Delta \bar{X}_3 \\
\Delta \bar{X}_4 \\
\Delta \bar{X}_5 \\
\Delta \bar{X}_6
\end{array}
$$

where B''_{MOD} is the part of B'' which becomes modified. Only the diagonal elements become modified by the presence of the converters.

Off-diagonal elements will be present in B''_{MOD} if there is any a.c. connection between converter terminal busbars. All off-diagonal elements of BB'' and AA'' are zero.

In addition, matrix A is block diagonal in 5×5 blocks with the exception of the d.c. interconnection equations.

Equation R(3) of (4.3.11) in each set of d.c. equations is derived from the d.c. interconnection. For the six-converter system shown in Fig. 4.4 the following equations are applicable:

$$V_{d1} + V_{d2} - I_{d1}(R_{d1} + R_{d2}) = 0$$
$$V_{d3} - I_{d3} \cdot R_{d3} = 0$$
$$I_{d1} - I_{d2} = 0$$
$$V_{d4} + V_{d6} - I_{d4}R_{d4} - I_{d6}R_{d6} = 0$$
$$V_{d5} - V_{d6} - I_{d5} \cdot R_{d5} + I_{d6}R_{d6} = 0$$
$$I_{d4} - I_{d5} - I_{d6} = 0.$$

This example indicates the ease of extension to the multiple-converter case.

4.6   D.C. CONVERGENCE TOLERANCE

The d.c. p.u. system is based upon the same power base as the a.c. system and on the nominal open-circuit a.c. voltage at the converter transformer secondary. The p.u. tolerances for d.c. powers, voltages and currents are therefore comparable with those adopted in the a.c. system.

In general, the control equations are of the form

$$\bar{X} - \bar{X}^{sp} = 0$$

where X may be the tap or cosine of the firing angle, i.e. they are linear and are thus solved in one d.c. iteration. The question of an appropriate tolerance for these mismatches is therefore irrelevant.

An acceptable tolerance for the d.c. residuals which is compatible with the a.c. system tolerance is typically 0.001 p.u. on a 100 MVA base, i.e. the same as that normally adopted for the a.c. system.

4.7   TEST SYSTEM AND RESULTS

The A.E.P. standard 14-bus test system is used to show the convergence properties of the a.c.–d.c. algorithm, with the a.c. transmission line between busbars 5 and 4 replaced by a h.v.d.c. link. As these two buses are not voltage controlled, the interaction between the a.c. and d.c. systems will therefore be considerable.

Various control strategies have been applied to the link and the convergence results are given in Table 4.1. The number of iteration (i, j) should be interpreted as follows.

Table 4.1
Convergence results

	Case specification	*Number of iterations to convergence (0.1 MW/MVAR)*			
	Specified d.c. link constraints m-rectifier end n-inverter end	*5 variables*		*4 variables*	
		$1_{P,Q,DC}$	$2_{P,DC,Q,DC}$	$1_{P,Q,DC}$	$2_{P,DC,Q,DC}$
1	$\alpha_m P_{dm} \gamma_n V_{dn}$	4.3	4.3	4.4	4.3
2	$\alpha_m P_{dm} a_n V_{dn}$	4.4	5.5	4.4	Failed
3	$a_m P_{dm} a_n V_{dn}$	4.4	5.5	4.4	Failed
4	$a_m P_{dm} \gamma_n V_{dn}$	4.4	4.4	4.4	4.4
5	$a_m P_{dm} \gamma_n a_n$	4.4	4.4	4.4	4.4
6	$a_m P_{dm} \alpha_m \gamma_n$	4.3	4.3	4.4	4.3
7	$\alpha_m I_{d} \gamma_n V_{dn}$	4.3	4.3	4.4	4.3
8	$a_m V_{dm} \gamma_n P_{dn}$	4.4	4.4	4.4	4.4
	Case 1 with initial condition errors				
9	50% error	4.4	4.3	4.4	4.3
10	80% error	7.6*	5.4*	4.4	4.3

*indicates a false solution.

Table 4.2
Characteristics of d.c. link

	Converter 1	*Converter 2*
A.C. busbar	Bus 5	Bus 4
D.C. voltage base	100 kV	100 kV
Transformer reactance	0.126	0.0728
Commutation reactance	0.126	0.0728
Filter admittance B_f^*	0.478	0.629
D.C. link resistance	0.334 Ω	
Control parameters for Case 1		
D.C. link power	58.6 MW	—
Rectifier firing angle (deg)	7	—
Inverter extinction angle (deg)	—	10
Inverter d.c. voltage	—	−128.87 kV

*Filters are connected to a.c. terminal busbar.
Note: All reactances are in p.u. on a 100 MVA base.

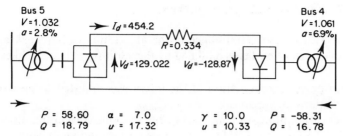

Figure 4.5
D.c. link operation for Case 1

- i is the number of reactive power-voltage updates required.
- j is the number of real power-angle updates.

Although the number of d.c. iterations varies for the different sequences, this is of secondary importance and may if required be assessed in each case from the number of a.c. iterations.

The d.c. link data and specified controls of Case 1 are given in Table 4.2 and the corresponding d.c. link operation is illustrated in Fig. 4.5. The specified conditions for all cases are derived from the results of Case 1. Under those conditions, the a.c. system in isolation, (with each converter terminal modelled as an equivalent a.c. load) requires (4, 3) iterations. The d.c. system in isolation (operating from fixed terminal voltages) requires two iterations under all control strategies.

The sequential method (P, Q, DC) produces fast and reliable convergence although the reactive power convergence is slower than for the a.c. system alone.

With the removal of the variable ϕ, Q_{term} (dc) converges faster but the convergence pattern is more oscillatory and an overall deterioration of a.c. voltage convergence results.

With the second sequential method, (P, DC, Q, DC) convergence is good in all cases except 2 and 3, i.e. the cases where the transformer tap and d.c. voltage are specified at the inverter end. However, this set of specifications is not likely to occur in practice.

4.7.1 Initial Conditions for D.C. System

Initial values for the d.c. variables $\bar{x}$ are assigned from estimates for the d.c. power and d.c. voltage and assuming a power factor of 0.9 at the converter terminal busbar. The terminal busbar voltage is set at 1.0 p.u. unless it is a voltage-controlled busbar.

This procedure gives adequate initial conditions in all practical cases as good estimates of P_{term} (dc) and V_d are normally obtainable.

With starting values for d.c. real and reactive powers within $\pm 50\%$, which are available in all practical situations, all algorithms converged rapidly and reliably (see Case 9).

4.7.2 Effect of A.C. System Strength

In order to investigate the performance of the algorithms with a weak a.c. system, the test system described earlier is modified by the addition of two a.c. lines as shown in Fig. 4.6.

The reactive power compensation of the filters was adjusted to give similar d.c. operating conditions as previously.

The number of iterations to convergence for the most promising algorithms are shown in Table 4.3 for the control specifications corresponding to cases 1 to 4 in the previous results.

In all other cases, where the control angle at one or both converters is free, an oscillatory relationship between converter a.c. terminal voltage and the reactive power of the converter is possible.

To illustrate the nature of the iteration, the convergence pattern of the converter reactive power demand and the a.c. system terminal voltage of the rectifier is plotted in Fig. 4.7

A measure of the strength of a system in a load-flow sense is the short-circuit-to-converter power ratio (SCR) calculated with all machine reactances set to zero. This short-circuit ratio is invariably much higher than the usual value.

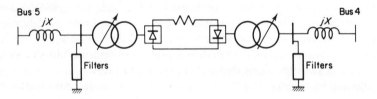

Figure 4.6
D.C. link operating from weak a.c. system

Table 4.3
Numbers of iterations of the P, Q, DC sequence for weak a.c. systems

Case specification m—rectifier n—inverter	$x_l = 0.3$		$x_l = 0.4$	
	(i)	(ii)	(i)	(ii)
11 $\alpha_m P_{dm} \gamma_n V_{dn}$	4.4	4.4	5.4	4.4
12 $\alpha_m P_{dm} a_n V_{dn}$	9.8	10.12	>30	Diverges
13 $a_m P_{dm} a_n V_{dn}$	9.8	10.12	>30	Diverges
14 $a_m P_{dm} \gamma_n V_{dn}$	6.5	7.7	28.27	>30

(i) using the five-variable formulation; (ii) using the four-variable formulation.

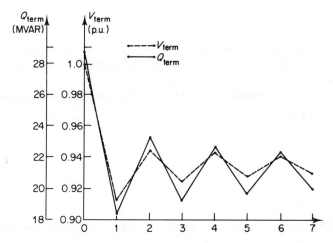

Figure 4.7
Convergence pattern for a.c.–d.c. load flow with weak a.c. system. Sequential method (P, Q, DC, five variables)

In practice, converter operation has been considered down to a SCR of 3. A survey of existing schemes shows that, almost invariably, with systems of very low SCR, some form of voltage control, often synchronous codensers, is an integral part of the converter installation. These schemes are therefore often very strong in a load-flow sense.

It may therefore be concluded that the sequential integration should converge in all practical situations although the convergence may become slow if the system is weak in a load-flow sense.

4.7.3 Discussion of convergence properties

The overall convergence rate of the a.c.–d.c. algorithms depends on the successful interaction of the two distinct parts. The a.c. system equations are solved using the well-behaved constant tangent fast-decoupled algorithm, whereas the d.c. system equations are solved using the more powerful, but somewhat more erratic, full Newton–Raphson approach.

The powerful convergence of the Newton–Raphson process for the d.c. equations can cause overall convergence difficulties. If the first d.c. iteration occurs before the reactive power-voltage update then the d.c. variables are converged to be compatible with the incorrect terminal voltage. This introduces an unnecessary discontinuity which may lead to convergence difficulties. The solution time of the d.c. equations is normally small compared to the solution time of the a.c. equations. The relative efficiencies of the alternative algorithms may therefore be assessed by comparing the total numbers of voltage and angle updates.

In general, those schemes which acknowledge the fact that the d.c. variables are strongly related to the terminal voltage give the fastest and most reliable performance.

4.8 NUMERICAL EXAMPLE

The complete New Zealand primary transmission system was used as a basis for a planning study which included an extra multiterminal h.v.d.c. scheme, i.e. involving six converter stations as illustrated in Fig. 4.4.

Representative input and output information obtained from the computer is given on the following pages.

AC DC LOAD FLOW PROGRAM

DEPARTMENT OF ELECTRICAL & ELECTRONIC ENGINEERING, UNIVERSITY OF CANTERBURY, NEW ZEALAND

SYSTEM NO.   3   23 MAR 90

MAXIMUM NUMBER OF ITERATIONS 10	
POWER TOLERANCE 0.00100	
PRINT OUT INDICATOR 000000000	NUMBER OF BUSES 114
SYSTEM MVA BASE 100.00	NUMBER OF LINES 206
D.C. LINK INDICATOR 6	NUMBER OF TRANSFORMERS 19
NUMBER OF A.C. SYSTEMS 2	
SLACK BUSBARS 80 218	

B U S   D A T A

BUS	NAME	TYPE	VOLTS	LOAD MW	LOAD MVAR	GENERATION MW	GENERATION MVAR	MINIMUM MVAR	MAXIMUM MVAR	SHUNT SUSCEPTANCE
104	AVIEMORE–220	1	1.0520	0.00	0.00	220.00	–34.40	–500.00	500.00	0.000
108	BENMORE—220	1	1.0520	97.20	0.00	540.00	46.60	–500.00	500.00	0.000
118	BRLY——220	0	1.0030	329.60	95.80	0.00	0.00	0.00	0.00	0.000
127	CROM1——220	0	1.0520	0.00	0.00	0.00	0.00	0.00	0.00	0.000
128	CROM2——220	0	1.0520	0.00	0.00	0.00	0.00	0.00	0.00	0.000
129	CLUTHA——220	1	1.0300	0.00	0.00	600.00	0.00	0.00	0.00	0.000
138	GERALDINE220	0	1.0210	0.00	0.00	0.00	0.00	0.00	0.00	0.000
143	HWBS——220	0	1.0270	95.30	80.40	0.00	0.00	0.00	0.00	0.000

L I N E   D A T A

BUS	NAME	BUS	NAME	RESISTANCE	REACTANCE	SUSCEPTANCE
104	AVIEMORE–220	108	BENMORE—220	0.00330	0.01530	0.02298
104	AVIEMORE–220	108	BENMORE—220	0.00330	0.01530	0.02298
104	AVIEMORE–220	268	WAITAKI—220	0.00150	0.00730	0.01052
108	BENMORE—220	255	TWIZEL—220	0.00370	0.02610	0.06954
118	BRLY——220	167	ISLINGTON220	0.00210	0.01651	0.05285
118	BRLY——220	181	LAND–T02–220	0.00110	0.00861	0.02751
127	CROM1——220	218	ROXBURGH–220	0.00770	0.04450	0.07251
127	CROM1——220	255	TWIZEL—220	0.00820	0.09260	0.16746
128	CROM2——220	218	ROXBURGH–220	0.00770	0.04450	0.07251
128	CROM2——220	255	TWIZEL—220	0.00820	0.09260	0.16746

TRANSFORMER DATA

BUS	NAME	BUS	NAME	RESISTANCE	REACTANCE	TAP	CODE
6	BUNTHORPE110	7	BUNTHORPE220	0.00400	0.09560	1.000	0
6	BUNTHORPE110	7	BUNTHORPE220	0.00400	0.09560	1.000	0
6	BUNTHORPE110	7	BUNTHORPE220	0.00170	0.04590	1.000	0
10	EDGECOMBE110	11	EDGECOMBE220	0.00400	0.09560	1.000	0
10	EDGECOMBE110	11	EDGECOMBE220	0.00400	0.09560	1.000	0
21	HAYWARDS-110	22	HAYWARDS-220	0.00170	0.05140	1.000	0
21	HAYWARDS-110	22	HAYWARDS-220	0.00170	0.05140	1.000	0
21	HAYWARDS-110	22	HAYWARDS-220	0.00410	0.10120	1.000	0
21	HAYWARDS-110	22	HAYWARDS-220	0.00410	0.10120	1.000	0
23	HENDERSON110	24	HENDERSON220	0.00090	0.01840	1.053	0
39	MARSDEN—110	40	MARSDEN—220	0.00000	0.05500	1.000	0
39	MARSDEN—110	40	MARSDEN—220	0.00000	0.05500	1.000	0
48	NEWPLYMTH110	49	NEWPLYMTH220	0.00080	0.02480	1.000	0
54	OTAHUHU—110	55	OTAHUHU—220	0.00700	0.04100	1.000	0
54	OTAHUHU—110	55	OTAHUHU—220	0.00700	0.04100	1.000	0
54	OTAHUHU—110	55	OTAHUHU—220	0.00160	0.04550	1.000	0
58	PENROSE—110	59	PENROSE—220	0.00090	0.02750	1.000	0
62	STRATSORD110	63	STRATFORD220	0.00200	0.05290	1.000	0
66	TARUKENGA110	67	TARUKENGA220	0.00080	0.02530	1.000	0

TEM EQUATIONS

```
2+VD3+VD4+VD5+VD6+VD7+VD8+VD9+VD10-ID1.RD1-ID2.RD2-ID3.RD3-ID4.RD4-ID5.RD5-ID6.RD6-ID7.RD7-ID8.RD8-ID9.RD9-ID10.RD10=0
4   0   0   0   0   0   0   0   0    25.5600  0.0000  0.0000  0.0000  0.0000  0.0000  0.0000  0.0000  0.0000  0.0000
)   1   0   0   0   0   0   0   0     0.0000  0.0000  0.0019  0.0000  0.0000  0.0000  0.0000  0.0000  0.0000  0.0000
)   0   1   0   1   0   0   0   0     0.0000  0.0000  0.0000 10.0000  0.0000 20.0000  0.0000  0.0000  0.0000  0.0000
)   0   0   1  -1   0   0   0   0     0.0000  0.0000  0.0000  0.0000  3.0000-20.0000  0.0000  0.0000  0.0000  0.0000

2+ID3+ID4+ID5+ID6+ID7+ID8+ID9+ID10=0
4   0   0   0   0   0   0   0

2+ID3+ID4+ID5+ID6+ID7+ID8+ID9+ID10=0
9   0   1  -1  -1   0   0   0   0
```

<table>
<tr><td>

DC CONVERTER NUMBER   1   INPUT DATA

CONVERTER ATTACHED TO BUS NUMBER 108

NOMINAL DC VOLTAGE	110.00000
MAXIMUM DC VOLTAGE	150.00000
MINIMUM DC VOLTAGE	0.00000
MAXIMUM DC CURRENT	0.0000
COMMUTATION REACTANCE (P.U.)	0.08970
TRANSFORMER REACTANCE (P.U.)	0.08970
FIRING ANGLE:MINIMUM (DEG)	10.00000
MAXIMUM (DEG)	110.00000
TRANSFORMER TAP:MINIMUM (P.C.)	0.00000
MAXIMUM (P.C.)	0.00000
INCREMENT	0.00000
FILTER REACTANCE (P.U.)	1.00000
NUMBER OF BRIDGES IN SERIES	4

SPECIFIED CONTROLS

CONVERTOR POWER FACTOR	
DC LINK VOLTAGE (KV)	
CONVERTOR CONTROL ANGLE	10.00000
TRANSFORMER TAP	
CONVERTER DC POWER (MW)	500.00000
TERMINAL REACTIVE POWER (MVAR)	
AC TERMINAL VOLTAGE (KV)	
CONVERTER DC CURRENT (KA)	

</td><td>

DC CONVERTER NUMBER   6   INPUT DATA

CONVERTER ATTACHED TO BUS NUMBER   7

NOMINAL DC VOLTAGE	90.00000
MAXIMUM DC VOLTAGE	140.00000
MINIMUM DC VOLTAGE	0.00000
MAXIMUM DC CURRENT	0.0000
COMMUTATION REACTANCE (P.U.)	0.07000
TRANSFORMER REACTANCE (P.U.)	0.07000
FIRING ANGLE:MINIMUM (DEG)	8.00000
MAXIMUM (DEG)	150.00000
TRANSFORMER TAP:MINIMUM (P.C.)	0.00000
MAXIMUM (P.C.)	0.00000
INCREMENT	0.00000
FILTER REACTANCE (P.U.)	0.70000
NUMBER OF BRIDGES IN SERIES	2

SPECIFIED CONTROLS

CONVERTER POWER FACTOR	
DC LINK VOLTAGE (KV)	-220.00000
CONVERTER CONTROL ANGLE	8.00000
TRANSFORMER TAP	
CONVERTER DC POWER (MW)	
TERMINAL REACTIVE POWER (MVAR)	
AC TERMINAL VOLTAGE (KV)	
CONVERTER DC CURRENT (KA)	

</td></tr>
</table>

SOLUTION CONVERGED IN 7 P–D AND 6 Q–V ITERATIONS

| LOAD | | GENERATION | | AC LOSSES | | MISMATCH | | SHUNTS |
MW	MVAR	MW	MVAR	MW	MVAR	MW	MVAR	MVAR
4496.80	1518.60	5226.58	791.80	194.91	−306.06	534.87	−182.89	37.85

OPERATING STATE OF CONVERTER 6 WHICH IS ATTACHED TO BUS 7 (BUNTHORPE220)

CONVERTER IS OPERATING IN THE INVERTION MODE
THE CONTROL ANGLE IS THE EXTINCTION ADVANCE ANGLE
DC POWER SUPPLIED TO THE AC SYSTEM = −309.23 MW

CONVERTER AC VOLTAGE (K–VOLTS)	TRANSFORMER TAP (PER CENT)	CONTROL ANGLE (DEGS)	COMMUTATION ANGLE (DEGS)	DC CURRENT (K–AMPS)	DC VOLTAGE (K–VOLTS)
87.94	−8.26	8.00	22.45	1.406	−220.00

POWER TRANSFERS

LINK TERMINAL POWER = −309.23 MW 207.61 MVAR
FROM TRANSFORMER TO CONVERTER = −309.23 MW 125.89 MVAR
REACTIVE POWER OF FILTERS = 142.86 MVAR

BUS DATA

BUS	NAME	VOLTS	ANGLE	GENERATION MW	MVAR	LOAD MW	MVAR	SHUNT MVAR	BUS	NAME	MW	MVAR
104	AVIEMORE–220	1.052	4.78	220.00	−33.87	0.00	0.00	0.00				
									108	BENMORE—220	41.89	−10.17
									108	BENMORE—220	41.89	−10.17
									268	WAITAKI—220	136.22	−13.53
									MISMATCH		0.000	0.000
108	BENMORE—220	1.052	4.43	540.00	−88.04	97.20	0.00	0.00				
									104	AVIEMORE–220	−41.83	7.88
									104	AVIEMORE–220	−41.83	7.88
									255	TWIZEL—220	26.47	6.87
									MISMATCH		500.000	−110.672
118	BRLY——220	0.968	−12.95	0.00	0.00	329.60	95.80	0.00				
									167	ISLINGTON220	−120.18	−76.16
									181	LAND–T02–220	−209.41	−19.64
									MISMATCH		−0.019	−0.002

THREE-PHASE ALGORITHM

4.9 INTRODUCTION

Any converter which is operating from an unbalanced a.c. system will itself operate with unbalanced power flows and unsymmetric valve conduction periods. In addition any unbalance present in the converter control equipment or any asymmetry in the converter transformer will introduce additional unbalance.

Considerable interaction exists between the unbalanced operation of the a.c. and d.c. systems. The exact nature of this interaction depends on features such as the converter transformer connection and the converter firing controller.

High-power converters often operate in systems of relatively low short-circuit ratios where unbalance effects are more likely to be significant and require additional consideration. The steady-state unbalance and its effect in converter harmonic currrent generation may also influence the need for transmission line transpositions and the means of reactive power compensation.

The converter model for unbalanced analysis is considerably more complex than those developed for the balanced case. The additional complexity arises from the need to include the effect of the three-phase converter transformer connection and of the different converter firing control modes. Early h.v.d.c. control schemes were based on phase angle control, where the firing of each valve is timed individually with respect to the appropriate crossing of the phase voltages. This control scheme has proved susceptible to harmonic stability problems when operating from weak a.c. systems. An alternative control, based on equidistant firings on the steady state, is generally accepted to provide more stable operation [5–7]. Under normal steady-state and perfectly balanced operating conditions, there is no difference between these two basic control strategies. However, their effect on the a.c. system and d.c. voltage and current waveshapes during normal, but not balanced, operation, is quite different. A three-phase converter model must be capable of representing the alternative control strategies.

The remainder of this chapter describes the development of a model for the unbalanced converter and its sequential integration with the three-phase fast-decoupled load flow described in Chapter 3.

4.10 FORMULATION OF THE THREE-PHASE A.C.–D.C. LOAD-FLOW PROBLEM

The operating state of the combined system is defined by

$$[\bar{V}_{int}, \bar{\theta}_{int}, \bar{V}, \bar{\theta}, \bar{x}]$$

where

$\bar{V}_{int}/\bar{\theta}_{int}$ are vectors of the balanced internal voltages at the generator internal busbars

$\bar{V}/\bar{\theta}$ are vectors of the three-phase voltages at every generator terminal busbar and every load busbar

$\bar{x}$ is a vector of the d.c. variables (as yet, undefined).

The significance of the three-phase a.c. variables was discussed in Chapter 3 and the selection of d.c. variables $\bar{x}$ is discussed in this section.

To enable a Newton–Raphson-based technique to be used, it is necessary to formulate a set of n independent equations in terms of the n variables describing the system. As explained in Chapter 3, the equations which relate to the a.c. system variables are derived from the specified a.c. system operating conditions. The only modification to these equations, which results from the presence of the d.c. system, occurs at the converter terminal busbars. These equations become

$$\Delta P^p_{term} = (P^p_{term})^{SP} - P^p_{term}(ac) - P^p_{term}(dc) \qquad (4.10.1)$$

$$\Delta Q^p_{term} = (Q^p_{term})^{SP} - Q^p_{term}(ac) - Q^p_{term}(dc) \qquad (4.10.2)$$

where $P^p_{term}(dc)$ and $Q^p_{term}(dc)$ are functions of the a.c. terminal conditions and the converter variables, i.e.

$$P^p_{term}(dc) = f(V^p_{term}, \theta^p_{term}, \bar{x}) \qquad (4.10.3)$$

$$Q^p_{term}(dc) = f(V^p_{term}, \theta^p_{term}, \bar{x}). \qquad (4.10.4)$$

The equations for the a.c. system may therefore be summarised as

$$
\begin{bmatrix}
\Delta \bar{P}(\bar{V}, \bar{\theta}) \\
\Delta \bar{P}_{\text{term}}(\bar{V}, \bar{\theta}, \bar{x}) \\
\Delta \bar{P}_{\text{gen}}(\bar{V}, \bar{\theta}) \\
\Delta \bar{Q}(\bar{V}, \bar{\theta}) \\
\Delta \bar{Q}_{\text{term}}(\bar{V}, \bar{\theta}, \bar{x}) \\
\Delta \bar{V}_{\text{reg}}(\bar{V})
\end{bmatrix} = 0
\tag{4.10.5}
$$

where the mismatches at the converter terminal busbars are indicated separately.

Further equations are derived from the d.c. system conditions. That is, for each converter k a set of equations

$$
\bar{R}(V^p_{\text{term}}, \theta^p_{\text{term}}, \bar{x})_k = 0
\tag{4.10.6}
$$

is derived in terms of the terminal conditions and the converter variables $\bar{x}$.

Equations (4.10.3), (4.10.4) and (4.10.6) form a mathematical model of the d.c. system suitable for inclusion into load-flow analysis. The three-phase a.c.–d.c. load-flow problem may therefore be formulated as the solution of

$$
\begin{bmatrix}
\Delta \bar{P}(\bar{V}, \bar{\theta}) \\
\Delta \bar{P}_{\text{term}}(\bar{V}, \bar{\theta}, \bar{x}) \\
\Delta \bar{P}_{\text{gen}}(\bar{V}, \bar{\theta}) \\
\Delta \bar{Q}(\bar{V}, \bar{\theta}) \\
\Delta \bar{Q}_{\text{term}}(\bar{V}, \bar{\theta}, \bar{x}) \\
\Delta \bar{V}_{\text{reg}}(\bar{V}) \\
R(\bar{V}_{\text{term}}, \bar{\theta}_{\text{term}}, \bar{x})
\end{bmatrix}
\tag{4.10.7}
$$

for the set of variables $(\bar{V}, \bar{\theta}, \bar{x})$.

4.11   D.C. SYSTEM MODELLING

The basic h.v.d.c. interconnection shown in Fig. 4.8 is used as a reference and its extension to other configurations is clarified throughout the development of the model. Under balanced conditions, the converter transformer modifies the source voltages applied to the converter and also affects the phase distribution of current and power. In addition, the a.c. system operation may be influenced (e.g. by a zero-sequence current flow to a star-g-delta transformer) by the transformer connection. Each bridge in Fig. 4.8 will thus operate with a different degree of unbalance, due to the influence of the converter transformer connections, and must be modelled independently. This feature is in contrast to the balanced d.c. model where it is possible to combine bridge in series and in parallel into an equivalent single bridge. The dimensions of the three-phase d.c. model will, therefore, be much greater than the balanced d.c. model.

All converters, whether rectifying or inverting, are represented by the same model (Fig. 4.9) and their equations are of the same form.

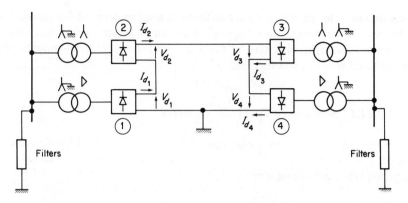

Figure 4.8
Basic h.v.d.c. interconnection

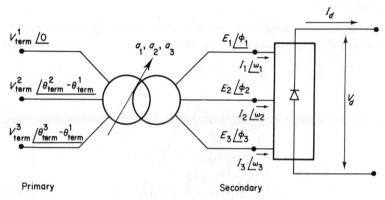

Figure 4.9
Basic converter unit

4.11.1 Basic Assumptions

To enable the formulation of equation (4.10.6) and to simplify the selection of variables $\bar{x}$ the following assumptions are made.

(i) The three a.c. phase voltages at the terminal busbar are sinusoidal.

(ii) The direct voltage an direct current are smooth.

(iii) The converter transformer is lossless and the magnetising admittance is ignored.

Assumptions (ii) and (iii) are equally as valid for unbalanced three-phase analysis as for single-phase analysis. Assumption (i) is commonly used in unbalanced converter studies [8,9] and appears to be backed from the experience of existing schemes. However, a general justification will require more critical examination of the problem.

Under balanced operation only characteristic harmonics are produced and, as filtering is normally provided at these frequencies, the level of harmonic voltages will be small. However, under even small amounts of unbalance, significant noncharacteristic harmonics may be produced and the voltage harmonic distortion at the terminal busbars will increase.

4.11.2   Selection of Converter Variables

The selection of converter variables has already been discussed with regard to the balanced converter model. The main considerations are also relevant to the unbalanced three-phase converter model.

(i) For computing efficiency, the smallest number of variables should be used. A minimum of six independent variables is required to define the operating state of an unbalanced converter, e.g. the three firing angles and the three transformer tap positions.

(ii) To enable the incorporation of a wide range of control specifications, all variables involved in their formulation should be retained. The following variables, defined

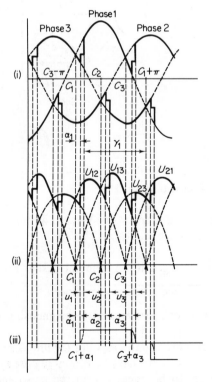

Figure 4.10
Unbalanced converter voltage and current waveform. (i) Phase voltages; (ii) D.C. voltage waveform; (iii) Assumed current waveshape for Phase 1 (actual waveform is indicated by dotted line)

with reference to Fig. 4.9 and 4.10, are required in the formulation of the control specifications for unbalanced converter operation.

- a_i off-nominal tap ratios on the primary side
- $U_{13} \underline{/C_1}$, $U_{23} \underline{/C_2}$, $U_{21} \underline{/C_3}$ phase-to-phase source voltages for the converter referred to the transformer secondary. C_i are therefore the zero crossings for the timing of firing pulses
- α_i Firing delay angle measured from the respective zero crossing
- V_d total average d.c. voltage from complete bridge
- I_d Average d.c. current.

where $i = 1, 2, 3$ for the three phases involved.

In contrast to the balanced case, the secondary phase-to-phase source voltages are included among the variables as they depend not only on the transformer taps but also on the transformer connection. Moreover, the zero crossings, C_i, are explicitly required in the formulation of the symmetrical firing controller and they are also included.

Although these fourteen variables do not constitute the final d.c. model it is convenient to formulate equation (4.10.6) in terms of these variables at this stage, i.e. vector $\bar{x}$ has the form $[U_i, C_i, \alpha_i, a_i, V_d, I_d]^T$. The necessary fourteen equations are derived in the following sections.

4.11.3 Converter Angle References

In the three-phase a.c. load flow described in Chapter 3 all angles are referred to the slack generators internal busbar. Similarly to the single-phase a.c.–d.c. load flow, the angle reference for each converter may be arbitrarily assigned. By using one of the converter angles (e.g. θ^1_{term} in Fig. 4.9) as a references, the mathematical coupling between the a.c. system and converter equations is weakened and the rate of convergence improved.

4.11.4 Per Unit System

Similarly to the single-phase case, computational simplicity is achieved by using common power and voltage bases on both sides of the converter.

In the three-phase case, however, the phase-neutral voltage is used as the base parameter and therefore

$$\text{MVA}_{\text{base}} = \text{base power per phase}$$
$$V_{\text{base}} = \text{phase-neutral voltage base.}$$

The current base on the a.c. and d.c. sides are also equal. Therefore the p.u. system does not change the form of any of the converter equations.

4.11.5 Converter Source Voltages

The phase-to-phase source voltages referred to the transformer secondary are found by a consideration of the transformer connection and off-nominal turns ratio. For example, consider the star–star transformer of Fig. 4.11.

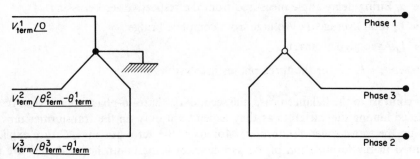

Figure 4.11
Star–star transformer connection

The phase-to-phase source voltages referred to the secondary are

$$U_{13}\!\bigg/ C_1 = \frac{1}{a_1} V^1_{\text{term}}\!\!\underline{/0} - \frac{1}{a_3} V^3_{\text{term}}\!\!\underline{/\theta^3_{\text{term}} - \theta^1_{\text{term}}} \tag{4.11.1}$$

$$U_{23}\!\bigg/ C_2 = \frac{1}{a_2} V^2_{\text{term}}\!\!\underline{/\theta^2_{\text{term}} - \theta^1_{\text{term}}} - \frac{1}{a_3} V^3_{\text{term}}\!\!\underline{/\theta^3_{\text{term}} - \theta^1_{\text{term}}} \tag{4.11.2}$$

$$U_{21}\!\bigg/ C_3 = \frac{1}{a_2} V^2_{\text{term}}\!\!\underline{/\theta^2_{\text{term}} - \theta^1_{\text{term}}} - \frac{1}{a_1} V^1_{\text{term}}\!\!\underline{/0} \tag{4.11.3}$$

which, in terms of real and imaginary parts, yield six equations.

4.11.6 D.C. Voltage

The d.c. voltage, found by integration of the waveforms in Fig. 4.10(ii), may be expressed in the form

$$V_d = \frac{\sqrt{2}}{\pi} \{ U_{21}[\cos(C_1 + \alpha_1 - C_3 + \pi) - \cos(C_2 + \alpha_2 - C_3 + \pi)]$$

$$+ U_{13}[\cos(C_2 + \alpha_2 - C_1) - \cos(C_3 + \alpha_3 - C_1)]$$

$$+ U_{23}[\cos(C_3 + \alpha_3 - C_2) - \cos(C_1 + \alpha_1 + \pi - C_2)]$$

$$- I_d(X_{c1} + X_{c2} + X_{c3})\} \tag{4.11.4}$$

where X_{ci} is the commutation reactance for phase i.

4.11.7 D.C. Interconnection

An equation is derived for each converter, from the d.c. system topology relating the d.c. voltages and currents, i.e.

$$f(V_d, I_d) = 0. \tag{4.11.5}$$

For example, the system shown in Fig. 4.8 provides the four equations

$$Vd_1 + Vd_2 + Vd_3 + Vd_4 - Id_1 Rd = 0$$
$$Id_1 - Id_2 = 0$$
$$Id_1 - Id_3 = 0$$
$$Id_1 - Id_4 = 0.$$

The apparent redundancy in the number of d.c. variables is due to the generality of the d.c. interconnection.

4.11.8 Incorporation of Control Strategies

Similarly to the single-phase case, any function of the variables is a (mathematically) valid control equation so long as the equation is independent of all the others.

Detailed consideration of the alternative firing controls is of particular interest in this respect. With reference to symmetrical firing control, one equation results from the specification of minimum firing angle control, i.e.

$$\alpha_i - \alpha_{min} = 0.$$

For a six-pulse unit, the interval between firing pulses in specified as $60°$. This provides two more equations.

In the equation above, phase (i) is selected during the solution procedure such that the other two phases will have, in the unbalanced case, firing angles greater than α_{min}.

With conventional phase angle control, the firing angle on each phase is specified as being equal to α_{min}, i.e.

$$\alpha_1 - \alpha_{min} = 0 \tag{4.11.6}$$

$$\alpha_2 - \alpha_{min} = 0 \tag{4.11.7}$$

$$\alpha_3 - \alpha_{min} = 0. \tag{4.11.8}$$

The remaining three-control equations required are derived from the operating conditions. Usually, the off-nominal taps are specified as being equal, i.e.

$$a_1 - a_2 = 0 \tag{4.11.9}$$

$$a_2 - a_3 = 0. \tag{4.11.10}$$

The final equation will normally relate to the constant current or constant power controller, e.g.

$$I_d - I_d^{sp} = 0 \tag{4.11.11}$$

or

$$V_d I_d - P_d^{sp} = 0. \tag{4.11.12}$$

4.11.9 Inverter Operation with Minimum Extinction Angle

In contrast to the single-phase load flow, for three-phase inverter operation it is necessary to retain the variable α in the formulation, as it is required in the specification of the symmetrical firing controller. Therefore, the restriction upon the extinction advance angle γ requires the implicit calculation of the commutation angle for each phase.

Using the specification for γ defined in Fig. 4.10, the following expression applies:

$$\cos \gamma_1^{sp} + \cos \alpha_1 - I_d \frac{(X_{c1} + X_{c3})}{\sqrt{2} U_{13}} = 0. \tag{4.11.13}$$

Similar equations apply to the other two phases with a cyclic change of suffixes.

4.11.10 Enlarged Converter Model

The three-phase equations so far developed are exact parallel of the four variable sequential version of the single-phase algorithm.

The mathematical model of the converter includes the formulation of equations (4.10.3) and (4.10.4) for the individual phase real and reactive power flows on the primary of the converter transformer. It is in connection with these equations that the three-phase model deviates significantly from the single-phase model.

The calculation of the individual phase, real and reactive powers at the terminal busbar requires the values of both the magnitude and angle of the fundamental components of the individual phase currents flowing into the converter transformer.

In the single-phase analysis, the magnitude of the fundamental current, obtained from the Fourier analysis of the current waveshape on the transformer secondary, was transferred across the converter transformer. This procedure is trivial and the relevant equations were not included in the d.c. solution. The angle of the fundamental component was calculated by simply equating the total real power on the a.c. and d.c. sides of the converter.

A similar procedure may be applied to the three-phase analysis of the unbalanced converter. In this case, however, the transfer of secondary currents to the primary is no longer a trivial procedure due to the influence of the three-phase transformer connection. In addition, the three-phase converter transformer may influence the a.c. system operation, for example, a star-g–delta connection provides a zero-sequence path for the a.c. system.

The simplest way of accounting for such influence is to include the converter transformer within the d.c. model. The three-phase converter transformer is represented by its nodal admittance model, i.e.

$$Y_{\text{node}} = \begin{array}{|c|c|} \hline Y_{pp} & Y_{ps} \\ \hline Y_{sp} & Y_{ss} \\ \hline \end{array} \tag{4.11.14}$$

where p indicates the primary and s the secondary side of the transformer. The 3×3

submatrices (Y_{pp}, etc.) for the various transformer connections, including modelling of the independent phase taps, were derived in Chapter 3.

The inclusion of the converter transformer within the d.c. model requires 12 extra variables, as follows:

- E_i/ϕ_i the fundamental component of the voltage waveshape at the transformer secondary busbar
- I_i/ω_i the fundamental component of the secondary current waveshapes; where $i = 1, 3$ for the three phases.

Thus a total set of 26 variables is required for each converter in the d.c. system model, fourteen of which have already been developed in previous sections.

4.11.11 Remaining Twelve Equations

With reference to equation (4.11.14), and assuming a lossless transformer (i.e. $Y_{pp} = jb_{pp}$, etc.), the currents at the converter side busbar are expressed as

$$I_i e^{j\omega_i} = - \sum_{k=1}^{3} [jb_{ss}^{ik}E_k e^{j\phi_k} + jb_{sp}^{ik}V_{term}^k e^{j(\theta_{term}^k - \theta_{term}^1)}]. \tag{4.11.15}$$

By subtracting θ_{term}^1 in the above equation, the terminal busbar angles are related to the converter angle reference.

Separating this equation into real and imaginary components, the following six equations result:

$$I_i \cos \omega_i = \sum_{k=1}^{3} [b_{ss}^{ik}E_k \sin \phi_k + b_{sp}^{ik}V_{term}^k \sin (\theta_{term}^k - \theta_{term}^1)] \tag{4.11.16}$$

$$I_i \sin \omega_i = \sum_{k=1}^{3} [- b_{ss}^{ik}E_k \cos \phi_k - b_{sp}^{ik}V_{term}^k \cos (\theta_{term}^k - \theta_{term}^1)]. \tag{4.11.17}$$

Three further equations are derived from approximate expressions for the fundamental r.m.s. components of the line current waveforms as shown in Fig. 4.10, i.e.

$$I_i = 0.995 \frac{4}{\pi} \frac{I_d}{\sqrt{2}} \sin (T_i/2) \tag{4.11.18}$$

where T_i is the assumed conduction period for phase i.

The sum of the real powers on the three phases of the transformer secondary may be equated to the total d.c. power, i.e.

$$\sum_{i=1}^{3} E_i I_i \cos (\phi_i - \omega_i) - V_d I_d = 0. \tag{4.11.19}$$

The derivation of the last two equations is influenced by the position of the fundamental frequency voltage reference for the secondary of the converter transformer.

The voltage reference for the a.c. system is earth, while in d.c. transmission the actual earth is placed on one of the converter d.c. terminals. This point is used as a reference to define the d.c. transmission voltages and the insulation levels of the converter transformer secondary windings.

In load-flow analysis, it is possible to use arbitrary references for each converter unit to simplify the mathematical model. The actual voltages to earth, if required, can then be obtained from knowledge of the particular configuration and earthing arrangements.

With a star-connected secondary winding an obvious reference is the star point itself. If the nodal admittance matrix is formed for a star-g-star-g connection then this reference is implicitly present through the admittance model of the transformer. In this case, however, the converter transformer does not restrict the flow of zero-sequence currents and the following two equations may be written:

$$\sum_{i=1}^{3} I_i \underline{/\omega_i} = 0. \tag{4.11.20}$$

These two equations (real and imaginary parts) complete the set of 12 independent equations in terms of 12 additional variables.

However, the above considerations do not apply to delta-connected secondary windings.

To obtain a reference which may be applied to all transformer secondary windings, an artificial reference node is created corresponding to the position of the zero-sequence secondary voltage. This choice of reference results in the following two equations:

$$\sum_{i=1}^{3} E_i \cos \phi_i = 0 \tag{4.11.21}$$

$$\sum_{i=1}^{3} E_i \sin \phi_i = 0. \tag{4.11.22}$$

The nodal admittance matrix for the star-connected transformer secondary is now formed for an unearthed star winding. The restriction on the zero-sequence current flowing on the secondary is therefore implicitly included in the transformer model for both star and delta connections.

For a star-connected secondary winding both alternatives yield exactly the same solution to the load-flow problem.

4.11.12 Summary of Equations and Variables

The 26 equations ($\bar{R}$) which define the operation of each converter are

$$R(1) = \sum_{i=1}^{3} E_i \cos \phi_i = 0$$

$$R(2) = \sum_{i=1}^{3} E_i \sin \phi_i = 0$$

$$R(3) = \sum_{i=1}^{3} E_i I_i \cos (\phi_i - \omega_i) - V_d I_d$$

$$R(4) = I_1 - \frac{4}{\pi} \frac{I_d}{\sqrt{2}} \sin(T_1/2)$$

$$R(5) = I_2 - \frac{4}{\pi} \frac{I_d}{\sqrt{2}} \sin(T_2/2)$$

$$R(6) = I_3 - \frac{4}{\pi} \frac{I}{\sqrt{2}} \sin(T_3/2)$$

$$R(7) = I_1 \cos \omega_1 - \sum_{k=1}^{3} [b_{ss}^{1k} E_k \sin \phi_k + b_{sp}^{1k} V_{\text{term}}^k \sin(\theta_{\text{term}}^k - \theta_{\text{term}}^1)]$$

$$R(8) = I_2 \cos \omega_2 - \sum_{k=1}^{3} [b_{ss}^{2k} E_k \sin \phi_k + b_{sp}^{2k} V_{\text{term}}^k \sin(\theta_{\text{term}}^k - \theta_{\text{term}}^1)]$$

$$R(9) = I_3 \cos \omega_3 - \sum_{k=1}^{3} [b_{ss}^{3k} E_k \sin \phi_k + b_{sp}^{3k} V_{\text{term}}^k \sin(\theta_{\text{term}}^k - \theta_{\text{term}}^1)]$$

$$R(10) = I_1 \sin \omega_1 + \sum_{k=1}^{3} [b_{ss}^{1k} E_k \cos \phi_k + b_{sp}^{1k} V_{\text{term}}^k \cos(\theta_{\text{term}}^k - \theta_{\text{term}}^1)]$$

$$R(11) = I_2 \sin \omega_2 \times \sum_{k=1}^{3} [b_{ss}^{2k} E_k \cos \phi_k + b_{sp}^{2k} V_{\text{term}}^k \cos(\theta_{\text{term}}^k - \theta_{\text{term}}^1)]$$

$$R(12) = I_3 \sin \omega_3 + \sum_{k=1}^{3} [b_{ss}^{3k} E_k \cos \phi_k + b_{sp}^{3k} V_{\text{term}}^k \cos(\theta_{\text{term}}^k - \theta_{\text{term}}^1)]$$

$R(13)$

$\vdots$ depend on transformer connection

$R(18)$

$R(19)$

$\vdots$ depend on the control specifications

$R(24)$

$$R(25) = V_d \pi - \sqrt{2} U_{21}[\cos(C_1 + \alpha_1 - C_3 + \pi) - \cos(C_2 + \alpha_2 - C_3 + \pi)]$$
$$- \sqrt{2} U_{13}[\cos(C_2 + \alpha_2 - C_1) - \cos(C_3 + \alpha_3 - C_1)]$$
$$- \sqrt{2} U_{23}[\cos(C_3 + \alpha_3 - C_2) - \cos(C_1 + \alpha_1 + \pi - C_2)]$$
$$+ I_d(X_{c1} + X_{c2} + X_{c3})$$

$R(26) = f(V_{di}, I_{di})$ from d.c. system topology.

The 26 variable vector $(\bar{x})$ is:

$$[E_1, E_2\ E_3, \phi_1, \phi_2, \phi_3, I_1, I_2, I_3, \omega_1, \omega_2, \omega_3, u_{12}, u_{13}, u_{23}, C_1, C_2, C_3, \alpha_1, \alpha_2, \alpha_3, a_1, a_2, a_3, V_d, I_d]^{\text{T}}.$$

4.12 LOAD-FLOW SOLUTION

A sequential technique, using the three-phase fast-decoupled a.c. algorithm and a full Newton–Raphson algorithm for the d.c. equations, involves the block successive iteration of the three equations

$$\begin{bmatrix} \Delta \bar{P}(\bar{V}, \bar{\theta})/\bar{V} \\ \Delta \bar{P}_{\mathrm{gen}}/\bar{V}_{\mathrm{int}} \end{bmatrix} = [B'] \begin{bmatrix} \Delta \bar{\theta} \\ \Delta \theta_{\mathrm{int}} \end{bmatrix} \tag{4.12.1}$$

$$\begin{bmatrix} \Delta \bar{Q}(\bar{V}, \bar{\theta})/\bar{V} \\ \Delta \bar{V}_{\mathrm{reg}}(\bar{V}) \end{bmatrix} = [B''] \begin{bmatrix} \Delta \bar{V} \\ \Delta \bar{V}_{\mathrm{int}} \end{bmatrix} \tag{4.12.2}$$

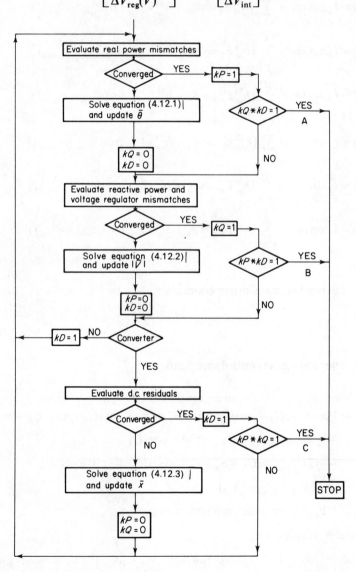

Figure 4.12
Flow chart for three-phase a.c.–d.c. load flow

$$[\bar{R}(\bar{x})] = [J][\Delta\bar{x}] \qquad (4.12.3)$$

where $[B']$ and $[B'']$ are the three-phase fast-decoupled a.c. Jacobian matrices as developed in Chapter 3, and $[J]$ is the d.c. Jacobian of first-order partial derivatives.

Equations (4.12.1) and (4.12.2) are the three-phase fast-decoupled algorithmic equations from Chapter 3. For the solution of the equations (4.12.1) and (4.12.2), the d.c. variables $\bar{x}$ are treated as constants and, in effect, the d.c. system is modelled simply by the appropriate real and reactive power injections at the converter terminal busbar.

These power injections are calculated from the latest solution of the d.c. system equations and are used to form the corresponding real and reactive power mismatches. For the d.c. iteration, the a.c. variables at the terminal busbars are considered to be constant.

The iteration sequence for the solution of equations (4.12.1), (4.12.2) and (4.12.3) is illustrated in Fig. 4.12. It is based on the P, Q, DC sequence described in Section 4.4 which proved the most sucessful sequential technique in the single-phase case.

This sequence acknowledges the fact that the converter operation is strongly related to the magnitude of the terminal voltages and more weakly dependent on their phase angles. Therefore, the converter solution follows the update of the a.c. terminal voltages. It should be noted, however, that in the three-phase case, final convergence is comparatively slow because the d.c. system behaviour is dependent on the phase-angle unbalance as much as on the voltage unbalance.

4.13 PROGRAM STRUCTURE AND COMPUTATIONAL ASPECTS

The main components of the computer program are illustrated in Fig. 4.13. The additional blocks and increase in size of the a.c.–d.c. program over the purely a.c. algorithm may be assesed by comparison with Fig. 3.21. The numbers in parenthesis are the approximate number of FORTRAN statements. The additional features are discussed in the following sections.

4.13.1 D.C. Input Data

The input data for the d.c. system consists of the parameters of each converter including maximum and minimum variable limits where appropriate. In addition, the d.c. network equations (4.11.5) must be formed from the d.c. system topology. As the d.c. system is relatively small and simple in its interconnection these equations are formed by inspection and effectively input directly by the user.

The d.c. system variables $\bar{x}$ are initialised as the balanced three-phase equivalent of the single-phase converter variables as discussed in Section 4.7.

4.13.2 Programmming Aspect of the Iterative Solution

The iterative solution (Fig. 4.12) for the a.c.–d.c. load flow is significantly enlarged over the purely a.c. case (Section 3.7). The basic reason is that the d.c. Jacobian must

124

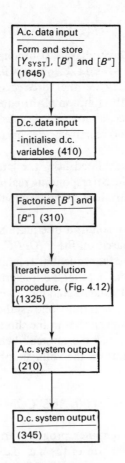

Figure 4.13

be reformed and refactorised at each iteration. In addition, because of the nonuniform nature of the d.c. Jacobians and residual equations, each term must be formulated separately in contrast to the a.c. case where compact program loops may be used.

Equations (4.12.1) and (4.12.2) are solved using sparsity techniques and near optimal ordering as described in Chapter 2. similarly to the single-phase case, the equations for each converter are separate except for those relating to the d.c. interconnection and the solution of equation (4.12.3) is carried out using a modified Gaussian elimination routine.

This feature may be utilised by appropriate ordering of variables to yield a block sparsity structure for the d.c. Jacobian. With this aim, the d.c. voltage variable is placed last for each block of converter equations and all the d.c. current variables are placed after all converter blocks. The d.c. Jacobian will then have a structure as illustrated in Fig. 4.14.

By using row pivoting only during the solution procedure, the block sparsity of Fig. 4.14 is preserved. Each block containing nonzero elements is stored in full, but only nonzero elements are processed.

This routine requires less storage than a normal sparsity program for nonsymmetrical matrices and the solution efficiency is improved.

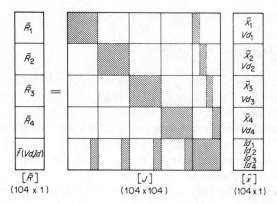

Figure 4.14
Jacobian structures for four-converter d.c. system (nonzero elements indicated)

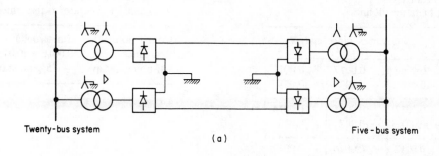

Twenty-bus system

Five-bus system

(a)

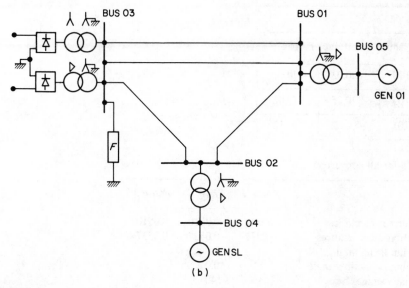

(b)

Figure 4.15
Three-phase a.c.–d.c. test system: (a) h.v.d.c. interconnection; (b) five-bus a.c. system

4.14 PERFORMANCE OF THE ALGORITHM

4.14.1 Test System

The performance of the algorithm is discussed with reference to the test system illustrated in Fig. 4.15. The system consists of two a.c. systems interconnected by a 600 kV, 600 MW h.v.d.c. link.

The 20-bus system is a representation of the 220 kV a.c network of the South Island of New Zealand. It includes mutually coupled parallel lines, synchronous generators and condensers, star–star and star–delta connected transformers and has a total generation in excess of 2000 MW.

At the other end of the link, a fictitious five-bus system represents 800 MW of remote hydrogeneration connected to a converter terminal and load busbar by long, untransposed high-voltage lines.

Table 4.4
System data
(a) Data for all lines.

(b) Data for generator transformers

Z_s series impedance matrix

0.0066 + j0.056	0.0017 + j0.027	0.0012 + j0.021
0.0017 + j0.027	0.0045 + j0.047	0.0014 + j0.022
0.0012 + j0.021	0.0014 + j0.0220	0.0062 + j0.061

Connection	Star-g/delta
Reactance	0.0016 + j0.015
Off-nominal tap	+ 2.5% on star

Y_s shunt admittance matrix

j0.15	− j0.03	− j0.01
− j0.03	j0.25	− j0.02
− j0.01	− j0.02	j0.125

(c) Data for all converters

	Phase 1	Phase 2	Phase 3
Transformer reactances	0.0510	0.0510	0.0510
Commutation reactances	0.0537	0.0537	0.0537
Minimum firing angle		7.0 deg	
Minimum extinction angle		10.0 deg	
Nominal voltage		140 kV	

D.C. link resistance = 25.0 ohms.

Table 4.4 *(continued)*

(d) Generator data

Name	Sequence reactances			Power (MW)	Voltage regulator V^a
	X_0	X_1	X_2		
GEN01	0.02	—	0.004	700.0	1.045
GENSL	0.02	—	0.004	Slack	1.061

(e) Busbar loadings

Bus name	Phase A		Phase B		Phase C	
	P-load	Q-load	P-load	Q-load	P-load	Q-load
BUS01	20.000	10.000	20.000	10.000	20.000	10.000
BUS02	66.667	26.667	66.667	26.667	66.667	26.667
BUS03	0.000	0.000	0.000	0.000	0.000	0.000
BUS04	0.000	0.000	0.000	0.000	0.000	0.000
BUS05	0.000	0.000	0.000	0.000	0.000	0.000

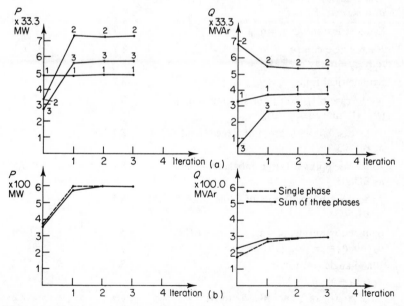

Figure 4.16
Convergence of terminal powers for three-phase converter model. (a) Unbalanced; (b) balanced

The small system is used to test the algorithm and to enable detailed discussion of results. The d.c. link should have considerable influence, as the link power rating is comparable to the total capacity of the small system. Relevant parameters for the a.c. system and d.c. link are given in Table 4.4.

4.14.2 Convergence of D.C. Model from Fixed Terminal Conditions

Typical convergence patterns for the terminal power flows for the three-phase model, under both balanced and unbalanced terminal conditions, are shown in Fig. 4.16. The convergence pattern of the single-phase algorithm is also illustrated. To enable a comparison to be made, the total three-phase powers are plotted for the balanced

Table 4.5
Case descriptions and convergence results

Case	Case description and rectifier specifications	Number of iterations to convergence (0.1 MW/MWAR)	
		20-bus system	5-bus system
a(i)	Converter modelled by equivalent balanced loads*	8.7	6.5
(ii)	Converter modelled by equivalent unbalanced loads*	8.7	6.5
b(i)	Phase-angle control; $\alpha_1 = \alpha_2 = \alpha_3 = \alpha_{min}$, $a_1 = a_2 = a_3$, $P_{dc} = P_d^{sp}$	8.7	6.5
(ii)	Symmetrical firing; $\alpha_i = \alpha_{min}$	8.7	6.5
(iii)	Phase-angle control; $\alpha_1 = \alpha_2 = \alpha_3 = \alpha_{min}$, $a_1 = a_2 = a_3$, $I_{dc} = I_{dc}^{sp}$, $V_{d1} = V_{d2}$	8.7	6.5
(iv)	Symmetrical firing; $\alpha_i = \alpha_{min}$	8.7	6.5
(v)	As for case b(1); with poor starting values. (P_{dc}, Q_{dc} in error by 70%)	8.7	8.7
(vi)	As for case b(i); with large unbalanced load at BUS03	8.7	7.6
(vii)	As for case b(ii); with large unbalanced load at BUS03	8.7	7.6
(viii)	As for case b(i); with loss of 1 line BUS01 to BUS03	8.7	9.9
(ix)	Symmetrical firing; $\alpha_i = \alpha_{min}$, $a_1 = -10\%$, $a_2 = 0$, $a_3 = +10\%$	8.7	7.6
(x)	Phase-angle control; $a_1 = a_2 = a_3 = a^{sp}$, $\alpha_1 = \alpha_2 = \alpha_3$, $p_{dc} = P_{dc}^{sp}$	8.7	7.6
(xi)	Case (x) loss of 1 line. BUS01 to BUS03	8.7	8.8

*Loading for case a(i) and a(ii) derived from results for case b(i). See Table 4.6

case. In all cases the d.c. starting values were selected to give large initial errors in the terminal powers to better illustrate the convergence pattern.

The d.c. equations require two iterations to converge for both the single- and three-phase models.

4.14.3 Performance of the Integrated A.C.–D.C. Load Flow

With reference to the test system illustrated in Fig. 4.15, the following control specifications are used at the inverting terminal for all test cases:

• symmetrical firing control with the reference phase on minimum extinction angle

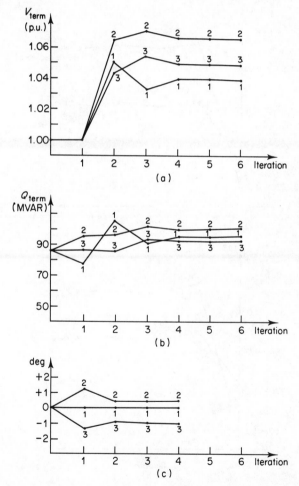

Figure 4.17
Convergence patterns of terminal conditions for a strong a.c. system: (a) a.c. terminal voltages; (b) terminal reactive power flows; (c) a.c. terminal angle unbalance (deviation from nominal)

- off-nominal tap ratios equal on all phases
- d.c. voltage specified.

A variety of different control strategies are considered at the rectifier terminal and the convergence results are given in Table 4.5. For comparison, the table includes cases with the converters modelled as equivalent a.c. loads.

It should be noted that the iteration scheme illustrated in Fig. 4.12 does not allow for each individual a.c. system to be converged independently, therefore, the number of iterations required is the larger of the two sets given in the table.

It is clear that the integration of the d.c. converter model does not cause any significant deterioration in performance. The only cases where convergence is slowed are (viii) and (xi) where the system is weakened by the loss of one transmission line.

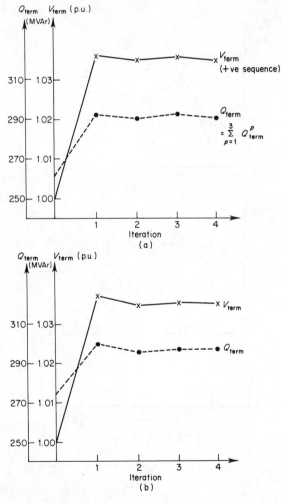

Figure 4.18
Comparison of single-phase and three-phase positive sequence convergence patterns. (a) Three-phase load flow; (b) single-phase load flow

This is to be expected from the discussion of single-phase sequential algorithms given in Section 4.7

To examine the effect of a weak system in the three-phase case, the convergence patterns for the terminal powers and voltages are shown for case (xi) in Fig. 4.17. The reactive power and voltage unbalance vary considerably over the first few iterations but this initial variation does not cause any convergence problems. With weaker systems, the unbalance increases and the convergence patterns become more oscillatory. The corresponding convergence pattern of the single-phase load flow for case (xi) is shown in Fig. 4.18(b) where a similar oscillatory pattern is observable. Moreover the sum of the three-phase reactive powers and the average phase voltage of Fig. 4.17, plotted in Fig. 4.18(a), shows an even closer similarity between the three-phase case and the single-phase behaviour.

4.14.4   Sample Results

The operating states of the two converters connected to BUS03 are listed for the most typical cases in Table 4.6. The corresponding a.c. system voltage profiles and generation results are given for cases a(i), b(i) and b(ii) in Table 4.7. The following discussion is with reference to these results.

Table 4.6(a)
Converter 1 results

				Commun-	Terminal powers		d.c. conditions	
		Firing	Tap	tation				
		angle	ratio	angle	Real	Reactive	Voltage	Current
Case	Phase	α_i (deg)	a_i (%)	u_i (deg)	P_i (MW)	Q_i (MVAr)	Vd_1 (kV)	Id_1 (kA)
b(i)	1	7.00	5.5	29.79	98.1	48.1	292.8	1.0246
	2	7.00	5.5	29.32	101.7	50.8	—	—
	3	7.00	5.5	29.61	100.3	48.3	—	—
b(ii)	1	7.00	5.3	29.78	98.6	49.0	292.8	1.0246
	2	7.20	5.3	29.14	100.9	51.3	—	—
	3	8.43	5.3	28.50	100.6	47.8	—	—
b(vi)	1	7.00	4.8	29.17	95.6	39.5	292.8	1.0246
	2	7.00	4.8	29.16	101.9	50.5	—	—
	3	7.00	4.8	30.43	102.44	57.2	—	—
b(vii)	1	7.00	3.9	29.03	97.6	39.1	292.8	1.0246
	2	11.64	3.9	25.63	101.8	54.7	—	—
	3	9.37	3.9	28.56	100.6	57.7	—	—
b(ix)	1	11.00	−10.0	24.32	104.6	49.4	314.1	0.9483
	2	7.00	0.0	27.76	101.1	45.4	—	—
	3	7.55	10.0	26.08	92.1	44.03	—	—

Table 4.6(b)
Converter 2 results

		Converter 2 (star–star)						
					Terminal powers		d.c. conditions	
Case	Phase	Firing angle α_i (deg)	Tap ratio a_i (%)	Commun-tation angle u_i (deg)	Real P_i (MW)	Reactive Q_i (MVAr)	Voltage Vd_1 (kV)	Current Id_1 (kA)
b(i)	1	7.00	5.5	29.80	97.3	49.2	292.8	1.0246
	2	7.00	5.5	29.60	102.6	53.2	—	—
	3	7.00	5.5	29.32	100.14	44.7	—	—
b(ii)	1	8.03	5.2	28.97	96.4	50.0	292.8	1.0246
	2	7.00	5.2	29.57	102.7	52.9	—	—
	3	8.55	5.2	28.08	100.87	45.66	—	—
b(vi)	1	7.00	4.3	30.63	67.9	13.0	292.8	1.0246
	2	7.00	4.3	28.92	95.5	89.4	—	—
	3	7.00	4.3	28.90	136.6	53.7	—	—
b(vii)	1	7.00	3.0	30.48	70.9	17.9	292.8	1.0246
	2	14.95	3.0	23.25	90.1	94.1	—	—
	3	13.41	3.0	24.25	138.9	52.2	—	—
b(ix)	1	8.08	−10.0	25.42	88.9	65.3	314.7	0.9483
	2	8.38	0.0	27.30	122.6	49.9	—	—
	3	7.00	10.0	26.96	86.9	24.2	—	—

The results of the realistic three-phase converter model (case b(i), although distinguishable from those of the balanced model a(i), are not significantly different as regards the a.c. system operation. They are definitely significant, however, as regards converter operation, particularly when consideration is given to the harmonic content.

A comparison of cases b(i) and b(ii) shows an increase in reactive power consumption in case b(ii) due to two phases having greater than minimum firing angles.

The results also show that the transformer connection modifies the converter source voltages and the phase distribution of power flows. Under balanced conditions, a zero-sequence voltage may appear at system busbars. As the converter has no zero-sequence path, zero-sequence current will only flow when the converter transformer provides a path, as in the case of the star-g–delta transformer. A typical example is illustrated in Fig. 4.19 where the zero-sequence voltages and currents are shown for case b(i). Accurate converter transformer models must therefore be included in the converter modelling.

4.14.5 Conclusions on Performance of the Algorithm

The fast-decoupled three-phase a.c.–d.c. load flow behaves in a very similar manner to the corresponding single-phase version. The following general conclusions can be made on its performance.

Table 4.7
Bus voltages and generation results
Case a(i)

Bus name	Phase A		Phase B		Phase C		Generation total	
	Voltage	Angle	Voltage	Angle	Voltage	Angle		
BUS01	1.067	27.294	1.067	−92.891	1.061	147.431	0.000	0.000
BUS02	1.054	25.190	1.065	−94.670	1.057	144.915	0.000	0.000
BUS03	1.038	23.185	1.071	−95.714	1.043	142.567	0.000	0.000
BUS04	1.045	−3.566	1.046	−123.479	1.047	116.436	173.621	74.723
BUS05	1.061	2.683	1.062	−117.367	1.061	122.628	700.000	113.920

Case b(i)

Bus name	Phase A		Phase B		Phase C		Generation total	
	Voltage	Angle	Voltage	Angle	Voltage	Angle		
BUS01	1.067	27.362	1.065	−92.955	1.062	147.437	0.000	0.000
BUS02	1.055	25.232	1.064	−94.717	1.057	144.925	0.000	0.000
BUS03	1.038	23.517	1.066	−95.965	1.049	142.543	0.000	0.000
BUS04	1.045	−3.552	1.046	−123.483	1.047	116.438	173.570	74.706
BUS05	1.061	2.690	1.062	−117.369	1.060	122.634	700.000	113.680

Case b(ii)

Bus name	Phase A		Phase B		Phase C		Generation total	
	Voltage	Angle	Voltage	Angle	Voltage	Angle		
BUS01	1.066	27.31	1.066	−92.942	1.062	147.421	0.000	0.000
BUS02	1.054	25.238	1.064	−94.705	1.057	144.913	0.000	0.000
BUS03	1.036	23.532	1.066	−95.947	1.049	142.506	0.000	0.000
BUS04	1.045	−3.563	1.046	−123.479	1.047	116.439	173.593	75.949
BUS05	1.061	2.690	1.062	−117.363	1.060	122.635	700.000	115.391

- The number of iterations to convergence is not significantly increased by the presence of the d.c. converters.
- D.C. convergence is not dependent on the specific control specifications applied to each converter.
- Wide errors in initial conditions may be tolerated.
- For very weak a.c. systems the ineraction of the converter with the a.c. system is increased and the convergence is slowed. Sucessful convergence can, however, be expected in all practical cases.
- The algorithm exibits good reliability even under extreme unbalance.

134

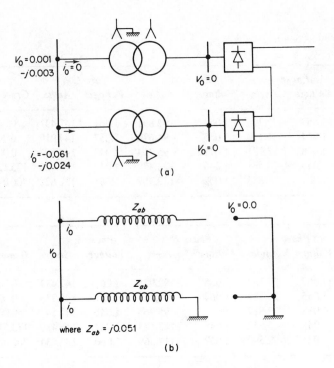

$V_0 = 0.001$
$-j0.003$ $i_0 = 0$

$V_0 = 0$

$i_0 = -0.061$
$-j0.024$

$V_0 = 0$

(a)

Z_{ab}

i_0

V_0

$V_0 = 0.0$

Z_{ab}

i_0

where $Z_{ab} = j0.051$

(b)

Figure 4.19
Sequence components and the converter transformer connection. (a) Zero sequence potentials for case *b* (i); (b) zero sequence network for converter transformers. (Note: Transformer secondary zero sequence reference is provided by equations (4.11.21) and (4.11.22).)

4.15   REFERENCES

[1] B. Stott and O. Alsac, 1974. Fast decoupled load flow, *IEEE Trans.* **PAS-93** 859–869.
[2] H. Sato and J. Arrillaga, 1969. Improved load-flow techniques for integrated a.c.–d.c. systems, *Proc. IEE* **116** 525–532.
[3] J. Reeve, G. Fahmy, and B. Stott, 1976. Versatile load-flow method for multi-terminal h.v.d.c. systems, *Paper F76-354-1* IEEE PES Summer Meeting, Portland.
[4] B. Stott, 1971. Load flow for a.c. and integrated a.c.–d.c. systems. *PhD Thesis* University of Manchester.
[5] J. D. Ainsworth, 1967. Harmonic instability between controlled static converters and a.c. networks, *Proc. IEE* **114** 949–957.
[6] J. D. Ainsworth, 1968. The phase-locked oscillator—A new control system for controlled static converters, *IEEE Trans.* **PAS-87** 859–865.
[7] J. Kauferle, R. Mey and Y. Rogowsky, 1970. H.V.D.C. stations connected to weak a.c. systems, *IEEE Trans.* **PAS-85** 1610–1617.
[8] A. G. Phadke and J. H. Harlow, 1966. Unbalanced converter operation, *IEEE Trans.* **PAS-85** 233–239.
[9] J. Arrillaga and A. E. Efthymiadis, 1968. Simulation of converter performance under unbalanced conditions, *Proc. IEE* **155** 1809–1818.

5. FAULTED SYSTEM STUDIES

5.1 INTRODUCTION

The main object of fault analysis is to calculate fault currents and voltages for the determination of circuit-breaker capacity and protective relay performance.

Early methods used in the calculation of fault levels involved the following approximations.

- All voltage sources assumed a one per unit magnitude and zero relative phase, which is equivalent to neglecting the prefault load current contribution.
- Transmission plant components included only inductive parameters.
- Transmission line shunt capacitance and transformer magnetising impedance were ignored.

Based on the above assumptions, simple equivalent sequence impedance networks were calculated and these were interconnected according to the fault specification. Conventional circuit analysis was then used to calculate the sequence voltage and currents and with them, by means of the inverse sequence component transformation, the phase components.

Although the basic procedure of the computer solution is still the same, the need for the various approximations has disappeared.

The three-phase models of transmission plant developed in Chapter 3, which included interphase and parallel line mutual effects, could be easily combined to produce the faulted system matrix admittance or matrix impedance and hence provide an accurate model for the analysis of a.c. system faults.

However, the main reasons given for the use of the phase frame of reference in load flows are less relevant here. Extra losses and harmonic content are less of a problem in the short period of time prior to fault clearance. Fault studies are normally performed on systems reasonably well balanced either at the operational or planning stage; in the latter case only after prospective system configurations have been proved acceptable through load-flow studies.

Moreover, faulted system studies constitute an integral part of multi-machine transient stability programs, the complexity of which will not normally permit the three-phase approach.

A single-phase representation, achieved with the help of the symmetrical components transformation [1] is used in this chapter as a basis for the development of a fault-study program [2–5].

5.2 ANALYSIS OF THREE-PHASE FAULTS

A preliminary stage to the analysis is the collection of appropriate data specifying the system to be analysed in terms of prefault voltage, loading and generating conditions. Such data is then processed to form a nodal equivalent network constituted by admittances and injected currents.

The equivalent circuits of loads, lines and transformers discussed in Chapter 3 are directly applicable here. The generators can be represented by a constant voltage E^M behind an approximate machine admitance y^M, the value of which depends on the time of the calculation from the instant of fault inception. This is illustrated in Fig. 5.1(a).

When analysing the first two or three cycles following the fault, the subtransient admittance of the machine is normally used, whilst for longer times, it is more appropriate to use the transient admittance. The machine model, illustrated in Fig. 5.1(a), is then converted to a nodal equivalent by means of Norton's Theorem which changes the voltage source into a current source injected at the bus j as shown in Fig. 5.1(b). This is most effective as otherwise a further node at j' is necessary to define the machine admittance y^M.

The injected nodal current is given by

$$I_j = y_j^M E_j^M \tag{5.2.1}$$

where

$$E_j^M = V_j + \frac{I_j^M}{y_j^M} \tag{5.2.2}$$

so that

$$I_j = y_j^M V_j + I_j^M. \tag{5.2.3}$$

I_j^M is the current required at the voltage V_j to produce the machine power $P_j^M + jQ_j^M$, so

$$(I_j^M)^* V_j = P_j^M + jQ_j^M. \tag{5.2.4}$$

Thus from the load-flow data of P^M, Q^M and V^M we may calculate the injected nodal current I_j as

$$I_j = y_j^M V_j + \frac{P_j^M - jQ_j^M}{V_j^*}. \tag{5.2.5}$$

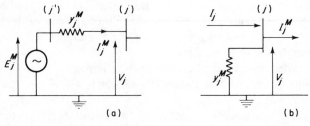

Figure 5.1
Generator representation

5.2.1 Admittance Matrix Equation

Let us take as a reference the small system of Fig. 5.2. Each element is converted to its nodal equivalent. These are connected together as shown in Fig. 5.3 and finally simplified to the equivalent circuit of Fig. 5.4.

The following equations may then be written for the network of Fig. 5.4:

$$I_1 = y_{11}V_1 + y_{12}(V_1 - V_2) \tag{5.2.6}$$

$$I_2 = y_{12}(V_2 - V_1) + y_{22}V_2 + y_{23}(V_2 - V_3) + y_{24}(V_2 - V_4) \tag{5.2.7}$$

$$I_3 = y_{23}(V_3 - V_2) + y_{33}V_3 + y_{34}(V_3 - V_4) \tag{5.2.8}$$

$$I_4 = y_{24}(V_4 - V_2) + y_{34}(V_4 - V_3) + y_{44}V_4 + y_{45}(V_4 - V_5) \tag{5.2.9}$$

$$I_5 = y_{45}(V_5 - V_4) + y_{55}V_5 \tag{5.2.10}$$

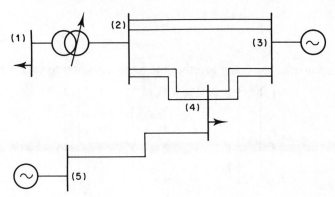

Figure 5.2
Example of small power system

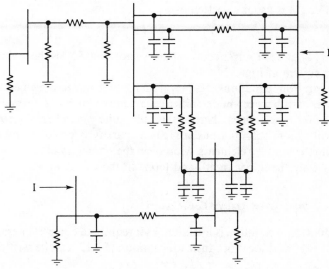

Figure 5.3
Model substitution

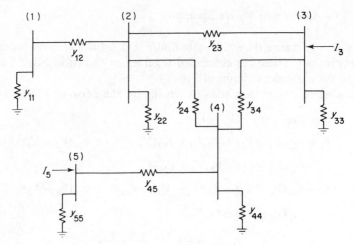

Figure 5.4
Final equivalent

or in matrix form after grouping together the terms common to each voltage

$$
\begin{bmatrix} I_1 \\ I_2 \\ I_3 \\ I_4 \\ I_5 \end{bmatrix} = \begin{bmatrix} Y_{11} & Y_{21} & Y_{31} & Y_{41} & Y_{51} \\ Y_{12} & Y_{22} & Y_{32} & Y_{42} & Y_{52} \\ Y_{13} & Y_{23} & Y_{33} & Y_{43} & Y_{53} \\ Y_{14} & Y_{24} & Y_{34} & Y_{44} & Y_{54} \\ Y_{15} & Y_{25} & Y_{35} & Y_{45} & Y_{55} \end{bmatrix} \begin{bmatrix} V_1 \\ V_2 \\ V_3 \\ V_4 \\ V_5 \end{bmatrix}
\tag{5.2.11}
$$

where

$$
Y_{ii} = \sum_j y_{ij} \qquad Y_{ij} = -y_{ij} \qquad i \neq j.
$$

Equation (5.2.11) is usually written as

$$
[I] = [Y] \cdot [V]
\tag{5.2.12}
$$

where $[I]$ and $[V]$ are the current and voltage vectors and $[Y]$ is the nodal admittance matrix of the system of Fig. 5.2.

It can be seen from equations (5.2.6) to (5.2.10) that nonzero elements only occur where branches exist between nodes. Since each node or busbar is normally connected to fewer than four other nodes, there are usually quite a number of zero elements in any system with more than ten busbars. Such sparsity is exploited by only storing and processing the nonzero elements. Moreover, the symmetry of the matrix ($Y_{ij} = Y_{ji}$) permits using only the upper right-hand terms in the calculations.

5.2.2 Impedance Matrix Equation

The nodal admittance equation is inefficient as it requires a complete iterative solution for each fault type and location. Instead, equation (5.2.12) can be written as

$$
[V] = [Y]^{-1} \cdot [I]
$$
$$
= [Z] \cdot [I].
\tag{5.2.13}
$$

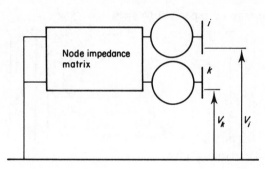

Figure 5.5
Thevenin equivalent of prefault system

This equation uses the bus nodal impedance matrix $[Z]$ and permits using the Thevenin equivalent circuit as illustrated in Fig. 5.5 which, as will be shown later, provides a direct solution of the fault conditions at any node. However, the use of conventional matrix inversion techniques results in an impedance matrix with nonzero terms in every position Z_{ij}.

The sparsity of the $[Y]$ matrix may be retained by using an efficient inversion technique [6, 7] and the nodal impedance matrix can then be calculated directly from the factorised admittance matrix.

5.2.3 Fault Calculations

From the initial machine data, the values of $[I]$ are first calculated from equation (5.2.5) using one per unit voltages. These may now be used to obtain a better estimate of $[V]$, the prefault voltage at every node from equation (5.2.13). If the initial data are supplied from a load flow, this calculation will not make any difference.

The program now has sufficient information to calculate the voltages and currents during a fault.

From Fig. 5.6 the voltage at the fault bus k is

$$V_k^f = Z^f I^f \qquad (5.2.14)$$

where k is the bus to be faulted, Z^f is the fault impedance and I^f is the fault current.

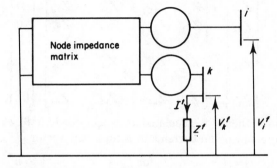

Figure 5.6
Thevenin equivalent of faulted system

Equation (5.2.13) may be expanded to yield

$$
\begin{bmatrix} V_1 \\ V_2 \\ \cdot \\ V_k \\ \cdot \\ V_n \end{bmatrix} = \begin{bmatrix} Z_{11} & Z_{12} & \cdot & Z_{1k} & \cdot & Z_{1n} \\ Z_{21} & Z_{22} & \cdot & Z_{2k} & \cdot & Z_{2n} \\ \cdot & & \cdot & & \cdot & \\ Z_{k1} & Z_{k2} & \cdot & Z_{kk} & \cdot & Z_{kn} \\ \cdot & & \cdot & & \cdot & \\ Z_{n1} & Z_{n2} & \cdot & Z_{nk} & \cdot & Z_{nn} \end{bmatrix} \cdot \begin{bmatrix} I_1 \\ I_2 \\ \cdot \\ I_k \\ \cdot \\ I_n \end{bmatrix}.
\tag{5.2.15}
$$

Selecting row k and expanding gives

$$
V_k = Z_{k1}I_1 + Z_{k2}I_2 + \ldots + Z_{kk}I_k + \ldots + Z_{kn}I_n.
\tag{5.2.16}
$$

This equation describes the voltage at bus k prior to the fault. During a fault a large fault current I^f flows out of bus k. Including this current in equation (5.2.16) and using equation (5.2.14) gives

$$
V_k^f = Z^f I^f = Z_{k1}I_1 + \ldots + Z_{kk}I_k + \ldots + Z_{kn}I_n - Z_{kk}I^f
\tag{5.2.17}
$$

or

$$
Z^f I^f = V_k - Z_{kk}I^f
\tag{5.2.18}
$$

and so the fault current is given directly by

$$
I^f = \frac{V_k}{Z_{kk} + Z^f}.
\tag{5.2.19}
$$

Also from equation (5.2.15) the prefault voltage at any other bus j is

$$
V_j = Z_{j1}I_1 + Z_{j2}I_2 + \ldots + Z_{jk}I_k + \ldots + Z_{jn}I_n
\tag{5.2.20}
$$

and during the fault

$$
V_j^f = Z_{j1}I_1 + Z_{j2}I_2 + \ldots + Z_{jk}I_k + \ldots + Z_{jn}I_n - Z_{jk}I^f
\tag{5.2.21}
$$

or

$$
V_j^f = V_j - Z_{jk}I^f.
\tag{5.2.22}
$$

From equations (5.2.19) and (5.2.22) the fault voltages at every bus in the system may be calculated, each calculation requiring only one column of the impedance matrix. The kth column can be obtained by multiplying the impedance matrix by a vector which has a '1' in the kth row and '0' elsewhere, i.e.

$$
\begin{bmatrix} Z_{1k} \\ Z_{2k} \\ \cdot \\ Z_{kk} \\ \cdot \\ Z_{nk} \end{bmatrix} = \begin{bmatrix} Z_{11} & Z_{12} & \cdot & Z_{1k} & \cdot & Z_{1n} \\ Z_{21} & Z_{22} & \cdot & Z_{2k} & \cdot & Z_{2n} \\ \cdot & & \cdot & & \cdot & \\ Z_{k1} & Z_{k2} & \cdot & Z_{kk} & \cdot & Z_{kn} \\ \cdot & & \cdot & & \cdot & \\ Z_{n1} & Z_{n2} & \cdot & Z_{nk} & \cdot & Z_{nn} \end{bmatrix} \cdot \begin{bmatrix} 0 \\ 0 \\ \cdot \\ 1 \\ \cdot \\ 0 \end{bmatrix}.
\tag{5.2.23}
$$

Once Z_{kk} is known then I^f is calculated from equation (5.2.19). I^f is then subtracted from the initial prefault nodal currents to form a new vector $[I^f]$ defined by

$$
\begin{aligned}
I_j^f &= I_j && \text{for } j \neq k, && j = 1 \text{ to } n \\
I_k^f &= I_k - I^f && \text{for } j = k.
\end{aligned}
$$

The voltages during the fault are given by the product of the impedance matrix and this new vector $[I^f]$, i.e.

$$[V^f] = [Z] \cdot [I^f].$$ (5.2.24)

Equation (5.2.24) is equivalent to (5.2.22) because of the expansion

$$[I^f] = [I] - [0, 0, 0, \ldots I^f, \ldots 0]^T$$

from which equation (5.2.24) expands as

$$[V^f] = [Z]\{[I] - [0, 0, 0, \ldots I^f, \ldots 0]^T\}$$

or

$$[V^f] = [V] - [Z] \cdot [0, 0, \ldots I^f, \ldots 0]^T$$

which is equivalent to equation (5.2.22).

Once the fault voltages are known the branch currents between buses can be calculated from the original branch admittances, i.e.

$$I_{ij}^f = y_{ij}\{V_i^f - V_j^f\}.$$ (5.2.25)

A correction is necessary for the sending end current of a tapped transformer, i.e.

$$I_{ij}^f = y_{ij}\{(1 - \tau)V_i^f - V_j^f\}.$$ (5.2.26)

With reference to Fig.5.7, a machine fault current contribution is

$$I_i^{Mf} = (E_i^M - V_i^f)y_i^M$$

or substituting $I_i = y_i^M E_i^M$ (from equation (5.2.1))

$$I_i^{Mf} = I_i^M - V_i^f y_i^M.$$ (5.2.27)

5.3 ANALYSIS OF UNBALANCED FAULTS

If the network is unsymmetrically faulted or loaded, neither the phase currents nor the phase voltage will possess three-phase symmetry. The analysis can no longer be

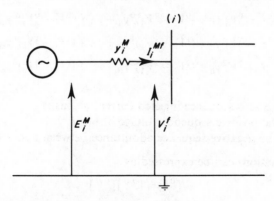

Figure 5.7
Machine representation showing fault current contribution

limited to one phase and the admittance of each element will consist of a 3×3 matrix which on the assumption of a reasonably balanced transmission system, will be symmetrical, i.e.

$$\begin{bmatrix} {}^{aa}Y & {}^{ab}Y & {}^{ac}Y \\ & {}^{aa}Y & {}^{ab}Y \\ & & {}^{aa}Y \end{bmatrix}. \tag{5.3.1}$$

Matrix (5.3.1) can be diagonalised by the symmetrical components transformation $(T^*)^t Y T$ into its sequence component equivalent, i.e.

$$\begin{bmatrix} {}^{0}Y & & \\ & {}^{1}Y & \\ & & {}^{2}Y \end{bmatrix} \tag{5.3.2}$$

where

$$\begin{bmatrix} {}^{0}Y = {}^{aa}Y + {}^{ab}Y + {}^{ac}Y \\ {}^{1}Y = {}^{aa}Y + a({}^{ab}Y) + a^2({}^{ac}Y) \\ {}^{2}Y = {}^{aa}Y + a^2({}^{ab}Y) + a({}^{ac}Y) \\ a = e^{j2\pi/3} \end{bmatrix}. \tag{5.3.3}$$

Moreover, for stationary balanced system elements the admittances ${}^{ab}Y$ and ${}^{ac}Y$ are equal and equations (5.3.3) show that the corresponding positive and negative sequence admittances are also equal. Further, the simplifying assumption is often made that the positive and negative sequence admittances of rotating machines are equal. This assumption is only reasonable when the subtransient admittances are being used and in such case the storage required by the program can be substantially reduced by deleting the negative sequence matrices.

5.3.1 Admittance Matrices

The data specifying each element of the system are then used to form the following three nodal equations.

$$^{0}I_i = {}^{0}V_i^{0}y_{ii} + ({}^{0}V_i + {}^{0}V_1)^{0}y_{ii} + \dots + ({}^{0}V_i - {}^{0}V_n)^{0}y_{ni} \tag{5.3.4}$$

$$^{1}I_i = {}^{1}V_i^{1}y_{ii} + ({}^{1}V_i - {}^{1}V_1)^{1}y_{ii} + \dots + ({}^{1}V_i - {}^{1}V_n)^{1}y_{ni} \tag{5.3.5}$$

$$^{2}I_i = {}^{2}V_i^{2}y_{ii} + ({}^{2}V_i - {}^{2}V_1)^{2}y_{ii} + \dots + ({}^{2}V_i - {}^{2}V_n)^{2}y_{ni} \tag{5.3.6}$$

where

$^{0}I_i$ is the zero-sequence injected current at bus i
$^{1}V_i$ is the positive-sequence voltage at bus i
$^{2}y_{ni}$ is the negative-sequence admittance between nodes n and i.

The above equations can be expressed as

$$[^{0}I] = [^{0}Y][^{0}V] \tag{5.3.7}$$

$$[^{1}I] = [^{1}Y][^{1}V] \tag{5.3.8}$$

$$[^{2}I] = [^{2}Y][^{2}V] \tag{5.3.9}$$

where

$$^{\gamma}Y_{ij} = -\,^{\gamma}Y_{ij} \qquad \text{for} \quad i=1,n;\,j=1,n;\,j \neq 1 \quad \text{and} \quad \gamma=0,1,\text{ or } 2$$

$$^{\gamma}Y_{ii} = \sum_{k=1}^{n}\,^{\gamma}y_{ik} \qquad \text{for} \quad i=1,n;\,\gamma=0,1,\text{ or } 2.$$

The sequence admittance matrices can now be triangularised by a factorisation method. Since the three admittance matrices have identical structure, this can be made more efficient by triangularising them simultaneously, i.e. in programming terms, only one set of vectors is needed to form pointers to the three arrays as they are stored by the factorisation routine.

5.3.2 Fault Calculations

As already explained for the three-phase fault, the nodal impedance matrices may now be calculated directly from the reduced admittance matrices and the following sequence impedance matrix equations result:

$$[^{0}V] = [^{0}Z][^{0}I] \tag{5.3.10}$$

$$[^{1}V] = [^{1}Z][^{1}I] \tag{5.3.11}$$

$$[^{2}V] = [^{2}Z][^{2}I]. \tag{5.3.12}$$

Because the system is assumed to be balanced prior to the fault, the vectors of negative- and zero-sequence currents are zero, i.e. there are no prefault negative- or zero-sequence voltages.

The positive-sequence network then models the prefault network condition and equation (5.3.11) is used to calculate the prefault voltages. If the original voltages used in the machine models were obtained from a load-flow calculation, then the use of equation (5.3.11) will make no difference to those results; however, if the voltages were assumed at one p.u. with zero angle then this calculation will provide more accurate prefault voltages.

The single-phase equivalent circuit is then set up by linking the three sequence network together according to the type of fault to be analysed [8].

5.3.3 Short-circuit Faults

A convenient way of simulating the fault location F for the analysis of short-circuit faults is illustrated in Fig. 5.8. It includes three fault impedances $^{a}Z, ^{b}Z$ and ^{c}Z and three injected currents $^{a}I^{f}, ^{b}I^{f}$ and $^{c}I^{f}$.

For each type of fault, it is possible to write 'boundary conditions' for the currents and voltages at the fault location. For example, Fig. 5.9 shows the case of a line-to-ground fault at bus k.

The boundary conditions are

$$^{b}I_{k}^{f} = \,^{c}I_{k}^{f} = 0 \tag{5.3.13}$$

and

$$^{a}V_{k}^{f} = \,^{a}Z^{f} \cdot {}^{a}I^{f}. \tag{5.3.14}$$

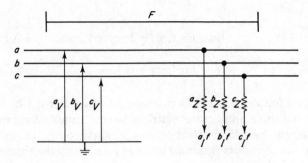

Figure 5.8
The fault location

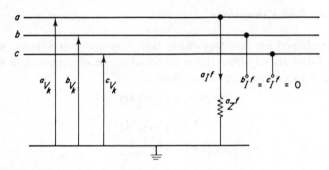

Figure 5.9
Single line-to-ground fault

Using equations (5.3.13) and (5.3.14) with the sequence components transformation the following relationships result:

$$^0I^f = {}^1I^f = {}^2I^f = {}^aI^f/3 \tag{5.3.15}$$

and

$$^0V_k^f + {}^1V_k^f + {}^2V_k^f = {}^aZ^f \cdot {}^aI^f = 3 \cdot (Z^f)^1 I^f. \tag{5.3.16}$$

Also, the sequence voltages at the fault location may be described by the equations

$$^0V_k^f = - {}^0Z_{kk} \cdot {}^0I^f \tag{5.3.17}$$

$$^1V_k^f = {}^1V_k - {}^1Z_{kk} \cdot {}^1I^f \tag{5.3.18}$$

$$^2V_k^f = - {}^2Z_{kk} \cdot {}^2I^f. \tag{5.3.19}$$

From equations (5.3.15) to (5.3.19), the following relationships are obtained.

$$^0I^f = {}^1I^f = {}^2I^f = \frac{{}^1V_k}{{}^0Z_{kk} + {}^1Z_{kk} + {}^2Z_{kk} + 3Z^f}. \tag{5.3.20}$$

Similar considerations yield the fault currents for other types of short-circuit fault. The results for line-to-ground, line-to-line, line-to-line-to-ground, and line-to-line-to-line faults are illustrated in Table 5.1.

These fault currents at the fault location are than added to the current vectors $[^0I]$, $[^1I]$ and $[^2I]$ to produce the fault current vectors $[^0I^f]$, $[^1I^f]$ and $[^2I^f]$. For a fault

Table 5.1.
Fault currents for short-circuit faults

Fault	$^1I^f$	$^2I^f$	$^0I^f$
$L-G$	$\dfrac{V_i}{^1Z_{ii} + {}^2Z_{ii} + {}^0Z_{ii} + 3Z^f}$	$^1I^f$	$^1I^f$
$L-L$	$\dfrac{V_i}{^1Z_{ii} + {}^2Z_{ii} + Z^f}$	$-{}^1I^f$	0
$L-L-G$	$\dfrac{V_i}{(^2Z'_{ii} \cdot {}^0Z'_{ii})/(^2Z'_{ii} + {}^0Z'_{ii}) + {}^1Z'_{ii}}$	$\dfrac{-{}^0Z'_{ii} \cdot {}^1I^f}{^2Z'_{ii} + {}^0Z'_{ii}}$	$\dfrac{-{}^2Z'_{ii} \cdot {}^1I^f}{^2Z'_{ii} + {}^0Z'_{ii}}$
$L-L-L-G$	$\dfrac{V_i}{^1Z_{ii} + Z^f}$	0	0

where

$$^1Z'_{ii} = {}^1Z_{ii} + 0.5Z^f$$
$$^2Z'_{ii} = {}^2Z_{ii} + 0.5Z^f$$
$$^0Z'_{ii} = {}^0Z_{ii} + 0.5Z^f.$$

at bus k these are

$$^0I^f_i = \begin{cases} 0 & \text{for } i \neq k \\ -{}^0I^f & \text{for } i = k \end{cases} \tag{5.3.21}$$

$$^1I^f_i = \begin{cases} ^1I_i & \text{for } i \neq k \\ ^1I_k - {}^1I^f & \text{for } i = k \end{cases} \tag{5.3.22}$$

$$^2I^f_i = \begin{cases} 0 & \text{for } i \neq k \\ -{}^2I^f & \text{for } i = k. \end{cases} \tag{5.3.23}$$

The fault voltages are then obtained from equations (5.3.10) to (5.3.12) by substituting the fault current vector for the prefault current vector, i.e.

$$[^0V^f] = [^0Z][^0I^f] \tag{5.3.24}$$

$$[^1V^f] = [^1Z][^1I^f] \tag{5.3.25}$$

$$[^2V^f] = [^2Z][^2I^f]. \tag{5.3.26}$$

5.3.4 Open-circuit Faults

The system is now represented by a two-port network across which the faulty line is connected as shown in Fig. 5.10. In this case, the prefault voltages have to be obtained from a load-flow study.

For an open-circuit fault on phase b and c the boundary conditions are

$$I_b = I_c = 0$$
$$(^aV_l - {}^aV_k) = {}^aZ^aI.$$

146

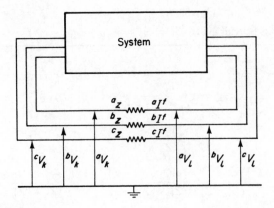

Figure 5.10
Two-port network with faulty line where ^aZ, ^bZ or ^cZ may be on open circuit

Using these equations with the sequence transformation, the following relationships result:

$$^0I^f = {}^1I^f = {}^2I^f = {}^aI^f/3 \tag{5.3.27}$$

and

$$^0V_{kl}^f + {}^1V_{kl}^f + {}^2V_{kl}^f = {}^aZ \cdot {}^aI^f = 3 \cdot ({}^aZ)^lI^f \tag{5.3.28}$$

where

$$V_{kl} = V_l - V_k.$$

Table 5.2.
Fault currents for open–circuit faults

Fault	$^1I^f$	$^2I^f$	$^0I^f$
1-o-c	$\dfrac{V_l - V_k}{({}^2Z' \cdot {}^0Z')/({}^2Z' + {}^0Z') + {}^1Z'}$	$\dfrac{{}^0Z' \cdot {}^1I^f}{{}^2Z' + {}^0Z'}$	$\dfrac{{}^2Z' \cdot {}^1I^f}{{}^2Z' + {}^0Z'}$
2-o-c	$\dfrac{V_l - V_k}{{}^1Z + {}^2Z + {}^0Z + Z^f}$	$^1I^f$	$^1I^f$

where

$$^1Z = {}^1Z_{kk} + {}^1Z_{ll} - {}^1Z_{kl} - {}^1Z_{lk}$$
$$^0Z = {}^0Z_{kk} + {}^0Z_{ll} - {}^0Z_{kl} - {}^0Z_{lk}$$
$$^2Z = {}^2Z_{kk} + {}^2Z_{ll} - {}^2Z_{kl} - {}^2Z_{lk}.$$

Z^f is the sum of the positive-, negative- and zero-sequence impedances of the faulty circuit, i.e.

$$Z^f = {}^1Z^f + {}^2Z^f + {}^0Z^f$$
$$^1Z' = {}^1Z + {}^1Z^f$$
$$^0Z' = {}^0Z + {}^0Z^f$$
$$^2Z' = {}^2Z + {}^2Z^f.$$

Equations (5.3.27) and (5.3.28) define the connection of the Thevenin equivalent sequence networks at the fault location to solve for the fault currents.

The equivalent Thevenin impedances are the sequence impedances of the system between the two buses k and l, i.e.

$$Z_{eqv} = Z_{kk} + Z_{ll} - Z_{lk} - Z_{kl} \tag{5.3.29}$$

and the equivalent Thevenin voltage is given by the difference between the voltage at buses l and k with the faulted line disconnected.

During the fault the sequence voltages $^{0}V_{kl}^{f}$, $^{1}V_{kl}^{f}$ and $^{2}V_{kl}^{f}$ have the same expressions as equations (5.3.17) to (5.3.19). Thus similar considerations, as in the case of the line-to-ground short circuit, lead to the following expression for the fault currents:

$$^{0}I^{f} = {}^{1}I^{f} = {}^{2}I^{f} = \frac{^{1}V_{kl}}{^{0}Z_{eqv} + {}^{1}Z_{eqv} + {}^{2}Z_{eqv} + 3(^{a}Z)}. \tag{5.3.30}$$

The case of a single open-circuit fault can be analysed in a similar manner and the final relevant equations are shown in Table 5.2.

The fault current vector is formed as follow:

$$^{0}I_{i}^{f} = \begin{cases} 0 & \text{for } i = 1, n \quad i \neq k \quad \text{or} \quad 1 \\ -{}^{0}I^{f} & \text{for } i = k \\ {}^{0}I^{f} & \text{for } i = l \end{cases}$$

$$^{1}I_{i}^{f} = \begin{cases} {}^{1}I_{i} & \text{for } i = 1, n \quad i \neq k \quad \text{or} \quad l \\ {}^{1}I_{k} - {}^{1}I^{f} & \text{for } i = k \\ {}^{1}I_{l} + {}^{1}I^{f} & \text{for } i = l \end{cases}$$

$$^{2}I_{i}^{f} = \begin{cases} 0 & \text{for } i = 1, n \quad i \neq k \quad \text{or} \quad l \\ -{}^{2}I^{f} & \text{for } i = k \\ {}^{2}I^{f} & \text{for } i = l \end{cases}$$

and the voltage vector is given by equations (5.3.24) to (5.3.26).

From the fault voltages the branch currents are obtained as follows:

$$^{0}I_{ij}^{f} = {}^{0}y_{ij}(^{0}V_{i}^{f} - {}^{0}V_{j}^{f}) \tag{5.3.31}$$

$$^{1}I_{ij}^{f} = {}^{1}y_{ij}(^{1}V_{i}^{f} - {}^{1}V_{j}^{f}) \tag{5.3.32}$$

$$^{2}I_{ij}^{f} = {}^{2}y_{ij}(^{2}V_{i}^{f} - {}^{2}V_{j}^{f}). \tag{5.3.33}$$

Where necessary the corrections for taps on the positive- and negative-sequence networks are

$$^{1}I_{ij}^{f} = {}^{1}y_{ij}\{^{1}V_{i}^{f}(1 - \tau_{ij}) - {}^{1}V_{j}^{f}\} \tag{5.3.34}$$

$$^{2}I_{ij}^{f} = {}^{2}y_{ij}\{^{2}V_{i}^{f}(1 - \tau_{ij}) - {}^{2}V_{j}^{f}\}. \tag{5.3.35}$$

Finally, the machine contributions may be calculated, i.e.

$$^{0}I_{i}^{Mf} = -{}^{0}y_{i}^{M} \cdot {}^{0}V_{i}^{f}$$

$$^{1}I_{i}^{Mf} = I_{i} - {}^{1}y_{i}^{M} \cdot {}^{1}V_{i}^{f}$$

$$^{2}I_{i}^{Mf} = -{}^{2}y_{i}^{M} \cdot {}^{2}V_{i}^{f}.$$

148

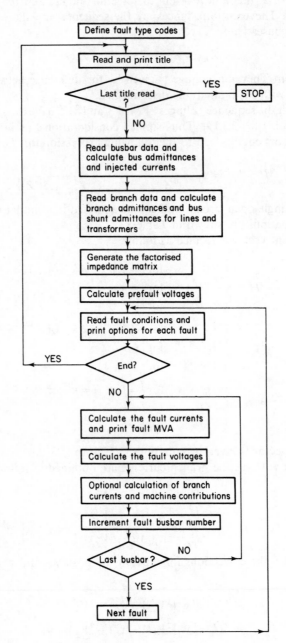

Figure 5.11
General flow diagram

5.4 PROGRAM DESCRIPTION AND TYPICAL SOLUTIONS

A fault analysis program must be capable of analysing the following a.c. system faults:

- line-to-ground short circuit
- line-to-line short circuit
- line-to-line-to-ground short circuit
- line-to-line-to-line to ground short circuit
- single open-circuit line
- double open-circuit line.

Basic to the fault study program is the determination of the impedance matrix of the system, the elements of which can be used, along with the conditions imposed by the type of fault, to directly solve for the fault currents and voltages.

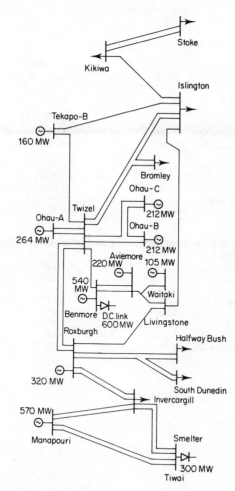

Figure 5.12
New Zealand South Island primary system

The main steps of a general-purpose fault program are indicated in Fig. 5.11. Prefault information and typical outputs for balanced and unbalanced fault conditions are illustrated in the following computer printouts. The printouts relate to studies carried out for the New Zealand South Island a.c. system illustrated in Fig. 5.12.

<u>AC FAULTS ANALYSIS PROGRAM</u>

<u>DEPARTMENT OF ELECTRICAL & ELECTRONIC ENGINEERING, UNIVERSITY OF CANTERBURY, NEW ZEALAND</u>

23 MAR 90

<u>SYSTEM NO. 21</u>

A.C. FAULTS ANALYSIS TEST PROGRAM – 20 BUSES, 34 BRANCHES

SYSTEM MVA BASE = 100.0 MVA

<u>BUSBAR DATA</u>

BUSBAR NAME	VOLTAGE MAG(PU)	VOLTAGE ANG(DEGS)	LOAD P (MW)	LOAD Q (MVAR)
AVIEMORE–220	1.05200	4.69	0.00	0.00
BENMORE—220	1.05000	4.39	597.20	180.00
BROMLEY—220	1.00100	−11.43	129.60	38.30
HALFWAYBU220	1.03800	−2.31	95.30	40.40
INVERCARG220	1.02900	−2.12	183.20	20.00
ISLINGTON220	1.00500	−12.11	504.10	124.30
KIKIWA—220	1.00400	−25.66	59.20	9.20
LIVINGSTN220	1.00000	0.00	0.00	0.00
MANAPOURI220	1.06000	2.75	0.00	0.00
OHAU–A—220	1.05000	4.40	0.00	0.00
OHAU–B—220	1.04900	4.15	0.00	0.00
OHAU–C—220	1.05000	4.35	0.00	0.00
ROXBURGH–220	1.05500	0.00	0.00	0.00
SOUTHDUNEDIN	1.04000	−2.18	34.20	12.90
STOKE——220	1.01000	−27.08	53.20	−20.30
TEKAPO–B–220	1.04600	3.72	0.00	0.00
TIWAI——220	1.02200	−2.45	0.00	0.00
TIWAI——76	0.99630	−12.11	288.00	105.72
TWIZEL—220	1.04900	4.03	0.00	0.00
WAITAKI—220	1.05100	4.16	0.00	0.00

BRANCH DATA

SENDING BUSBAR	RECEIVING BUSBAR	R1	SERIES IMPEDANCES (PU OR OHMS) X1	R0	X0	R2	X2	SUSCEPTANCE (PU OR OHMS)	TRANSFORMER TAP(%)	TYPE	NOMINAL VOLTS(KV)
AVIEMORE-220	BENMORE-220	0.00325	0.01509	0.00858	0.03767	0.00325	0.01509	0.02304	0.00	1	0.00
AVIEMORE-220	BENMORE-220	0.00330	0.01530	0.00870	0.03810	0.00330	0.01530	0.02298	0.00	1	0.00
AVIEMORE-220	WAITAKI-220	0.00153	0.00723	0.00404	0.02045	0.00153	0.00723	0.01062	0.00	1	0.00
BENMORE-220	TWIZEL-220	0.00429	0.02935	0.01839	0.08604	0.00429	0.02935	0.08201	0.00	1	0.00
BROMLEY-220	ISLINGTON220	0.00203	0.01651	0.01056	0.06501	0.00203	0.01651	0.05364	0.00	1	0.00
BROMLEY-220	TWIZEL-220	0.01714	0.13990	0.08952	0.55097	0.01714	0.13990	0.45460	0.00	1	0.00
HALFWAYBU220	ROXBURGH-220	0.00768	0.06592	0.03966	0.24614	0.00768	0.06592	0.19082	0.00	1	0.00
HALFWAYBU220	SOUTHDUNEDIN	0.00175	0.01010	0.00546	0.02565	0.00175	0.01010	0.01665	0.00	1	0.00
INVERCARG220	MANAPOURI220	0.01338	0.09178	0.05742	0.30410	0.01338	0.09178	0.25996	0.00	1	0.00
INVERCARG220	MANAPOURI220	0.01338	0.09178	0.05742	0.30410	0.01338	0.09178	0.25996	0.00	1	0.00
INVERCARG220	ROXBURGH-220	0.01880	0.11223	0.05851	0.32322	0.01880	0.11223	0.17208	0.00	1	0.00
INVERCARG220	ROXBURGH-220	0.01915	0.11252	0.05959	0.27184	0.01915	0.11252	0.17814	0.00	1	0.00
INVERCARG220	TIWAI-220	0.00226	0.01456	0.00970	0.04070	0.00226	0.01456	0.04596	0.00	1	0.00
INVERCARG220	TIWAI-220	0.00226	0.01456	0.00970	0.04070	0.00226	0.01456	0.04596	0.00	1	0.00
ISLINGTON220	KIKIWA-220	0.03326	0.20031	0.10355	0.57070	0.03326	0.20031	0.30182	0.00	1	0.00
ISLINGTON220	LIVINGSTN220	0.03230	0.17662	0.10359	0.55184	0.03230	0.17662	0.35841	0.00	1	0.00
ISLINGTON220	TEKAPO-B-220	0.02112	0.14576	0.09055	0.41498	0.02112	0.14576	0.39973	0.00	1	0.00
ISLINGTON220	TWIZEL-220	0.01630	0.13037	0.08517	0.52963	0.01630	0.13037	0.44180	0.00	1	0.00
KIKIWA-220	STOKE-220	0.00762	0.04370	0.02373	0.11286	0.00762	0.04370	0.07278	0.00	1	0.00
LIVINGSTN220	ROXBURGH-220	0.02649	0.12551	0.07003	0.35496	0.02649	0.12551	0.18426	0.00	1	0.00
LIVINGSTN220	WAITAKI-220	0.00588	0.02787	0.01555	0.07883	0.00588	0.02787	0.04092	0.00	1	0.00
MANAPOURI220	TIWAI-220	0.01549	0.10734	0.06648	0.35551	0.01549	0.10734	0.29780	0.00	1	0.00
MANAPOURI220	TIWAI-220	0.01549	0.10734	0.06648	0.35551	0.01549	0.10734	0.29780	0.00	1	0.00
OHAU-A-220	TWIZEL-220	0.00115	0.00662	0.00357	0.02030	0.00115	0.00662	0.00109	0.00	1	0.00
OHAU-A-220	TWIZEL-220	0.00115	0.00662	0.00357	0.02030	0.00115	0.00662	0.00109	0.00	1	0.00
OHAU-B-220	TWIZEL-220	0.00024	0.00179	0.00166	0.00580	0.00024	0.00179	0.00057	0.00	1	0.00
OHAU-B-220	TWIZEL-220	0.00024	0.00179	0.00166	0.00580	0.00024	0.00179	0.00057	0.00	1	0.00
OHAU-B-220	OHAU-C-220	0.00064	0.00477	0.00309	0.01546	0.00064	0.00477	0.00152	0.00	1	0.00
OHAU-C-220	TWIZEL-220	0.00088	0.00656	0.00425	0.02126	0.00088	0.00656	0.00209	0.00	1	0.00
ROXBURGH-220	SOUTHDUNEDIN	0.00849	0.07059	0.04219	0.25660	0.00849	0.07059	0.19854	0.00	1	0.00
ROXBURGH-220	TWIZEL-220	0.01590	0.13710	0.24001	0.06800	0.01590	0.13710	0.43180	0.00	1	0.00
ROXBURGH-220	TWIZEL-220	0.01590	0.13710	0.24001	0.06800	0.01590	0.13710	0.43180	0.00	1	0.00
TEKAPO-B-220	TWIZEL-220	0.00230	0.01554	0.00990	0.04732	0.00230	0.01554	0.04860	0.00	1	0.00

MACHINE DATA

| BUSBAR | GENERATION | | IMPEDANCE (PU OR OHMS) | | | | | | VOLTAGE |
NAME	MW	MVAR	R POS	X POS	R ZERO	X ZERO	R NEG	X NEG	BASE
AVIEMORE–220	220.00	–6.90	0.00182	0.09900	0.00182	0.04500	0.00182	0.07980	0.00
BENMORE—220	540.00	163.21	0.00066	0.07800	0.00066	0.02268	0.00066	0.05396	0.00
ISLINGTON220	0.00	142.80	0.00490	0.33500	0.00490	0.13830	0.00490	0.32200	0.00
MANAPOURI220	400.00	42.10	0.00050	0.03930	0.00050	0.01310	0.00050	0.02630	0.00
OHAU–A—220	214.00	–3.90	0.00153	0.09390	0.00153	0.02901	0.00153	0.06310	0.00
OHAU–B—220	175.00	–41.40	0.00165	0.09250	0.00165	0.03395	0.00165	0.06790	0.00
OHAU–C—220	175.00	15.30	0.00165	0.09250	0.00165	0.03395	0.00165	0.06790	0.00
ROXBURGH–220	74.30	10.30	0.00169	0.08580	0.00169	0.03190	0.00169	0.06900	0.00
TEKAPO–B–220	160.00	–10.70	0.00112	0.09900	0.00112	0.05621	0.00112	0.11250	0.00
WAITAKI—220	30.00	–6.60	0.00555	0.19200	0.05550	0.64200	0.00555	0.12840	0.00

SYSTEM NO. 21

L–G FAULT AT 'MANAPOURI220'

FAULT IMPEDANCE POSITIVE SEQUENCE   0.000 +J   0.000
ZERO SEQUENCE   0.000 +J   0.000

FAULT MVA =   5344.91 MVA

FAULT PHASE CURRENTS =   50.425   0.000   0.000 PU

B U S B A R   V O L T A G E S

PHASE VOLTAGES

| BUSBAR | PHASE A | | PHASE B | | PHASE C | |
NAME	MAGN	ANGLE	MAGN	ANGLE	MAGN	ANGLE
AVIEMORE–220	1.00769	4.96	1.03252	–114.45	1.03060	124.20
BENMORE—220	1.00856	4.66	1.03150	–114.78	1.02921	123.93
BROMLEY—220	0.95296	–11.33	0.98057	–130.49	0.97951	107.85
HALFWAYBU220	0.78102	–1.55	0.95290	–118.18	0.95810	114.03
INVERCARG220	0.24881	–1.22	0.87243	–114.44	0.86906	111.47
ISLINGTON220	0.95686	–12.02	0.98422	–131.18	0.98348	107.17
KIKIWA—220	0.95600	–25.55	0.98295	–144.74	0.98276	93.61
LIVINGSTN220	0.96254	2.01	1.01611	–116.92	1.01461	120.61
MANAPOURI220	0.00000	0.00	0.87259	–107.85	0.85324	115.38
OHAU–A—220	0.99886	4.65	1.02894	–114.55	1.02469	123.80
OHAU–B—220	0.99642	4.40	1.02789	–114.77	1.02314	123.54
OHAU–C—220	0.99896	4.60	1.02906	–114.61	1.02475	123.75
ROXBURGH–220	0.79211	0.62	0.97053	–115.90	0.97241	116.46
SOUTHDUNEDIN	0.78219	–1.42	0.95450	–118.05	0.95960	114.16
STOKE——220	0.96178	–26.96	0.98887	–146.15	0.98873	92.19
TEKAPO–B–220	0.99968	3.95	1.02713	–115.32	1.02359	123.13
TIWAI——220	0.21519	–2.52	0.86477	–114.63	0.85684	111.14
TIWAI——76	0.35752	–11.33	0.79432	–114.52	0.79316	91.51
TWIZEL—220	0.99542	4.27	1.02778	–114.88	1.02275	123.42
WAITAKI—220	1.00007	4.44	1.02952	–114.90	1.02776	123.59

BRANCH CURRENTS

SEQUENCE CURRENTS

SENDING BUSBAR	RECEIVING BUSBAR	POSITIVE		ZERO		NEGATIVE	
		REAL	IMAG	REAL	IMAG	REAL	IMAG
AVIEMORE—220	BENMORE—220	0.341664	0.069312	0.000397	0.009594	−0.029174	0.061505
AVIEMORE—220	BENMORE—220	0.336932	0.068457	0.000387	0.009484	−0.028790	0.060649
AVIEMORE—220	WAITAKI—220	1.393619	−0.219927	0.003232	−0.038128	0.065216	−0.309849
BENMORE—220	TWIZEL—220	0.252046	−0.212731	0.015295	−0.002452	0.020987	−0.159284
BROMLEY—220	ISLINGTON220	0.719796	0.173058	−0.001264	0.000020	−0.005469	0.003947
BROMLEY—220	TWIZEL—220	−1.926780	0.195419	0.002387	0.000115	0.026733	−0.010789
HALFWAYBU220	ROXBURGH—220	−0.567284	0.159465	0.011693	−0.005925	0.059888	−0.017658
HALFWAYBU220	SOUTHDUNEDIN	−0.216475	0.110046	0.004546	−0.003581	0.022787	−0.011950
INVERCARG220	MANAPOURI220	−0.443749	−1.056205	0.113502	−0.201024	0.424027	−0.919897
INVERCARG220	MANAPOURI220	−0.443749	−1.056205	0.113502	−0.201024	0.424027	−0.919897
INVERCARG220	ROXBURGH—220	−0.560268	2.145992	−0.062925	0.365157	−0.181347	1.486348
INVERCARG220	ROXBURGH—220	−0.564148	2.138058	−0.090041	0.427981	−0.184659	1.481390
INVERCARG220	TIWAI——220	0.433583	−1.167772	0.087504	−0.177243	−0.009138	−0.532094
INVERCARG220	TIWAI——220	0.433583	−1.167772	0.087504	−0.177243	−0.009138	−0.532094
ISLINGTON220	KIKIWA——220	1.116270	−0.188627	−0.000823	0.000196	−0.019587	0.003483
ISLINGTON220	LIVINGSTN220	−1.336068	0.002491	0.000639	−0.007278	0.031891	−0.070338
ISLINGTON220	TEKAPO—B—220	−1.889161	0.144159	0.001479	0.001140	0.023890	0.004622
ISLINGTON220	TWIZEL——220	−2.158592	0.182427	0.002640	0.000113	0.029400	−0.012001
KIKIWA——220	STOKE——220	0.559962	−0.026245	−0.000416	0.000056	−0.009836	0.000546
LIVINGSTN220	ROXBURGH—220	0.346169	−0.618743	0.004268	−0.046484	0.104859	−0.505846
LIVINGSTN220	WAITAKI——220	−1.672586	0.330301	−0.003542	0.040641	−0.072866	0.444313
MANAPOURI220	TIWAI——220	0.438018	0.743464	−0.087714	0.151205	−0.363512	0.714964
MANAPOURI220	TIWAI——220	0.438018	0.743464	−0.087714	0.151205	−0.363512	0.714964
OHAU—A—220	TWIZEL——220	1.008814	−0.078604	0.020271	−0.011051	0.003941	−0.141165
OHAU—A—220	TWIZEL——220	1.008814	−0.078604	0.020271	−0.011051	0.003941	−0.141165
OHAU—B—220	TWIZEL——220	1.197638	−0.030557	0.028701	−0.015632	0.004662	−0.196022
OHAU—B—220	TWIZEL——220	1.197638	−0.030557	0.028701	−0.015632	0.004662	−0.196022
OHAU—B—220	OHAU—C—220	−0.772727	0.212014	−0.015800	0.007908	−0.003212	0.121007
OHAU—C—220	TWIZEL——220	0.888666	−0.162493	0.019096	−0.010735	0.003609	−0.141476
ROXBURGH—220	SOUTHDUNEDIN	0.499726	−0.129419	−0.010788	0.005262	−0.052770	0.014384
ROXBURGH—220	TWIZEL——220	−0.586252	0.677463	−0.076022	0.036327	−0.061123	0.559474
ROXBURGH—220	TWIZEL——220	−0.586252	0.677463	−0.076022	0.036327	−0.061123	0.559474
TEKAPO—B—220	TWIZEL——220	−0.367431	−0.179452	0.016119	−0.007277	0.024243	−0.138814
TIWAI——220	TIWAI——76	1.736261	−1.068914	0.000000	0.000000	−0.742030	0.460065

SYSTEM NO. 21

LL–L FAULT AT 'MANAPOURI220'

FAULT IMPEDANCE	POSITIVE SEQUENCE	0.000 +J	0.000
	ZERO SEQUENCE	0.000 +J	0.000

FAULT MVA =   3845.61 MVA

FAULT PHASE CURRENTS =   36.280   36.280   36.280 PU

B U S B A R   V O L T A G E S

PHASE VOLTAGES

BUSBAR	PHASE A		PHASE B		PHASE C	
NAME	MAGN	ANGLE	MAGN	ANGLE	MAGN	ANGLE
AVIEMORE–220	0.99045	5.08	0.99045	–114.92	0.99045	125.08
BENMORE—220	0.99166	4.80	0.99166	–115.20	0.99166	124.80
BROMLEY—220	0.93568	–11.27	0.93568	–131.27	0.93568	108.73
HALFWAYBU220	0.73059	–1.84	0.73059	–121.84	0.73059	118.16
INVERCARG220	0.22966	–2.38	0.22966	–122.38	0.22966	117.62
ISLINGTON220	0.93973	–11.97	0.93973	–131.97	0.93973	108.03
KIKIWA——220	0.93882	–25.52	0.93882	–145.52	0.93882	94.48
LIVINGSTN220	0.93941	2.08	0.93941	–117.92	0.93941	122.08
MANAPOURI220	0.00000	0.00	0.00000	0.00	0.00000	0.00
OHAU-A——220	0.97954	4.83	0.97953	–115.17	0.97953	124.83
OHAU-B——220	0.97705	4.59	0.97705	–115.41	0.97705	124.59
OHAU-C——220	0.97974	4.78	0.97974	–115.22	0.97974	124.78
ROXBURGH–220	0.74232	0.46	0.74232	–119.54	0.74232	120.46
SOUTHDUNEDIN	0.73176	–1.71	0.73176	–121.71	0.73176	118.29
STOKE——220	0.94449	–26.93	0.94449	–146.93	0.94449	93.06
TEKAPO-B–220	0.98271	4.09	0.98271	–115.91	0.98271	124.09
TIWAI——220	0.20004	–3.24	0.20004	–123.24	0.20004	116.76
TIWAI——76	0.19507	–12.89	0.19507	–132.89	0.19507	107.10
TWIZEL—220	0.97601	4.48	0.97601	–115.52	0.97601	124.47
WAITAKI—220	0.98190	4.55	0.98190	–115.44	0.98190	124.55

5.5 REFERENCES

[1] Wagner and Evans, 1933. *Symmetrical Components* McGraw-Hill, New York.

[2] A. Brameller, 1972. Three-phase short-circuit calculation using digital computer. *Internal Report* University of Manchester Institute of Science and Technology.

[3] A. Brameller, 1972. User manual for unbalanced fault analysis. *Internal Report* University of Manchester Institute of Science and Technology.

[4] J. Preece, 1975. Symmetrical components fault studies. *MSc Dissertation* University of Manchester Institute of Science and Technology.

[5] C. B. Lake, 1978. A computer program for a.c. fault studies in a.c.–d.c. systems. *M.E. Thesis* University of Canterbury, New Zealand.

[6] B. Stott and E. Hobson, 1971. Solution of large power-system networks by ordered elimination: a comparison of ordering schemes, *Proc. IEE* **118** 125–134.

[7] K. Zollenkopf, 1970. Bifactorization—basic computational algorithm and programming techniques. *Conference on Large Sets of Sparse Linear Equations, Oxford* 75–96.

[8] P. M. Anderson, 1973. *Analysis of Faulted Power Systems* Iowa State University Press.

6. POWER SYSTEM STABILITY— BASIC MODEL

6.1 INTRODUCTION

The stability of a power system following some predetermined operating condition is a dynamic problem and requires more elaborate plant component models than the ones discussed in previous chapters. It is normally assumed that prior to the dynamic analysis, the system is operating in the steady state and that a load-flow solution is available.

Two types of stability studies are normally carried out. The subsequent recovery from a sudden large disturbance is referred to as 'transient stability' and the solution is obtained in the time domain. The period under investigation can vary from a fraction of a second, when first swing stability is being determined, to over ten seconds when multiple swing stability must be examined.

The term 'dynamic stability' is used to describe the long-time response of a system to small disturbances or badly set automatic controls. The problem can be solved either in the time domain or in the frequency domain. In this book, dynamic stability is treated as an extension of transient stability and is thus solved in the time domain. Such extension normally requires modification of some plant component models and often the introduction of new models, but because of the smaller perturbations and longer study duration the small-time constant effects can be ignored.

Consideration is given in this chapter to the dynamic modelling of a power system containing synchronous machines and basic loads. More advanced synchronous machine models as well as other power system components, such as induction motors and a.c.–d.c. converters, are considered in Chapter 7.

6.1.1 The Form of the Equations

To a greater or lesser extent, all system variables require time to respond to any change in operating conditions and a large set of differential equations can be written to determine this response. This is impractical, however, and many assumptions must be made to simplify the system model. The assumptions made depend on the problem being investigated and no clear definitive model exists.

A major problem with a time domain solution is the 'stiffness' of the system (Appendix IV). That is, the time constants associated with the system variables vary enormously. When only synchronous machines are being considered, rotor swing stability is the principal concern. The main time constants associated with the rotor are of the order of 1 to 10 s. The form of the solution is dominated by time constants of this order and smaller or greater time constants have less significance.

The whole of the a.c. transmission network responds rapidly to configurational changes as well as loading changes. The time constants associated with the network variables are extremely small and can be considered to be zero without significant loss of accuracy. Similarly the synchronous machine stator time constants may be taken as zero. The relevant differential equations for these rapidly changing variables are transformed into algebraic equations.

When the time constant is large or the disturbance is such that the variable will not change greatly, the time constant may be regarded as infinite, that is the variable becomes a constant. Excitation voltage or mechanical power to the synchronous machine may often be treated as constant in short-duration studies without appreciable loss of accuracy. Depending on how the computer program is written variables which become constant may be treated by either:

(i) retaining the differential equation but assigning a very large value to the relevant time constant;

(ii) removing the differential equation.

For flexibility both methods are usually incorporated into a program.

A system which after these initial simplifications contains α differential variables, contained in the vector Y_α, and β algebraic variables, contained in the vector X_β, may be described by the matrix equations

$$pY_\alpha = F_\alpha(Y_\alpha, X_\beta) \tag{6.1.1}$$

$$0 = G_\beta(Y_\alpha, X_\beta) \tag{6.1.2}$$

where p denotes the differential operator $\mathrm{d}/\mathrm{d}t$.

6.1.2 Frames of Reference

The choice of axes, or frame of reference, in which the system equations are formulated is of great importance as it infuences the analysis.

For synchronous machines, the most appropriate frame of reference is one which is attached to the rotor, i.e. it rotates at the same speed as the rotor. The main advantage of this choice is that the coefficients of the equations developed for the synchronous machine are not time-dependent. The major axis of this frame of reference is taken as the rotor pole or 'direct axis'. The second axis lies 90° (electrical) from each pole and is referred to as the 'quadrature axis'.

In the dynamic state, each synchronous machine is rotating independently and transforming between synchronous machine frames through the network is difficult. This is overcome by choosing an independent frame of reference for the network and transforming between this frame and the synchronous machine frames at the machine terminals. The most obvious choice for the network is a frame of reference which rotates at synchronous speed. The two axes are obtained from the initial steady-state load-flow slack busbar. Although the network frame is rotating synchronously, this does not stop each nodal voltage or branch current from having an independent frequency during the dynamic analysis.

6.2 SYNCHRONOUS MACHINES—BASIC MODELS

6.2.1 Mechanical Equations

The mechanical equations of a synchronous machine are very well established [1, 2] and need be only briefly outlined. Three basic assumptions are made in deriving the equations.

(i) Machine rotor speed does not vary greatly from synchronous speed (1.0 p.u.).

(ii) Machine rotational power losses due to windage and friction are ignored.

(iii) Mechanical shaft power is smooth, that is the shaft power is constant except for the results of speed governor action.

Assumption (i) allows per unit power to be equated with per unit torque. From Assumption (ii), the accelerating power of the machine (Pa) is the difference between the shaft power (Pm) as supplied by the prime mover or absorbed by the load and the electrical power (Pe). The acceleration (α) is thus

$$\alpha = \frac{Pa}{Mg} = \frac{(Pm - Pe)}{Mg} \tag{6.2.1}$$

where Mg is the angular momentum.

The acceleration is independent of any constant speed frame of reference and it is convenient to choose a synchronously rotating frame to define the rotor angle (δ). Thus

$$\frac{d^2\delta}{dt^2} = \frac{(Pm - Pe)}{Mg}. \tag{6.2.2}$$

The angular momentum may be further defined by the inertia constant Hg (measured in MWs/MVA) which is relatively constant regardless of the size of the machine, i.e.

$$Mg = \frac{Hg}{\pi f_0} \tag{6.2.3}$$

where f_0 is the system base frequency.

Eddy currents induced in the rotor iron or in the damping windings produce torques which oppose the motion of the rotor relative to the synchronous speed. A deceleration power can be introduced into the mechanical equations to account for this damping, giving

$$\frac{d^2\delta}{dt^2} = \frac{1}{Mg}\left(Pm - Pe - Da\frac{d\delta}{dt}\right). \tag{6.2.4}$$

The damping coefficient (Da), measured in Watts/rad/sec, has been largely superseded by a synchronous machine model which includes the subtransient effect of the damper windings in the electrical equations, but it is still used in some programs.

Two single-order ordinary differential equations may now be written to describe the mechanical motion of the synchronous machine, i.e.

$$p\omega = \frac{1}{Mg}(Pm - Pe - Da(\omega - 2\pi f_0)) \qquad (6.2.5)$$

$$p\delta = \omega - 2\pi f_0. \qquad (6.2.6)$$

6.2.2 Electrical Equations

The derivation of equations to account for flux changes in a synchronous machine has been given by Concordia [3] and Kimbark [4]. A brief outline only will be given in this section, so that various electrical quantities may be defined and phasor diagrams constructed. The approximations made in the derivation are as follows.

(i) The rotor speed is always sufficiently near 1.0 p.u. that it may be considered a constant.

(ii) All inductances defined in this section are independent of current. The effects due to saturation of iron are considered in Chapter 7.

(iii) Machine winding inductances can be represented as constants plus sinusoidal harmonics of rotor angle.

(iv) Distributed windings may be represented as concentrated windings.

(v) The machine may be represented by a voltage behind an impedance.

(vi) There are no hysteresis losses in the iron, and eddy currents are only accounted for by equivalent windings on the rotor.

(vii) Leakage reactance only exists in the stator.

Using these assumptions, classical theory permits the construction of a model for the synchronous machine in the steady-state, transient and subtransient states.

The per unit system adopted is normalised to eliminate factors of $\sqrt{2}$, $\sqrt{3}$, π and turns ratio, although the term 'proportional' should be used instead of 'equal' when comparing quantities. Note that one p.u. field voltage produces 1.0 p.u. field current and 1.0 p.u. open-circuit terminal voltage at rated speed.

6.2.2.1 Steady State Equations

Figure 6.1 shows the flux and voltage phasor diagram for a cylindrical rotor synchronous machine in which all saturation effects are ignored. The flux Ff is proportional to the field current If and the applied field voltage and it acts in the direct axis of the machine. The stator open-circuit terminal voltage Ei is proportional

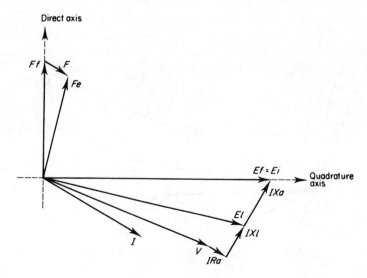

Figure 6.1
Phasor diagram of a cylindrical rotor synchronous machine in the steady state

to Ff but lies on the quadrature axis. The voltage Ei is also proportional to the applied field voltage and may be referred to as Ef.

When the synchronous machine is loaded, a flux F proportional to and in phase with the stator current I is produced which, when added vectorially to the field flux Ff, gives an effective flux Fe. The effective internal stator voltage El is due to Fe and lags it by 90°. The terminal voltage V is found from this voltage El by considering the voltage drops due to the leakage reactance Xl and armature resistance Ra. By similar triangles, the difference between Ef and El is in phase with the IXl voltage drop and is proportional to I. Therefore the voltage difference may be treated as a voltage drop across an armature reactance Xa. The sum of Xa and Xl is termed the synchronous reactance.

For the salient pole synchronous machine the phasor diagram is more complex. Because the rotor is symmetrical about both the d and q axes it is convenient to resolve many phasor quantities into components in these axes. The stator current may be treated in this manner. Although F_d will be proportional to I_d and F_q will be proportional to I_q, because the iron paths in the two axes are different, the total armature reaction flux F will not be proportional to I nor necessarily be in phase with it. Retaining our earlier normalising assumptions, it may be assumed that the proportionality between I_d and F_d is unity but the proportionality between I_q and F_q is less than unity and is a function of the saliency.

In Fig. 6.2 the phasor diagram of the salient pole synchronous machine is shown. Note that the d and q axes armature reactances have been developed as in the cylindrical rotor case. From these, direct and quadrature synchronous reactances (X_d and X_q) can be established, i.e.

$$X_d = Xl + Xa_d \tag{6.2.7}$$

$$X_q = Xl + Xa_q \tag{6.2.8}$$

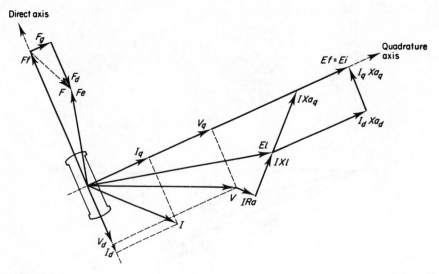

Figure 6.2
Phasor diagram of a salient pole synchronous machine in the steady state

$$Ei - V_q = Ra\,I_q - X_d I_d \tag{6.2.9}$$
$$- V_d = Ra\,I_d + X_q I_q \tag{6.2.10}$$

where V_d and V_q are the axial components of the terminal voltage V.

In steady-state conditions it is quite acceptable to use as the machine model, the field voltage Ef or the voltage equivalent to field current Ei behind the synchronous reactances. In these circumstances the rotor position (quadrature axis) with respect to the synchronously rotating frame of reference is given by the angular position of Ef.

Only the salient pole machine will now be considered, as the cylindrical rotor model may be regarded as a special case of a salient machine ($X_d = X_q$).

6.2.2.2 Transient Equations

For faster changes in the conditions external to the synchronous machine, the above model is no longer suitable. Due to the 'inertia' of the flux linkages these changes cannot be reflected throughout the whole of the model immediately. It is therefore necessary to create new fictitious voltages E'_d and E'_q which represent the flux linkages of the rotor windings. These transient voltages can be shown to exist behind the transient reactances X'_d and X'_q:

$$E'_q - V_q = Ra\,I_q - X'_d I_d \tag{6.2.11}$$

$$E'_d - V_d = Ra\,I_d + X'_q I_q. \tag{6.2.12}$$

The voltage E_i should now be considered as the sum of two voltages, E_d and E_q, and is the voltage behind synchronous reactance. In the prevous section, where steady state was considered, current flowed only in the field winding and, hence, in that case, $E_d = 0$ and $E_q = E_i$.

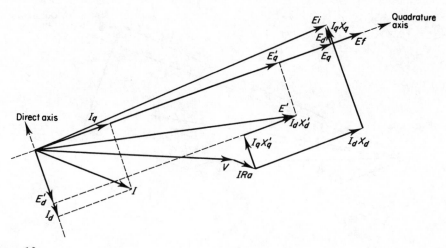

Figure 6.3
Phasor diagram of a synchronous machine in the transient state

Where it is necessary to allow the rotor flux linkages to change with time, the following ordinary differential equations are used:

$$pE'_q = (Ef - E_q)/T'_{d0} = (Ef + (X_d - X'_d)I_d - E'_q)/T'_{d0} \tag{6.2.13}$$

$$pE'_d = -E_d/T'_{q0} = (-(X_q - X'_q)I_q - E'_d)/T'_{q0}. \tag{6.2.14}$$

The phasor diagram of the machine operating in the transient state is shown in Fig. 6.3.

6.2.2.3 Subtransient Equations

Either deliberately, as in the case of damper windings, or unavoidably, other circuits exist in the rotor. These circuits are taken into account if a more exact model is required. The reactances and time constants involved are small and can often be justifiably ignored. When required, the development of these equations is identical to that for transients and yields

$$E''_q - V_q = Ra\,I_q - X''_d I_d \tag{6.2.15}$$

$$E''_d - V_d = Ra I_d + X''_q I_q \tag{6.2.16}$$

$$pE''_q = (E'_q + (X'_d - X''_d)I_d - E''_q)/T''_{d0} \tag{6.2.17}$$

$$pE''_d = (E'_d - (X'_q - X''_q)I_q - E''_d)/T''_{q0}. \tag{6.2.18}$$

The equations are developed assuming that the transient time constants are large compared with the subtransient time constants. A phasor diagram of the synchronous machine operating in the subtransient state is shown in Fig. 6.4. It should be noted that equations (6.2.11) and (6.2.12) are now true only in the steady-state mode of operation, although once subtransient effects have decayed, the error will be small.

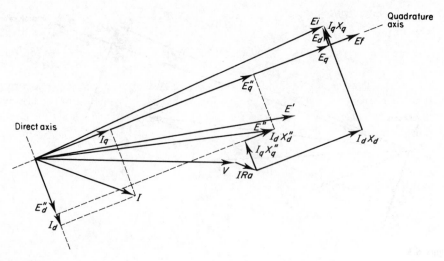

Figure 6.4
Phasor diagram of a synchronous machine in the subtransient state

6.2.2.4   Machine Models

It is possible to extend the model beyond subtransient level but this is seldom done in multi-machine programs. Investigations [5] using a generator model with up to seven rotor windings have shown that using the standard machine data the more complex models do not necessarily given more accurate results. However, improved results can be obtained if the data, especially the time constants, are suitably modified.

The most convenient method of treating synchronous machines of differing complexity is to allow each machine the maximum possible number of equations and then let the actual model used be determined automatically according to the data presented.

Five models are thus possible for a four-winding rotor.

Model 1—constant voltage magnitude behind d-axis transient reactance (X'_d) requiring no differential equations. Only the algebraic equations (6.2.11) and (6.2.12) are used.
Model 2—d-axis transient effects requiring one differential equation (pE'_q). Equations (6.2.11) to (6.2.13) are used.
Model 3—d- and q-axis transient effects requiring two differential equations $(pE'_q$ and $pE'_d)$. Equations (6.2.11) to (6.2.14) are used.
Model 4—d- and q-axis subtransient effects requiring three differential equations $(pE'_q, pE''_q$ and $pE''_d)$. Equations (6.2.13), (6.2.15) to (6.2.17) and

$$pE''_d = \frac{(-(X_q - X''_q)I_q - E''_d)}{T''_{qo}} \tag{6.2.19}$$

are used. This last equation is merely equation (6.2.14) with modified primes. Whether it is a subtransient or transient equation is open to argument.

Model 5—d- and q-axis subtransient effects requiring four differential equations (pE_q', pE_d', pE_q'' and pE_d''). Equations (6.2.13) to (6.2.18) are used.

Thus mechanical equations (6.2.5) and (6.2.6) must also be solved for all these models.

Groups of synchronous machines or parts of the system may be represented by a single synchronous machine model. An infinite busbar, representing a large stiff system, may be similarly modelled as a single machine represented by model 1, with the simplification that the mechanical equations (6.2.5) and (6.2.6) are not required. This sixth model is thus defined as:

Model 0—Infinite machine-constant voltage (phase and magnitude) behind d-axis transient reactance (X_d'). Only equations (6.2.11) and (6.2.12) are used.

6.3 SYNCHRONOUS MACHINE AUTOMATIC CONTROLLERS

For dynamic power system simulations of 1 s or longer duration, it is necessary to include the effects of the machine controllers, at least for the machine most affected by the disturbance. Moreover, controller representation is becoming necessary, even for first swing stability, with systems being operated at their limits with near critical fault clearing times.

The two principal controllers of a turbine generator set are the automatic voltage regulator (AVR) and the speed governor. The AVR model consists of voltage sensing equipment, comparators and amplifiers controlling a synchronous machine which can be generating or motoring. The speed governor may be considered to have similar equipment but in addition it is necessary to take the turbine into account.

6.3.1 Automatic Voltage Regulators

Many different AVR models have been developed to represent the various types used in a power system. The application of such models is difficult and a better approach is to develop a single general purpose AVR model, on a similar basis to the synchronous machine model. The model can then revert to any desired type by using the correct data. The IEEE defined several AVR types [6], the main two of which (Type 1 and Type 2) are shown in Fig. 6.5.

A composite model of these two AVR types can be constructed. This model may also include a secondary signal which can be taken from any source, but usually either machine rotor speed deviation from synchronous speed or rate of change of machine output power. This model is shown in Fig. 6.6 and has been found to be satisfactory for all the systems studied so far. It is acknowledged that other AVR models may be necessary for specific studies.

In many systems studied, the amount of data available for an AVR model is quite small. The composite model can degenerate into a very simple model easily by defaulting time constants to zero and gains to either zero, unity or an extremely large value depending on their position.

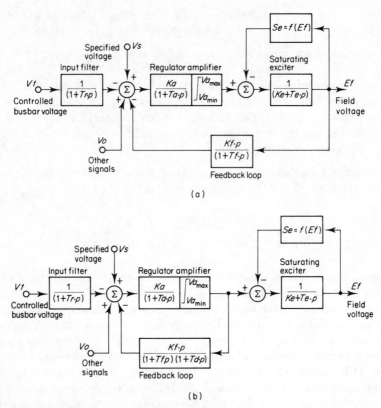

Figure 6.5
Block diagrams for two commonly used AVR models [6]. (a) IEEE Type 1 AVR model; (b) IEEE Type 2 AVR model (© 1982 IEEE)

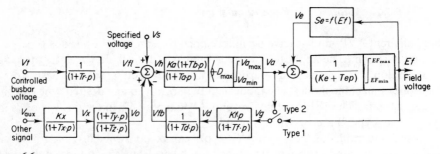

Figure 6.6
Block diagram of a composite automatic voltage regulator model

The equations for the AVR model shown in Fig. 6.6 are as follows:

$$pVfl = (Vt - Vfl)/Tr \tag{6.3.1}$$

$$pVa = (Ka(1 + T_b \cdot p)Vh - Va)/Ta \tag{6.3.2}$$

subject to

$$|pVa| \leqslant D_{\max}$$

and

$$Va_{\max} \geqslant Va \geqslant Va_{\min}$$
$$pEf = (Va - Ve - Ke \cdot Ef)/Te \tag{6.3.3}$$

subject to

$$Ef_{\max} \geqslant Ef \geqslant Ef_{\min}$$
$$pVd = (Kf \cdot pVg - Vd)/Tf \tag{6.3.4}$$
$$pVfb = (Vd - Vfb)/T_d \tag{6.3.5}$$
$$pVx = (KxV_{\text{aux}} - Vx)/Tx \tag{6.3.6}$$
$$pVo = ((1 + Ty \cdot p)Vx - Vo)/Tz \tag{6.3.7}$$
$$Vh = Vs - Vfb - Vfl + Vo \tag{6.3.8}$$
$$Ve = SeEf \tag{6.3.9}$$

where $Se = f(Ef)$ and

$$Vg = Ef \text{ [unless IEEE Type 2 when } Vg = Va] \tag{6.3.10}$$
$$V_{\text{aux}} = \text{a predefined signal.}$$

The IEEE [6] recommends that Se be specified at maximum field voltage ($Se_{\max}$) and at 0.75 of maximum field voltage ($Se_{0.75\max}$). From this Se may be determined for any value of field voltage by either linear interpolation or by fitting a quadratic. Where linear interpolation is used, equation (6.3.9) may be transformed to

$$Ve = (k_1 Ef - k_2)Ef \tag{6.3.11}$$

where

$$\left. \begin{aligned} k_1 &= (4Se_{0.75\max})/(3Ef_{\max}) \\ k_2 &= 0 \end{aligned} \right\} \quad \text{if } Ef \leqslant 0.75 Ef_{\max}$$

or

$$\left. \begin{aligned} k_1 &= 4(Se_{\max} - Se_{0.75\max})/Ef_{\max} \\ k_2 &= 4Se_{0.75\max} - 3Se_{\max} \end{aligned} \right\} \quad \text{if } Ef > 0.75 Ef_{\max}.$$

A means of modelling lead–lag circuits such as those in the regulator amplifier, the stabilising loop and the auxiliary signal circuits is given at the end of this section.

Despite the advantages of one composite AVR model, if there are a great many AVRs to be modelled most of which have simple characteristics then it is better to make two models. One model, which contains only the commonly used parts of the composite model can then be dimensioned for all AVRs. The other model, which contains only the less commonly used parts of the composite model can be quite small dimensionally. A connection vector is all that is necessary to interconnect the two models whenever necessary.

6.3.2 Speed Governors

For speed governors, as with AVRs, a composite model which can be reduced to any desired level is the most satisfactory. The speed governor models recommended

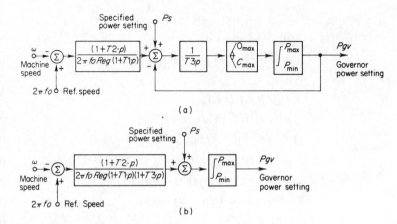

Figure 6.7
Typical models of speed governors and valves [7]. (a) Thermal governor and valve; (b) hydro
governor and valve (© 1982 IEEE)

by the IEEE [7] are shown in Fig. 6.7. Notice that if limits are not exceeded, the
two models are identical. The difference is due to the assumption that, in a hydro
governor, gate servo and gate positions are the same. One model can be used for the
governors of both turbines provided that the limits are either internal or external to
the second transfer function block of Fig. 6.8. Also, very little extra effort is required
to divorce the governor from the actual turbine power and keep it instead as a
function of valve position.

The equations of the speed governor shown in Fig. 6.8 are

$$pG_1 = [R(1 + T_2p)(2\pi f_0 - \omega) - G_1]/\!/T_1 \tag{6.3.12}$$

$$pG_2 = (G_1 - G_2)/T_3) \tag{6.3.13}$$

$$Gv = G_2 + Gs. \tag{6.3.14}$$

The valve/gate position setting (Gv) is subject to opening and closing rate limits (o_{max}
and c_{max} respectively) and to physical travel limits so that

$$-c_{max} < pGv < o_{max}$$
$$0 < Gv < 1. \tag{6.3.15}$$

The valve equation is

$$Pgv = Gv \cdot Pb. \tag{6.3.16}$$

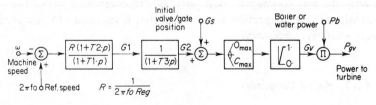

Figure 6.8
Generalised model of a speed governor and valve

For thermal turbines, where a boiler is modelled, in the steady state Pb will be the actual power delivered and Gs will be unity, i.e. the valve will be fully open. If a boiler is not modelled or a hydro turbine is being controlled then, in the steady state, Pb will be the maximum output from the boiler or water system (i.e. maximum turbine mechanical power output) and Gv, and hence Gs, will be such that Pgv is the actual mechanical power output of the turbine.

This method of modelling a valve has the advantage that nonlinearities between valve position and power can be easily included and also the operation of the governor and valve can be readily interpreted.

For a hydro governor where the limits are external, the model is as given in equations (6.3.12) to (6.3.16) but for a thermal governor, G_2 is reset after the valve limits are applied to be

$$G_{2(\text{lim})} = Gv - Gs \qquad \text{(thermal governor only)}. \qquad (6.3.17)$$

6.3.3 Hydro and Thermal Turbines

This section is restricted to the modelling of simple turbines only. Compound thermal turbines may require a detailed model, as given in Chapter 7, but, for stability studies of only 1 or 2 s duration, the effect of all but the high pressure (HP) turbine can usually be ignored. The time constant associated with the steam entrained between the HP turbine outlet and the IP or LP turbine inlet is usually very large (greater than 5 s) and the output from all turbines other than the HP turbine may be treated as constant.

Simple linear models of hydro and thermal turbines are shown in Fig. 6.9. The hydro turbine model includes the penstock which gives the characteristic lead–lag response of this type of turbine. The model is generally sufficient for all hydro turbines and, from Fig. 6.9, the differential equation for the mechanical power output (Pm) of the turbine is

$$pPm = ((1 - Tw \cdot p)Pgv - Pm)/T_4 \qquad (6.3.18)$$

with $T_4 = 0.5\,Tw$ as a further close approximation.

For the thermal turbine using Fig. 6.9(b) this equation is

$$pPm = (K_1 \cdot Pgv - Pm)/T_4 \qquad (6.3.19)$$

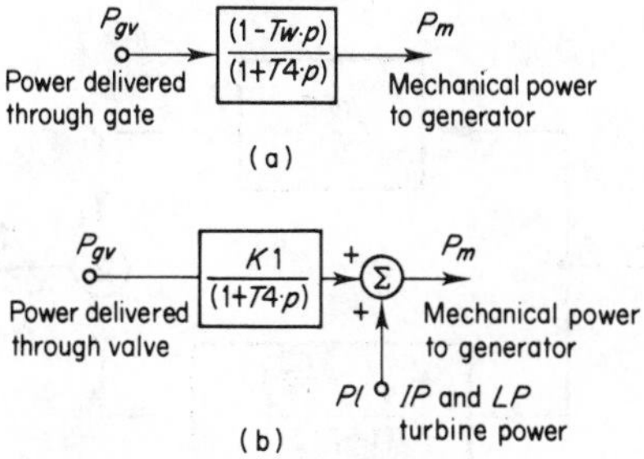

Figure 6.9
Simple linear models of turbines: (a) hydro turbine; (b) thermal turbine

with K_1 representing the fraction of power delivered by the HP turbine. For simple turbines K_1 is thus unity. For compound turbines, the power (Pl) from the IP and LP turbines is obtained from

$$Pl = (1 - K_1)Pm_0. \tag{6.3.20}$$

Here Pm_0 is the initial steady-state mechanical power. Note that for this simple model, the initial value of Pgv is Pm/K_1, even though all the steam passes through the valve.

Provided that the HP valve does not close fully, then, rather than inject the power from the IP and LP turbines as shown, it is easier to treat it as a simple turbine ($Pl = 0$) but with the speed regulation modified by

$$Reg_{(mod)} = \frac{Reg}{K_1}. \tag{6.3.21}$$

6.3.4 Modelling Lead–Lag Circuits

Lead–lag circuits may present a problem depending on the integration scheme adopted. Where the differential equations are not used directly and the derivatives are not explicitly calculated, the following can be used to convert the model into a more acceptable form.

For the circuit shown in block diagram form in Fig. 6.10, the equation is

$$V_{out} = \frac{K(1 + T_2 p)}{(1 + T_1 p)} V_{in}. \tag{6.3.22}$$

This can be transformed to

$$V_{out} = \frac{K \cdot T_2}{T_1} \left(\frac{(T_1/T_2 + T_1 \cdot p)}{1 + T_1 \cdot p} \right) V_{in}$$

$$V_{in} \quad \boxed{\dfrac{K(1 + T2 \cdot p)}{(1 + T1 \cdot p)}} \quad V_{out}$$

Figure 6.10
Typical lead–lag circuit block diagram

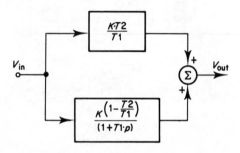

Figure 6.11
Modified block diagram of a lead–lag circuit

and then to

$$V_{\text{out}} = \left(\frac{K \cdot T_2}{T_1} + \frac{K(1 - T_2/T_1)}{(1 + T_1 \cdot p)} \right) V_{\text{in}} \tag{6.3.23}$$

which can be represented by the block diagram in Fig. 6.11, and is a lag circuit in parallel with a gain.

It is important to remember that the time constant T_1, must be nonzero even if the integration method can accommodate zero time constants.

6.4   LOADS

Early transient stability studies were concerned primarily with generator stability, and little importance was attached to loads. In the two-machine problem for example, the remainder of the system, generators and loads were represented by an infinite busbar. A great deal of attention has been given to load modelling since then.

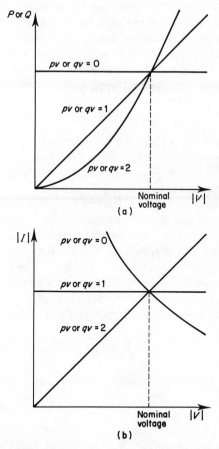

Figure 6.12
Characteristics of different load models: (a) active and reactive power against voltage; (b) current against voltage

Table 6.1
Typical values of characteristic load parameters [9]

Load	pv	qv	pf	qf
Filament lamp	1.6	0	0	0
Fluorescent lamp	1.2	3.0	−1.0	2.8
Heater	2.0	0	0	0
Induction motor half load	0.2	1.6	1.5	−0.3
Induction motor full load	0.1	0.6	2.8	1.8
Reduction furnace	1.9	2.1	−0.5	0
Aluminium plant	1.8	2.2	−0.3	0.6

Much of the domestic load and some industrial load consist of heating and lighting, especially in the winter, and in early load models these were considered as constant impedances. Rotating equipment was often modelled as a simple form of synchronous machine and composite loads were simulated by a mixture of these two types of load.

A lot of work has gone into the development of more accurate load models. These include some complex models of specific large loads which are considered in the next chapter. Most loads, however, consist of a large quantity of diverse equipment of varying levels and composition and some equivalent model is necessary.

A general load characteristic [8] may be adopted such that the MVA loading at a particular busbar is a function of voltage (V) and frequency (f):

$$P = Kp(V)^{pv} \cdot (f)^{pf} \tag{6.4.1}$$

$$Q = Kq(V)^{qv} \cdot (f)^{qf} \tag{6.4.2}$$

where Kp and Kq are constants which depend upon the nominal value of the variables P and Q.

Static loads are relatively unaffected by frequency changes, i.e. $pf = qf = 0$, and with constant impedance loads $pv = qv = 2$.

The importance of accurate load models has been demonstrated by Dandeno and Kundur [8] when considering voltage-sensitive loads. Figure 6.12 demonstrates the power and current characteristics of constant power, constant current and constant impedance loads. Berg [9] has identified the characteristic load parameters for various homogeneous loads and these are given in Table 6.1. These characteristics may be combined to give the overall load characteristic at a busbar. For example, a group of n homogeneous loads, each with a characteristic of pv_j and a nominal power of P_j may be combined to give an overall characteristic of

$$pv_{(overall)} = \sum_{j=1}^{n} (pv_j \cdot P_j) \bigg/ \sum_{j=1}^{n} (P_j). \tag{6.4.3}$$

The other three overall characteristics may be similarly determined.

6.4.1 Low-voltage Problems

When the load parameters pv and qv are less than or equal to unity, a problem can occur when the voltage drops to a low value. As the voltage magnitude decreases, the current magnitude does not decrease. In the limiting case with zero voltage magnitude, a load current flows which is clearly irrational, given the nondynamic nature of the load model. From a purely practical point of view, the load characteristics are only valid for a small voltage deviation from nominal. Further, if the voltage is small, small errors in magnitude and phase produce large errors in current magnitude and phase. This results in loss of accuracy and with iterative solution methods poor convergence or divergence.

These effects can be overcome by using a constant impedance characteristic to represent loads where the voltage is below some predefined value, for example 0.8 p.u.

6.5 THE TRANSMISSION NETWORK

It is usual to represent the static equipment which constitutes the transmission system by lumped 'equivalent-π' parameters independent of the changes occurring in the generating and load equipment. This representation is used for multi-machine stability programs because the inclusion of time-varying parameters would cause enormous computational problems. Moreover, frequency, which is the most obvious variable in the network, usually varies by only a small amount and thus the errors involved are small. Also, the rates of change of network variables are assumed to be infinite which avoids the introduction of differential equations into the network solution.

The transmission network can thus be represented in the same manner as in the load-flow or short-circuit programs, that is by a square complex admittance matrix.

The behaviour of the network is described by the matrix equation

$$[I_{\text{inj}}] = [Y][V] \tag{6.5.1}$$

where $[I_{\text{inj}}]$ is the vector of injected currents into the network due to generators and loads and $[V]$ is the vector of nodal voltages.

Any loads represented by constant impedances may be directly included in the network admittance matrix with the injected currents due to these loads set to zero. Their effect is thus accounted for directly by the network solution.

6.6 OVERALL SYSTEM REPRESENTATION

Two alternative solution methods are possible. The preferred method uses the nodal matrix approach, while the alternative is the mesh matrix method.

Matrix reduction techniques can be used with both methods if specific network information is not required, but this gives little advantage as the sparsity of the reduced matrix is usually very much less.

6.6.1 Mesh Matrix Method

In this method, the system-loading components are treated as Thevenin equivalents of voltages behind impedances. The network is increased in size to include these impedances and the mesh impedance matrix of the increased network is created. This is then inverted or the factorised form of the inverse determined.

The solution process is as follows.

(i) Calculate the Thevenin voltages of the system loading components by solving the relevant differential and algebraic equations.

(ii) Determine the network currents using the Y matrix or factors. As the network current around a mesh containing the Thevenin voltage is the loading current this may affect the Thevenin voltage in which case an iterative process will be required.

6.6.2 Nodal Matrix Method

In this method, all network loading components are converted into Norton equivalents of injected currents in parallel with admittance. The admittances can be included in the network admittance matrix to form a modified admittance matrix which is then inverted, or preferably factorised by some technique so that solution at each stage is straightforward.

The following solution process applies.

(i) For each network-loading component, determine the injected currents into the modified admittance matrix by solving the relevant differential and algebraic equations.

(ii) Detemine network voltages from the injected currents using the Z matrix or factors.

As the network voltages affect the loading components, an iterative process is often required, although good approximations [8] can be used to avoid this.

With the nodal matrix method, busbar voltages are available directly and branch currents can be calculated if necessary while with the mesh matrix method, mesh currents are available directly and busbar voltages and branch currents must be calculated if necessary.

Although much work has been spent on the systematic construction of the mesh impedance matrix, the nodal admittance matrix is easier to construct and has gained wide acceptance in load-flow and fault analysis. For this reason, the remainder of this section will consider the nodal matrix method.

6.6.3 Synchronous Machine Representation in the Network

The equations representing a synchronous machine, as defined in Section 6.2, are given in the form of Thevenin voltages behind impedances. This must be modified to a current in parallel with an admittance by use of Norton's theorem. The admittance of the machine thus formed may be added to the shunt admittance of the machine busbar and treated as a network parameter. The vector $[I_{inj}]$ in equation (6.5.1) thus contains the Norton equivalent currents of the synchronous machines.

The synchronous machine equations are written in a frame of reference rotating

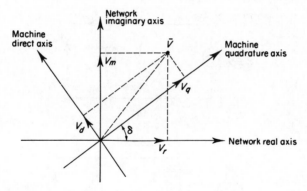

Figure 6.13
Synchronous machine and network frames of reference

with its own rotor. The real and imaginary components of the network equations, as given in Fig. 6.13, are obtained from the following transformation:

$$
\begin{vmatrix} V_r \\ V_m \end{vmatrix} = \begin{vmatrix} \cos\delta & -\sin\delta \\ \sin\delta & \cos\delta \end{vmatrix} \cdot \begin{vmatrix} V_q \\ V_d \end{vmatrix} \tag{6.6.1}
$$

This transformation is equally valid for currents as is the reverse transformation:

$$
\begin{vmatrix} V_q \\ V_d \end{vmatrix} = \begin{vmatrix} \cos\delta & \sin\delta \\ -\sin\delta & \cos\delta \end{vmatrix} \cdot \begin{vmatrix} V_r \\ V_m \end{vmatrix} \tag{6.6.2}
$$

When saliency exists, the values of X_d'' and X_q'' used in equations (6.2.15) and (6.2.16) and/or X_d' and X_q' used in equations (6.2.11) and (6.2.12) are different. Therefore, the Norton shunt admittance will have a different value in each axis and when transformed into the network frame of reference, will have time-varying components. However, a constant admittance can be used, provided that the injected current is suitably modified to retain the accuracy of the Norton equivalent [10]. This approach can be justified by comparing the two circuits of Fig. 6.14 in which $\bar{Y}_i$ is a time-varying admittance, whereas $\bar{Y}_0$ is fixed.

At any time t, the Norton equivalent of the machine is illustrated in Fig. 6.14(a), but the use of a fixed admittance results in the modified circuit of Fig. 6.14(b).

The machine current is

$$
\bar{I} = \bar{Y}_t(\bar{E}'' - \bar{V}) = \bar{Y}_0(\bar{E}'' - \bar{V}) + \bar{I}_{adj}
$$

and hence

$$
\bar{I}_{adj} = (\bar{Y}_t - \bar{Y}_0)(\bar{E}'' - \bar{V}) \tag{6.6.3}
$$

where $\bar{I}_{adj}$ accounts for the fact that the apparent current source is not accurate in this case.

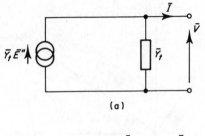

(a)

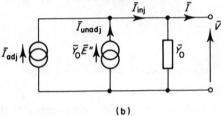

(b)

Figure 6.14
Method of representing synchronous machines in the network: (a) Norton equivalent circuit; (b) modified equivalent circuit

The injected current into the network which includes $\bar{Y}_0$ is given by

$$\bar{I}_{inj} = \bar{I}_{unadj} + \bar{I}_{adj} \qquad (6.6.4)$$

where

$$\bar{I}_{unadj} = \bar{Y}_0 \bar{E}''.$$

A suitable value for $\bar{Y}_0$ is found by using the mean of direct and quadrature admittances, i.e.

$$\bar{Y}_0 = \frac{(Ra - jX_{dq})}{(Ra^2 + X_d'' \cdot X_q'')} \qquad (6.6.5)$$

where

$$X_{dq} = \tfrac{1}{2}(X_d'' + X_q'').$$

The unadjusted value of current injected into the busbar is

$$
\begin{vmatrix} I_{unadj_r} \\ I_{unadj_m} \end{vmatrix}
= \frac{1}{(Ra^2 + X_d'' \cdot X_q'')}
\begin{vmatrix} Ra & X_{dq} \\ -X_{dq} & Ra \end{vmatrix}
\cdot
\begin{vmatrix} E_r'' \\ E_m'' \end{vmatrix}
\qquad (6.6.6)
$$

The adjusting current is not affected by rotor position in the machine frame of reference but it is when considered in the network frame. From equation (6.6.3) and also equations (6.2.15) and (6.2.16)

$$
\begin{vmatrix} I_{adj_q} \\ I_{adj_d} \end{vmatrix}
= \frac{\tfrac{1}{2}(X_d'' - X_q'')}{(Ra^2 + X_d'' X_q'')}
\begin{vmatrix} 0 & 1 \\ 1 & 0 \end{vmatrix}
\cdot
\begin{vmatrix} E_q'' - V_q \\ E_d'' - V_d \end{vmatrix}
\qquad (6.6.7)
$$

and transforming

$$
\begin{array}{|c|}
\hline I_{adj_r} \\
\hline I_{adj_m} \\
\hline
\end{array}
=
\frac{\frac{1}{2}(X_d'' - X_q'')}{(Ra^2 + X_d''X_q'')}
\begin{array}{|c|c|}
\hline -\sin 2\delta & \cos 2\delta \\
\hline \cos 2\delta & \sin 2\delta \\
\hline
\end{array}
\cdot
\begin{array}{|c|}
\hline E_r'' - V_r \\
\hline E_m'' - V_m \\
\hline
\end{array}
\tag{6.6.8}
$$

The total nodal injected current is therefore

$$
\begin{array}{|c|}
\hline I_{inj_r} \\
\hline I_{inj_m} \\
\hline
\end{array}
=
\begin{array}{|c|}
\hline I_{unadj_r} \\
\hline I_{unadj_m} \\
\hline
\end{array}
+
\begin{array}{|c|}
\hline I_{adj_r} \\
\hline I_{adj_m} \\
\hline
\end{array}
\tag{6.6.9}
$$

6.6.4 Load Representation in the Network

To be suitable for representation in the overall solution method, loads must be transformed into currents injected into the transmission network from which the terminal voltages can be calculated. A Norton equivalent model of each load must therefore be created. In a similar way to that adopted for synchronous machines, the Norton admittance may be included directly in the network admittance matrix.

A constant impedance load is therefore totally included in the network admittance matrix and its injected current is zero. This representation is extremely simple to implement, causes no computational problems and improves the accuracy of the network solution by strengthening the diagonal elements in the admittance matrix.

Nonimpedance loads may be treated similarly. In this case, the steady-state values of voltage and complex power obtained from the load flow are used to obtain a steady-state equivalent admittance ($\bar{Y}_0$) which is included in the network admittance matrix $[Y]$. During the stability run, each load is solved sequentially along with the generators, etc. to obtain a new admittance ($\bar{Y}$):

$$
\bar{Y} = \frac{\bar{S}^*}{|V|^2}.
\tag{6.6.10}
$$

The current injected into the network thus represents the deviation of the load characteristic from an impedance characteristic:

$$
\bar{I}_{inj} = (\bar{Y}_0 - \bar{Y})\bar{V}.
\tag{6.6.11}
$$

By converting the load characteristic to that of a constant impedance when the voltage drops below some predetermined value (V_{low}), as described in Section 6.4, the injected current is kept relatively small. An example of a load characteristic and its corresponding injected current is shown in Fig. 6.15.

In an alternative model the low-voltage impedance is added to the network and the injected current compensates for the deviation from the actual characteristic. In this case, there is a nonzero injected current in the initial steady-state operating condition.

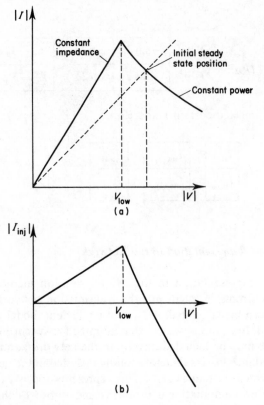

Figure 6.15
Load and injected currents for a constant power type load with low-voltage adjustment: (a) load
current; (b) injected current

6.6.5 System Faults and Switching

In general most power system disturbances to be studied will be caused by changes
in the network. These changes will normally be caused by faults and subsequent
switching action but occasionally the effect of branch or machine switching will be
considered.

6.6.5.1 Faults

Although faults can occur anywhere in the system, it is much easier computationally
to apply a fault to a busbar. In this case, only the shunt admittance at the busbar
need be changed, that is, a modification to the relevant self-admittance of the Y
matrix. Faults on branches require the construction of a dummy busbar at the fault
location and suitable modification of the branch data unless the distance between
the fault position and the nearest busbar is small enough to be ignored.

The worst case is a three-phase zero-impedance fault and this involves placing an infinite admittance in parallel with the existing shunt admittance. In practice, a nonzero but sufficiently low fault impedance is used so that the busbar voltage is effectively brought to zero. This is necessary to meet the requirements of the numerical solution method.

The application or removal of a fault at an existing busbar does not affect the topology of the network and where the solution method is based on sparsity exploiting ordered elimination, the ordering remains unchanged and only the factors required for the forward and backward substitution need be modified. Alternatively the factors can remain constant and diakoptical techniques [11] can be used to account for the network change.

6.6.5.2 *Branch Switching*

Branch switching can easily be carried out by either modifying the relevant mutual and self-admittances of the Y matrix or using diakoptical techniques. In either case, the topology of the network can remain unchanged as an open branch is merely one with zero admittance. While this does not fully exploit sparsity, in almost all cases the gain in computation time by not reordering exceeds the loss of retaining zero elements.

The only exception is the case of a branch switched into a network where no interconnection existed prior to that event. In this case, either diakoptical or reordering techniques become necessary. To avoid this problem, a dummy branch of sufficiently high impedance that the power flow is negligible under all conditions may be included with the steady-state data, or alternatively, the branch resistance may be set negative to represent an initial open circuit. A negative branch reactance should not be used as this is a valid parameter where a branch contains series capacitors.

Where a fault occurs on a branch but very close to a busbar, nonunit protection at the near busbar will normally operate before that at the remote end. Therefore, there will be a period when the fault is still being supplied from the remote end. There are two methods of accounting for this type of fault.

The simplest method only requires data manipulation. The fault is initially assumed to exist at the local busbar rather than on the branch. When the specified time for the protection and local circuit breaker to operate has elapsed, the fault is removed and the branch on which the fault is assumed to exist is opened. Simultaneously, the fault is applied at the remote busbar, but in this case, with the fault impedance increased by the faulted branch impedance, similarly the fault is maintained until the time specified for the protection and remote circuit breaker to operate has elapsed.

The second method is generally more involved but it is better when protection schemes are modelled. In this case, a dummy busbar is located at the fault position, even though it is close to the local busbar, and a branch with a very small impedance is inserted between the dummy busbar and the local busbar. The faulted branch then connects the dummy busbar to the remote busbar and the branch shunt susceptance originally associated with the local busbar is tranferred to the dummy busbar. This may all be done computationally at the time when the fault is being specified. The two branches can now be controlled independently by suitable protection systems. An advantage of this scheme is that the fault duration need not be specified as part of the input data. Opening both branches effectively isolates the fault, which can remain permanently attached to the dummy busbar, or if auto-reclosing is required, it can be removed automatically after a suitable deionisation period.

The second method will give problems if the network is not being solved by a direct method. During the iterative solution of the network, slight voltage errors will cause large currents to flow through a branch with a very small impedance. This will slow convergence and in extreme cases will cause divergence. With a direct method, based on ordered elimination, an exact solution of the busbar voltages is obtained for the injected currents specified at that particular iteration. Thus, provided that the impedance is not so small that numerical problems occur when calculating the admittance, and the subsequent factors for the forward and backward substitution, then convergence of the overall solution between machines and network will be unaffected. The value of the low-impedance branch between the dummy and local busbars may be set at a fraction of the total branch impedance, subject to a minimum value. If this fraction is under 1/100, the change in branch impedance is very small compared to the accuracy of the network data input and it is unnecessary to modify the impedance of the branch from the remote to the dummy busbar.

6.6.5.3 Machine Switching

Machine switching may be considered, either as a network or as a machine operation. It is a network operation if a dummy busbar is created to which the machine is connected. The dummy busbar is then connected to the original machine busbar by a low-impedance branch.

Alternatively, it may be treated as a machine operation by retaining the original network topology. When a machine is switched out, it is necessary to remove its injected current from the network solution. Also, any shunt admittance included in the network Y matrix, which is due to the machine, must be removed.

Although a disconnected machine can play no direct part in system stability, its response should still be calculated as before, with the machine stator current set to zero. Thus machine speed, terminal voltage, etc., can be observed even when disconnected from the system and in the event of reconnection, sensible results are obtained.

Where an industrial system is being studied many machines may be disconnected and reconnected at different times as the voltage level changes. This process will require many recalculations of the factors involved in the forward and backward substitution solution method of the network. However, these can be avoided by using the method adopted earlier to account for synchronous machine saliency. That is, an appropriate current is injected at the relevant busbar, which cancels out the effect of the shunt admittance.

6.7 INTEGRATION

Many integration methods have been applied to the power system transient stability problem and the principal methods are discussed in Appendix IV. Of these, only three are considered in this section. They are simple and easily applied methods which have gained wide acceptance. The purpose of the third method is not to provide another alternative but to clarify the differences between the other two methods.

Explicit Runge–Kutta methods have been used extensively in transient stability studies. They have the advantage that a 'packaged' integration method is usually available or quite readily constructed and the differential equations are incorporated with the method explicitly. It has only been with the introduction of more detailed system component models with very small time constants that the problems of stability have caused interest in other methods.

Fourth-order methods ($p = 4$) have probably been the most popular and among these the Runge–Kutta Gill method has the advantage that round-off error is minimised. With reference to equations (IV.4.1) to (IV.4.3), for this method the number of function substitutions is four ($v = 4$) and

$$
\begin{aligned}
w_1 &= 1/6 \\
w_2 &= (2 - \sqrt{2})/6 \\
w_3 &= (2 + \sqrt{2})/6 \\
w_4 &= 1/6
\end{aligned}
\tag{6.7.1}
$$

$$
\begin{aligned}
k_1 &= hf(t_n, y_n) \\
k_2 &= hf(t_n + h/2, y_n + k_1/2) \\
k_3 &= hf(t_n + h/2, y_n + (\sqrt{2} - 1)k_1/2 + (2 - \sqrt{2})k_2/2) \\
k_4 &= hf(t_n + h, y_n - \sqrt{2}k_2/2 + (2 + \sqrt{2})k_3/2).
\end{aligned}
\tag{6.7.2}
$$

The characteristic root of this fourth-order method, when applied to equation (IV.3.3), is

$$
z_1 = 1 + h\lambda + \tfrac{1}{2}h^2\lambda^2 + \tfrac{1}{6}h^3\lambda^3 + \tfrac{1}{24}h^4\lambda^4
\tag{6.7.3}
$$

and to ensure stability, the step length h must be sufficiently small that z_1 is less than unity.

The basic trapezoidal method is very well known, having been established as a useful method of integration before digital computers made hand calculation redundant.

More recently an implicit trapezoidal integration method has been developed for solving the multimachine transient stability problem [10], and has gained recognition as being very powerful, having great advantages over the more traditional methods.

The method is derived from the general multistep equation given by equation (IV.3.2) with k equal to unity and is thus a single-step method. The solution at the end of $n + 1$ steps is given by

$$
y_{n+1} = y_n + \frac{h_{n+1}}{2}(py_{n+1} + py_n).
\tag{6.7.4}
$$

It has second-order accuracy with the major term in the truncation error being $-\tfrac{1}{12}h^3$.

The characteristic root when applied to equation (IV.3.3) is

$$
z_1 = 1 - 2b_{n+1}
\tag{6.7.5}
$$

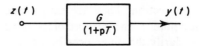

Figure 6.16
Simple transfer function

where

$$b_{n+1} = \frac{h_{n+1}}{(h_{n+1} - 2/\lambda)}. \tag{6.7.6}$$

If $\text{Re}(\lambda) < 0$ then $0 \leqslant b_{n+1} \leqslant 1.0$ and $|z_1| \leqslant 1.0$. The trapezoidal method is therefore A-stable, a property which is shown in Appendix IV to be more important in the solution process than accuracy. The trapezoidal method is linear and thus in a multivariable problem, like power system stability, the method is Σ-stable.

It can be shown that an A-stable linear multistep method cannot have an order of accuracy greater than two, and that the smallest truncation error is achieved by the trapezoidal method. The trapezoidal method is thus the most accurate Σ-stable finite difference method possible.

The method, as expressed by equation (6.7.4), is implicit and requires an iterative solution. However, the solution can be made direct by incorporating the differential equations into equation (6.7.4). Rearranging forms algebraic equations as described in Appendix IV.

For example, consider the trivial transfer function shown in Fig. 6.16. The differential equation for this system is given by

$$py(t) = (G \cdot z(t) - y(t))/T \tag{6.7.7}$$

with the input variable being denoted by 'z' to indicate that it may be either integrable or nonintegrable.

The algebraic form of equation (6.7.7) has a solution at the end of the $(n + 1)$th step of

$$y_{n+1} = c_{n+1} + m_{n+1} \cdot z_{n+1} \tag{6.7.8}$$

where

$$c_{n+1} = (1 - 2b_{n+1})y_n + b_{n+1} \cdot G \cdot z_n \tag{6.7.9}$$

$$m_{n+1} = b_{n+1} \cdot G \tag{6.7.10}$$

and

$$b_{n+1} = h_{n+1}/(2T + h_{n+1}). \tag{6.7.11}$$

Provided that the step length h remains constant it is unnecessary to reevaluate b or m at each step, i.e.

$$\left. \begin{matrix} b_{n+1} = b_n \\ m_{n+1} = m_n \end{matrix} \right\} \quad \text{if } h_{n+1} = h_n. \tag{6.7.12}$$

There is little to be gained by this, however, as it is a simple process and it is often desirable to change h during a study.

A comparison between the Runge–Kutta Gill and the trapezoidal methods when used to solve two power system transient stability problems is given in Tables 6.2

Table 6.2

Step length (ms)	Runge–Kutta Gill Max. error (degs)	CPU time (s)	Trapezoidal Max. error (degs)	CPU time (s)	Backward Euler Max. error (degs)	CPU time (s)
100.0	—	—	2.2	0.26	—	—
50.0	—	—	0.7	0.27	—	—
25.0	21.0	0.43	0.1	0.29	5.7	0.41
10.0	13.0	0.72	—	0.49	2.4	0.47
5.0	7.8	1.18	—	0.69	1.3	0.67
2.0	3.7	2.57	—	1.34	0.5	1.31
1.0	1.9	4.88	—	2.42	0.2	2.35
0.5	1.0	9.52	—	4.60	—	4.42
0.2	0.4	24.19	—	—	—	10.58
0.1	0.2	47.95	—	—	—	—

Table 6.3

Step length (ms)	Runge–Kutta Gill Max. error (degs)	CPU time (s)	Trapezoidal Max. error (degs)	CPU time (s)	Backward Euler Max. error (degs)	CPU time (s)
10.0	8.6	1.67	0.5	2.37	—	—
5.0	4.4	3.06	0.1	2.31	8.5	2.76
2.0	1.7	7.24	—	3.74	3.8	3.64
1.0	1.2	14.19	—	7.12	1.8	6.80
0.5	0.9	28.00	—	13.88	0.6	13.24

and 6.3. The comparison is made in terms of maximum error (based on results using very small step lengths) and central processor unit (CPU) execution time.

The advantages of the Σ-stable trapezoidal method are apparent from both tables, but the results are sufficiently different to show that an absolute comparison between methods cannot be made. The nonlinearity of the equations in any system also effect the errors obtained. CPU time using the Runge–Kutta Gill method is a function of the step length but this is not so with the trapezoidal method. For very small step lengths, only one iteration per step is needed using the trapezoidal method but as the step length increases so does the number of iterations. The relationship between step length and iterations is nonlinear, with the result that there is an optimum step length in which the iterations per step are small but greater than one.

For comparison, the backward Euler method is also included. This is a first-order method with the solution given by

$$y_{n+1} = y_n + h \cdot p y_{n+1} \tag{6.7.13}$$

and the characteristic root when applied to equation (IV.3.3) is

$$z_1 = 1/(1 - h\lambda). \tag{6.7.14}$$

Despite the three orders of accuracy difference between it and the Runge–Kutta Gill, the backward Euler method compares well.

The results for the trapezoidal and backward Euler methods were obtained using linear extrapolation of the nonintegrable variables at the beginning of each step. This required the storing of machine terminal voltages and currents together with other nonintegrable variables obtained at the end of the previous step.

6.7.1 Problems with the Trapezoidal Method

Although the trapezoidal method is Σ-stable and the step length is not constrained by the largest negative eigenvalue, the accuracy of the solution corresponding to the largest negative eigenvalues will be poor if a reasonable step length is not chosen.

With the backward Euler method, the larger the step length the smaller the characteristic root, i.e.

$$z_{1(BE)} \to 0 \quad \text{as } h\lambda \to -\infty \tag{6.7.15}$$

whereas for the trapezoidal method

$$z_{1(TRAP)} \to -1 \quad \text{as } h\lambda \to -\infty. \tag{6.7.16}$$

For small step lengths the characteristic roots of both methods tend towards, but never exceed, unity (positive), i.e.

$$z_{1(BE)} \text{ and } z_{1(TRAP)} \to +1 \quad \text{as } h\lambda \to 0. \tag{6.7.17}$$

The effect of too large a step length can be shown in a trivial but extreme example. The system shown earlier in Fig. 6.16 and equation (6.7.7) with a zero time constant T, and unity gain G, is such an example.

If the input $z(t)$ is a unit step function from an initial value of zero, then with a zero time constant, the output $y(t)$ should follow the input exactly, that is a constant output of unity. In fact, the output oscillates with $y_1 = 2$, $y_2 = 0$, $y_3 = 2$, etc.

Table 6.4 shows the effect of different step lengths on this simple system with a

Table 6.4
The effect of different step lengths on the solution of a simple system (Fig. 6.16) by the trapezoidal method

Step No.	$h\lambda = -0.5$		$h\lambda = -2.0$		$h\lambda = -8.0$		$h\lambda = -32.0$	
	Trap method	Exact solution	Trap method	Exact solution	Trap method	Exact solution	Trap method	Exact solution
0	0	0	0	0	0	0	0	0
1	0.4000	0.3935	1.0000	0.8647	1.6000	0.9997	1.8824	1.0000
2	0.6400	0.6321		0.9817	0.6400	1.0000	0.2215	
3	0.7840	0.7769		0.9975	1.2160		1.6870	
4	0.8704	0.8647		0.9997	0.8704		0.3938	
5	0.9222	0.9179	1.0000	1.0000	1.0778	1.0000	1.5349	1.0000

nonzero time constant T. This table shows that oscillations occur when $h\lambda$ is smaller than -2, that is, when the characteristic root z_1 is negative. The oscillations decay with a rate dependent on $h\lambda$, that is, the rate is dependent on the magnitude of z_1. It can also be seen that accuracy is good provided that $h\lambda$ is greater than or equal to -0.5.

Oscillations are only initiated at a discontinuity. Provided that there is no step function input, the output of a transfer function with zero time constant duplicates the input.

The example given is an extreme case and for the power system stability problem this usually only occurs in the input circuit of the AVR.

For the mechanical equation of the synchronous machine, the speed is given by

$$pw = \frac{1}{Mg}(Pa) \tag{6.7.18}$$

where Pa is the accelerating power given by $Pa = P_m - Pe$, and the damping factor Da is zero. Therefore, in this case

$$\omega_{n+1} = c_{n+1} + m_{m+1}Pa_{n+1} \tag{6.7.19}$$

where

$$c_{n+1} = \omega_n + m_{n+1}Pa_n \tag{6.7.20}$$

and

$$m_{n+1} = h_{n+1}/2Mg \tag{6.7.21}$$

and oscillations do not occur. Da, when it does exist, is usually very small and any oscillations will similarly be very small.

For the electrical equations of the synchronous machine, only the current can change instantaneously, and the effect is not as pronounced as for a unit step function.

Techniques [12] are available to remove the oscillations but they require a lot of storage and it is simpler to reduce the step length.

6.7.2 Programming the Trapezoidal Method

There is no means of estimating the value of errors in the trapezoidal method but the number of iterations required to converge at each step may be used as a very good indication of the errors. As previously mentioned, the number of iterations increases more rapidly than the step length and thus the number of iterations is a good reference for the control of the step length. It is suggested [13] to double the step length if the number of iterations per step is less than 3 and to halve it if the number of iterations per step exceeds 12. The resulting bandwidth (3–12) is necessary to stop constant changes in the step length.

To avoid problems of step length chattering a factor of about 1.5 (instead of 2) may be used. Unfortunately it is difficult to maintain a regular print out interval if a noninteger factor is used.

Even using a step length changing factor of 2, it is difficult to maintain a regular print out interval. Step halving can be carried out at any time but indiscriminate step doubling may mean that there is no solution at the desired print out time.

Doubling the step length thus should only be done immediately after a print out and the step length should not be allowed to exceed the print out interval.

On rare occasions, it is possible that the number of iterations at a particular step greatly exceeds the upper desired limit. It can be shown that the convergence pattern is geometric and usually oscillatory [13] after the first five or six iterations. Even when diverging, the geometric and oscillatory pattern can be observed. Schemes can thus be devised which estimate the correct solution. However, these schemes are relatively costly to implement in terms of programming, storage and execution and a more practical method is to stop iterating after a fixed number of iterations and start again with a half-step. It is not necessary to store all the information obtained at the end of the previous step, in anticipation of a restart, as this information is already available for the nonintegrable variables if an extrapolation method is being used at the beginning of each step. Further, much information is available in the C and M constants of the algebraic form of the integration method.

For example, with the two-variable problem given by equation (6.7.7), if $z(t)$ is a nonintegrable variable, then its value at the end of the nth step z_n is stored. The value of the integrable variable $y(t)$ at the end of the nth step y_n can be reevaluated from equations (6.7.9) and (6.7.10) to be

$$y_n = \frac{(c_{n+1} - m_{n+1}z_n)}{(1 - 2m_{n+1}/G)}. \tag{6.7.22}$$

In only a few cases where the differential equation is complex need the value of y_n be stored at the beginning of each step. While the method requires programming effort it is very economical on storage and the few instances where it is used do not affect the overall execution time appreciably.

Linear extrapolation of nonintegrable variables at the beginning of each step is a very worthwhile addition to the trapezoidal method. Although not essential, the number of iterations per step is reduced and the storage is not prohibitive. Higher orders of extrapolation give very little extra improvement and as they are not effective until some steps after a discontinuity their value is further reduced.

It is only at the first step after a discontinuity that linear extrapolation cannot be used. As this often coincides with a large rate of change of integrable variables, the number of iterations to convergence can be excessive. This is overcome by automatic step length reduction after a discontinuity. Two half-step lengths, before returning to the normal step length, has been found to be satisfactory in almost all cases [13].

6.7.3 Application of the Trapezoidal Method

The differential equations developed in this chapter have all been associated with the synchronous machine and its controllers. These equations can be transformed into the algebraic form of the trapezoidal method given by equation (6.7.8). While these algebraic equations can be combined to make a matrix equation this has little merit and makes discontinuities such as regulator limits more difficult to apply.

In order to simplify the following equations, the subscripts on the variables have been removed. It is rarely necessary to retain old values of variables and, where this is necessary, it is noted. The variable values are thus overwritten by new information as soon as they are available. The constants C and M associated with the algebraic form are evaluated at the beginning of a new integration step and hence use the information obtained at the end of the previous step.

6.7.3.1 Synchronous Machine

The two mechanical differential equations are given by equations (6.2.5) and (6.2.6) and the algebraic form is

$$\omega = C_\omega + M_\omega(Pm - Pe) \tag{6.7.23}$$

where

$$C_\omega = (1 - 2 \cdot M_\omega \cdot Da)\omega + M_\omega(Pm - Pe + 4\pi \cdot fo \cdot Da)$$
$$M_\omega = h/(2Mg + hDa)$$

and also

$$\delta = C_\delta + M_\delta(\omega) \tag{6.7.24}$$

where

$$C_\delta = \delta + M_\delta(\omega - 4\pi \cdot fo)$$
$$M_\delta = 0.5h.$$

It would be possible to combine these equations to form a single simultaneous solution of the form

$$\delta = C'_\delta + M'_\delta(Pm - Pe) \tag{6.7.25}$$

where

$$C'_\delta = C_\delta + M_\delta \cdot C_\omega$$
$$M'_\delta = M_\delta \cdot M_\omega$$

but machine speed ω is a useful piece of information and would still require evaluation in most problems.

It is also more convenient to retain the electrical power (Pe) as a variable rather than attempt to reduce it to its constituent parts:

$$Pe = I_d \cdot V_d + I_q \cdot V_q + (I_d^2 + I_q^2)Ra. \tag{6.7.26}$$

Thus Pe is extrapolated after C_ω and M_ω have been evaluated.

The mechanical power Pm is an integrable variable which, in the absence of a speed governor model for the machine, is constant.

There are four electrical equations associated with the change in flux in the synchronous machine and these are given by equations (6.2.13), (6.2.14), (6.2.17) and (6.2.18). The algebraic form of these equations is as follows:

$$E'_q = C_q + M_q(Ef + (X_d - X'_d)I_d) \tag{6.7.27}$$

where

$$C_q = (1 - 2M_q)E'_q + M_q(Ef + (X_d - X'_d)I_d)$$
$$M_q = h/(2T'_{do} + h)$$

also

$$E'_d = C_d - M_d(X_q - X'_q)(I_d) \tag{6.7.28}$$

where

$$C_d = (1 - 2M_d)E'_d - M_d(X_q - X'_q)I_q$$
$$M_d = h/(2T'_{q0} + h)$$

also

$$E''_q = C_{qq} + M_{qq}(E'_q + (X'_d - X''_d)I_d) \tag{6.7.29}$$

where

$$C_{qq} = (1 - 2M_{qq})E''_q + M_{qq}(E'_q + (X'_d - X''_d)I_d)$$
$$M_{qq} = h/(2T''_{d0} + h)$$

and also

$$E''_d = C_{dd} + M_d(E'_d - (X'_q - X''_q)I_q) \tag{6.7.30}$$

where

$$C_{dd} = (1 - 2M_{dd})E''_d + M_{dd}(E'_d - (X'_q - X''_q)I_q)$$
$$M_{dd} = h/(2T''_{q0} + h).$$

6.7.3.2 Synchronous Machine Controller Limits

There are usually limits associated with AVRs and speed governors and these require special consideration when applying the algebraic form of the trapezoidal rule. It is best to ignore the limits at first and develop the whole set of 'limitless' equations. Rather than confuse this discussion, it is easier to consider a simple AVR system as shown in Fig. 6.17, for which can be written

$$pV_{\text{out}} = (G_1 \cdot (V_{\text{in}} - Vfb) - V_{\text{out}})/T_1 \tag{6.7.31}$$

subject to

$$V_{\text{max}} \geqslant V_{\text{out}} \geqslant V_{\text{min}}$$
$$pVfb = (G_2 \cdot pV_{\text{out}} - Vfb)/T_2. \tag{6.7.32}$$

The feedback loop can be rearranged to avoid the derivative of V_{out} being explicitly required as described in Section 6.3 and this is shown in Fig. 6.18. Equation (6.7.32)

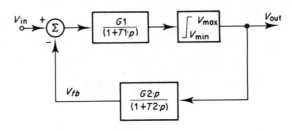

Figure 6.17
Block diagram of a simple controller

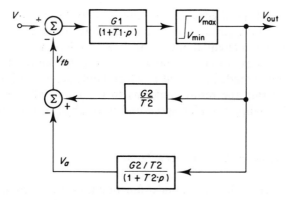

Figure 6.18
Modified block diagram of a simple controller of Fig. 6.17

is now replaced by

$$Vfb = \frac{G_2}{T_2} V_{out} - Va \tag{6.7.33}$$

$$pVa = \left(\frac{G_2}{T_2} V_{out} - Va \right) \bigg/ T_2. \tag{6.7.34}$$

Equations (6.7.31) and (6.7.34) can be transformed into the algebraic form

$$V_{out} = C_1 + M_1(V_{in} - Vfb) \tag{6.7.35}$$

$$V_a = C_2 + M_2(V_{out}) \tag{6.7.36}$$

where C_1, C_2, M_1 and M_2 may be determined in the usual way.

A simultaneous solution for the whole system is now possible by combining equations (6.7.33), (6.7.35) and (6.7.36) to give

$$V_{out} = C_3 + M_3(V_{in}) \tag{6.7.37}$$

where

$$C_3 = \frac{C_1 + M_1 C_2}{1 + M_1((G_2/T_2) - M_2)}$$

and

$$M_3 = \frac{M_1}{1 + M_1((G_2/T_2) - M_2)}.$$

After a solution of V_{out} is obtained it may be subjected to the limits of equation (6.7.31). If it is necessary Vfb can now be evaluated from Equations (6.7.33) and (6.7.36) using the limited value of V_{out}.

Where this simple controller model represents an AVR the input V_{in} may well be the (negated) deviation of terminal voltage V_t from its specified value (V_s). It is simpler to treat V_t as an extra nonintegrable variable rather than incorporate

$$V_t = \sqrt{(V_r^2 + V_m^2)}$$

in the model.

The usual models of speed governors do not have feedback loops associated with them, but the input to the governor (machine speed) is related to the turbine output (mechanical power) by differential equations. It is therefore necessary to solve a set of simultaneous equations in a similar manner to the example above. The simultaneous solution should be first made at a point at which limits are applied (i.e. at the valve) and then, after ensuring the result conforms to the limits, all the other variables around the loop (including machine speed and rotor angle) can be evaluated.

6.7.3.3 *Solution for Saturating AVR Exciter*

Another problem occurs when a nonlinear function is encountered. Equations (6.3.3) and (6.3.11) may be combined to form a single differential equation but this, then, involves a term in Ef^2 which complicates the evaluation of Ef. As the saturation function is approximate, it may be further simplified to give

$$Ve = (k_1^* Ef^* - k_2^*)Ef \qquad (6.7.38)$$

where Ef^* is the value of Ef at the previous iteration and k_1^* and k_2^* are determined from Ef^*. The equation describing the saturating exciter is thus

$$pEf = [Vam - (Ke - k_2^*)Ef - k_1^* Ef^* Ef]/Te \qquad (6.7.39)$$

and applying the trapezoidal rule the algebraic form of solution is:

$$Ef = [C_{ef} + M_{ef}(Vam)]/[1 + M_{ef}(k_1^* Ef^* - k_2^* - K_{ef})] \qquad (6.7.40)$$

where

$$C_{ef} = (1 - 2(Ke + K_{ef})M_{ef})Ef + M_{ef}Vam$$
$$M_{ef} = h/[2Te + h(Ke + K_{ef})]$$
$$K_{ef} = k_1^* Ef - k_2^*.$$

C_{ef}, M_{ef} and K_{ef} are evaluated once at the beginning of the step and, hence, only contain information obtained at the end of the previous step.

6.8 STRUCTURE OF A TRANSIENT STABILITY PROGRAM

6.8.1 Overall Structure

An overview of the structure of a transient stability program is given in Fig. 6.19. Only the main parts of the program have been included. The structure is such that the program can easily be modified to allow changes to be made or results displayed while a study is in progress, thus making the program interactive.

With care, the program can be divided into packages of subroutines each concerned with only one aspect of the system [13]. This permits the removal of component models when not required and the easy addition of new models whenever necessary. Thus for example, the subroutines associated with the synchronous machine, the AVRs, speed governors, etc., can be segregated from the network. Figure 6.20 shows

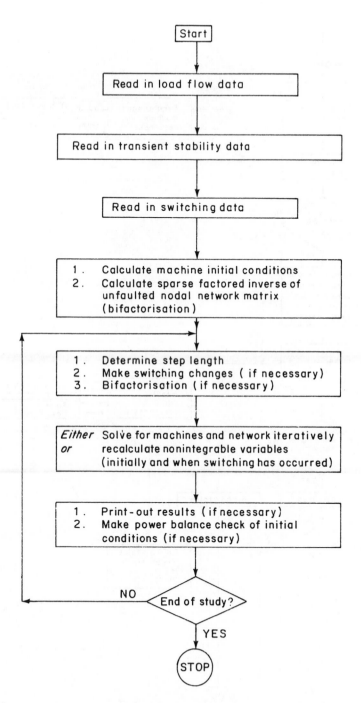

Figure 6.19
Conceptual overview of a transient stability program structure

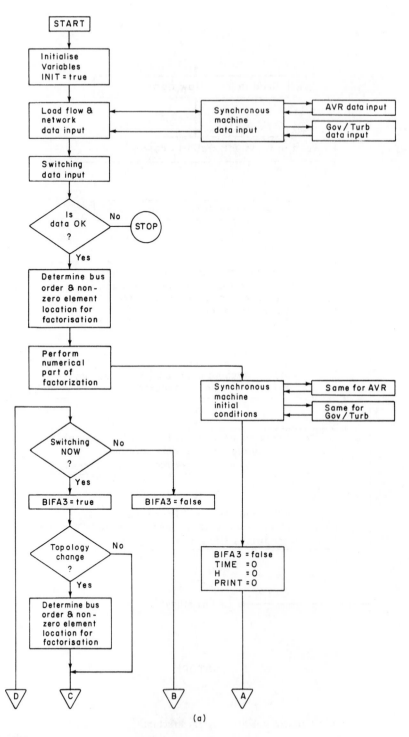

Figure 6.20
Structure of transient stability program: (a) Section 1; (b) Section 2

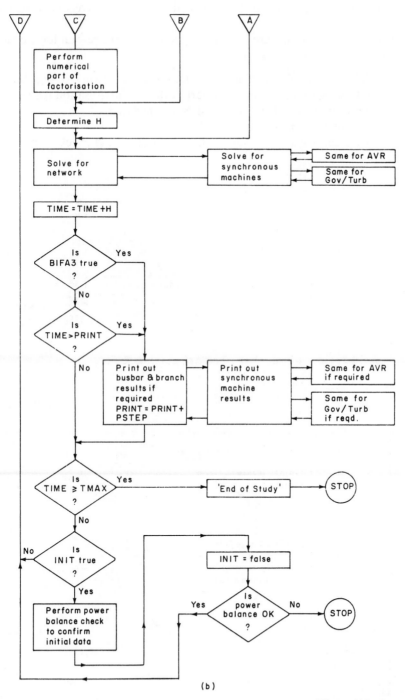

Figure 6.20 (*continued*)

192

a more detailed block diagram of the overall structure where this segregation is indicated.

While the block diagrams are intended to be self-evident several logic codes need to be explained.

BIFA3—This is a logical flag which is set true when a network change takes place indicating that the numerical part of the (bi-)factorisation (performed in a subroutine named BIFA3) must be recalculated.

   H—The integration step length.

INIT—A logical flag set true during the initialisation and checking period only.

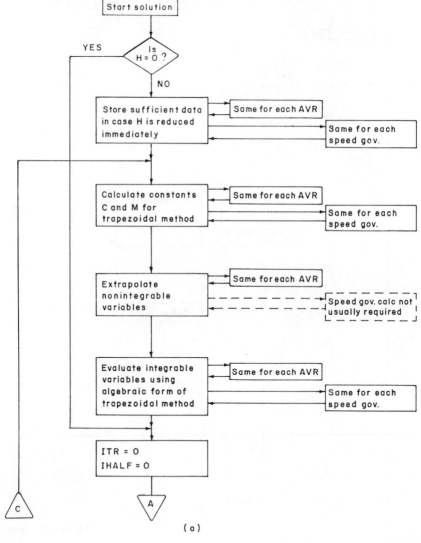

(a)

Figure 6.21
Structure of machine and network iterative solution: (a) Section 1; (b) Section 2; (c) Section 3

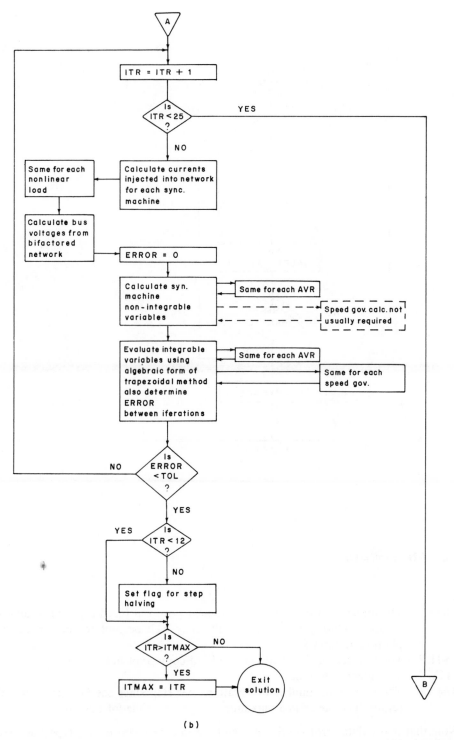

(b)

Figure 6.21 (*continued*)

194

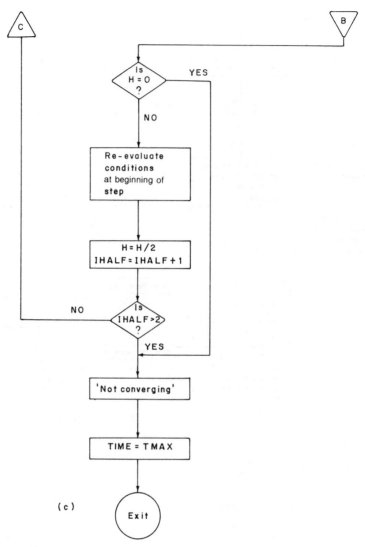

Figure 6.21 *(continued)*

PRINT—The integration time at which the next print out (hard copy) of results is required. During the study, results may well be sent to the screen for plotting at every step.

PSTEP—The integration time between the hard copy print out of results.

TIME—The integration time.

TMAX—The maximum number of iterations per step since the last print out of results. The predefined maximum integration time for the study.

Note that many data error checks are required in a program of this type but they have been omitted from the block diagram for clarity.

6.8.2 Structure of Machine and Network Iterative Solution

The structure of this part of the program requires further description. Two forms of solution are possible depending on whether an integration step is being evaluated or the nonintegrable variables are being recalculated after a discontinuity. A block diagram is given in Fig. 6.21.

The additional logic codes used in this part of the program are as follows.

ERROR—The maximum difference between any integrable variable from one iteration to another.

  ITR—Number of iterations required for solution.

IHALF—Number of immediate step halving required or the solution.

  TOL—Specified maximum value of ERROR for convergence.

ITMAX—The maximum number of iterations per step since the last print out of results. If this variable is sufficiently small (e.g. 3), when TIME = PRINT the step length (H) is doubled. PRINT is used for doubling H so that the change occurs at logical times. Also ITMAX ensures that the number of iterations is consistently small before initiating the change which prevents H chattering.

If convergence has not been achieved after a specified number of iterations, the study is terminated. This is done by setting the integration time equal to the maximum integration time.

6.9 GENERAL CONCLUSIONS

The transient stability program described in this chapter is sufficient for many basic stability studies. It is more than adequate when first swing stability is being evaluated and the machine detail and controllers will allow second and subsequent swing stability to be examined also.

However, if synchronous machine saturation or compound thermal turbines have to be modelled, it will be necessary to incorporate parts of Chapter 7 into the program. The structure set out at the end of this chapter should allow changes of this sort to be made quite easily. Similarly, if other system components are to be included this can be done without difficulty.

6.10 REFERENCES

[1] O. I. Elgerd, 1971. *Electrical Energy Systems Theory: An Introduction* McGraw-Hill, New York.

[2] B. M. Weedy, 1979. *Electric Power Systems* Wiley and Sons, London.

[3] C. Concordia, 1951. *Synchronous Machines. Theory and Performance* Wiley and Sons, New York.

[4] E. W. Kimbark, 1956. *Power System Stability: Synchronous Machines* (vol. 3) Wiley and Sons, New York.

[5] P. L. Dandeno, *et al.* 1973. Effects of synchronous machine modeling in large-scale system studies, *IEEE Trans.* **PAS-92** 574–582.

[6] IEEE Committee Report, 1968. Computer representation of exciter systems, *IEEE Trans* **PAS-87** 1460–1464.

[7] IEEE Committee Report, 1973. Dynamic models for steam and hydro turbines in power-system studies, *IEEE Trans* **PAS-92** 1904–1915.

[8] P. L. Dandeno and P. Kundur, 1973. A noniterative transient stability program including the effects of variable load-voltage characteristics, *IEEE Trans.* **PAS-92** 1478–1484.

[9] G. L. Berg, 1973. Power system load representation, *Proc. IEE* **120** 344–348.

[10] H. W. Dommell and N. Sato, 1972. Fast transient stability solutions, *IEEE Trans* **PAS-91** 1643–1650.

[11] A. Brameller *et al.* 1969. *Practical Diakoptics for Electrical Networks* Chapman and Hall, London.

[12] L. Lapidus and J. H. Seinfeld, 1971. *Numerical Solution of Ordinary Differential Equations* Academic Press, New York.

[13] C. P. Arnold, 1976. Solutions of the multi-machine power-system stability problem. *PhD Thesis* Victoria University of Manchester, UK.

7. POWER SYSTEM STABILITY— ADVANCED COMPONENT MODELLING

7.1 INTRODUCTION

This chapter develops further some of the component models described in Chapter 6 and introduces new models needed to investigate the effects of other a.c. system plant components. Turbine generator models are extended by considering the effects of saturation in the synchronous machine and the response of compound thermal turbines. Detailed consideration is also given to the modelling of induction motors and static power converters. The chapter also deals with protective gear modelling and unbalanced faults.

The induction motor model allows for a good representation over the whole speed range so that motor starting can be investigated. The model can be created in three ways depending upon the induction motor data available.

The basic formulation of three-phase bridge rectification and inversion is described in Appendix II and here it is extended so that the dynamic model can include abnormal operating conditions encountered during stability studies. It must be clarified, however, that the controllability of h.v.d.c. links during large disturbances in either the a.c. or d.c. system cannot be determined by transient stability programs. These and other problems associated with transient stability analysis involving h.v.d.c. links require the use of transient convertor simulation [1] or electromagnetic transient [2] programs.

The grouping of subroutines relevant to a particular component of the power system or aspect of the study, as developed in Chapter 6, should be retained for the models produced in this chapter. This ensures that additional models can be incorporated easily and models removed when not necessary.

7.2 SYNCHRONOUS MACHINE SATURATION

The relationship between mutual flux and the exciting MMF within a machine is not linear and some means of representing this nonlinearity is necessary if the results obtained from a stability study are to be accurate.

In most multimachine stability programs, each machine is represented by a voltage behind an impedance. As explained in Chapter 6 the impedance consists of armature resistance plus either transient or subtransient reactance. Also the voltage magnitude may be fixed or time-varying, depending on the complexity of the model.

Saturation may thus be taken into account by modifying the value of the reactance used in representing the machines. However, as explained for the model developed in Chapter 6, it is more convenient to fix the reactance and adjust the voltage accordingly.

Saturation is a part of synchronous machine modelling where there is still uncertainty as to the best method of simulation. The degree of saturation is not the same throughout the machine because the flux varies by the amount of leakage flux. Also, the saturation in the direct and quadrature axes are different, although this difference is small in the case of a cylindrical rotor.

Various methods have been adopted to account for saturation which differ not only in the model modification technique but also in the representation of the saturation characteristic of the machine.

7.2.1 Classical Saturation Model

Classical theory [3] for a cylindrical rotor machine assumes that the saturation is due to the total MMF produced in the iron and is the same in each axis.

It is necessary to make further assumptions in order to simplify the model.

(i) The magnetic reluctance in each axis is equal. Thus, the synchronous reactances are equal, i.e. $X_d = X_q$.

(ii) Saturation does not distort the sinusoidal variations assumed for rotor and stator inductances.

(iii) Because load-test data is not usually available, saturation is determined using the open-circuit saturation curve.

(iv) Potier reactance X_p may be used in calculating saturation.

(v) The total iron MMF (Fe) may be determined from

$$Fe = SIf \qquad (7.2.1)$$

where S is the saturation factor, defined as

$$S = 1 + \frac{\text{iron MMF}}{\text{air gap MMF}}. \qquad (7.2.2)$$

(vi) With reference to Fig. 7.1, the saturation factor S may be determined from

$$S = \frac{AC}{AB}. \qquad (7.2.3)$$

Figure 7.2 shows a typical voltage and MMF diagram for a round rotor synchronous machine. Potier voltage Ep, the voltage behind Potier reactance, may be determined readily from the terminal voltage Vt and the terminal current I. The MMF required to produce this voltage is found from the open-circuit saturation

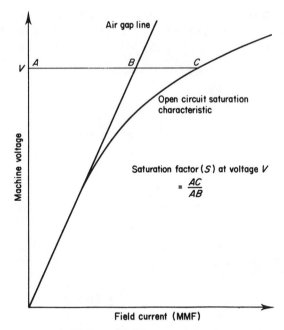

Figure 7.1
Open-circuit saturation characteristic of a synchronous machine

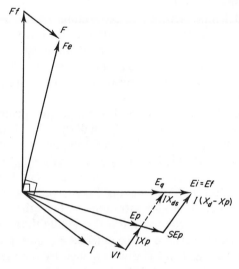

Figure 7.2
Vector diagram of cylindrical rotor synchronous machine saturation

curve of Fig. 7.1. Armature reaction F is found using assumption (iv) from which the field MMF (Ff) is calculated. The voltage equivalent to Ff referred to the stator is Ef. It is readily apparent that rotating the MMF diagram through $90°$ gives

$$Ff \propto Ei \text{ (in the steady state)}$$
$$Fe \propto SEp \qquad (7.2.4)$$
$$F \propto I(X_d - Xp).$$

In Fig. 7.2, the reactance X_{ds} is the saturated value of X_d. This produces an internal machine voltage E_q which lies on the quadrature axis. As $I(X_d - Xp)$ is parallel with IX_{ds}, then

$$\frac{Ep}{SEp} = \frac{I(X_{ds} - Xp)}{I(X_d - Xp)} = \frac{E_q}{Ei} \qquad (7.2.5)$$

and from this

$$Ei = SEq \qquad (7.2.6)$$

and

$$X_{ds} = \frac{(X_d - X_p)}{S} + Xp. \qquad (7.2.7)$$

All machine reactances subject to saturation are similarly modified.

7.2.2 Salient Machine Saturation

In the case of a salient synchronous machine, it may be assumed that the direct and quadrature axis armature reaction MMFs (F_d and F_q respectively) are proportional to the reactive voltage drops $I_d \cdot Xa_d$ and $I_q \cdot Xa_q$ respectively. Assumptions (iii) and (iv) for the classical model also apply.

There are many different methods of accounting for the saturation effect. The methods considered here assume that saturation in the d-axis is due at least in part to the component of flux in the d-axis. The first method ignores saturation in the q-axis, the second method accounts for quadrature axis saturation by the component of flux in the q-axis, the third method considers that the total flux contributes to the saturation in both axes.

In the first method, the density of the flux due to the quadrature axis armature reaction MMF (F_q) is considered sufficiently small that saturation effects on voltages are thus neglected in the direct axis. The other component of the armature reaction MMF (F_d) adds directly to the field MMF (Ff) to produce a main flux which in turn produces a quadrature axis voltage subject to saturation. The saturation level is determined by the quadrature component of Potier voltage (Ep_q).

The second method [4] allows for saturation in both the direct and quadrature axis components of the Potier voltage. It is assumed that the reluctances of the d-axis and q-axis paths differ only because of the different air gaps in each axis. The d-axis component of Potier voltage (Ep_d) is thus modified by the ratio X_q/X_d before the q-axis saturation factor is determined. Provided that it is assumed that the vector sum of the two saturated main flux components (Fe_q and Fe_d) is in phase with the MMF proportional to Potier voltage, then the saturated d- and q-axis synchronous reactances (X_{ds} and X_{qs}) are

$$X_{ds} = \frac{(X_d - Xp)}{S_d} + Xp \tag{7.2.8}$$

$$X_{qs} = \frac{(X_q - Xp)}{S_q} + Xp. \tag{7.2.9}$$

A third method [3] distinguishes between the saturation in the rotor and stator, and saturation factors based on Ep and Ep_q are obtained. This method is difficult to implement because it is necessary to ensure that saturation is not applied twice to any part of the machine. That is, the saturation in the field poles must be isolated from that of the armature, giving two saturation curves. The two saturation factors for this case may be defined as

$$S_{dq} = 1 + \frac{\text{iron MMF in the stator}}{\text{total air gap MMF}} \tag{7.2.10}$$

$$S_d = 1 + \frac{\text{iron MMF in the rotor}}{\text{direct axis air gap MMF}} \tag{7.2.11}$$

where S_{dq} acts equally on both d and q axes and S_d acts on the direct axis only.

Figures 7.3 to 7.5 demonstrate the differences between the three methods of saturation representation.

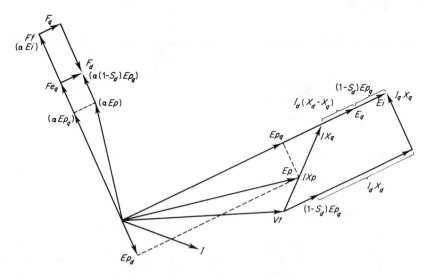

Figure 7.3
Salient pole synchronous machine with direct axis saturation only

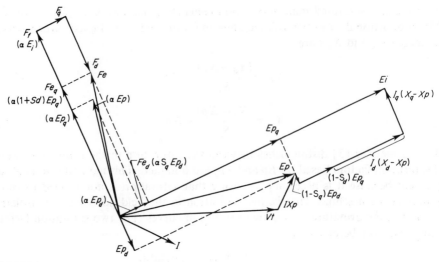

Figure 7.4
Salient pole synchronous machine with direct and quadrature axis saturation

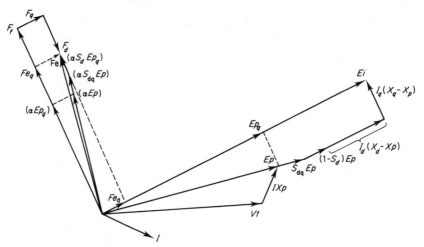

Figure 7.5
Salient pole synchronous machine with separate stator and rotor saturation

7.2.3 Simple Saturation Representation

An even simpler method of including the effect of saturation is to calculate the saturation initially (by some means) after which it is either held constant or varied according to the slope of the saturation curve at the initial point. This method is suitable for small perturbation studies where Potier voltage and machine angle do not vary greatly.

7.2.4 Saturation Curve Representation

The open-circuit saturation curve must be stored within the computer so that a new saturation factor can be determined at every stage of the study.

The most accurate method of storing this curve is to fit a polynomial of the form

$$If = C_0 + C_1 V + C_2 V^2 + C_3 V^3 \cdots + C_n V^n \qquad (7.2.12)$$

by taking $n + 1$ points on the curve. Normally n would be 5, 7 or 9. This is a clumsy method of both entering the data and storing it. In multimachine transient stability studies, where the machines are represented at best by subtransient parameters and an approximation to Potier reactance is made, nothing is achieved by such an elaborate method of representing the saturation curve.

The problem can be simplified by assuming that most of the coefficients of the polynomial are zero. A sufficiently good approximation is achieved with the equation [5]

$$If = V + C_n V^n \qquad (7.2.13)$$

where n is normally either 7 or 9. Only one point is needed to specify the curve and if If is always specified at a predetermined voltage, the data entries required per curve are reduced to one, from which C_n may be readily determined.

7.2.5 Potier Reactance

The Potier reactance of a machine is rarely quoted, although the open-circuit saturation curve is normally available. In order to model the saturation effects, it is thus necessary to estimate this reactance.

From knowledge of the leakage reactance Xl, Beckwith [6] calculated that

$$Xp = Xl + 0.63(X'_d - Xl) \qquad (7.2.14)$$

and if Xl is not available, then

$$Xp = 0.8 X'_d. \qquad (7.2.15)$$

Equation (7.2.15) may be modified to account for the type of synchronous machine [7] i.e.

$$Xp = 0.9 X'_d \qquad (7.2.16)$$

for a salient pole machine, as most of the saturation occurs in the poles, and

$$Xp = 0.7 X'_d \qquad (7.2.17)$$

for a round rotor machine, as most of the saturation occurs in the rotor teeth and the Potier and leakage reactances have similar values.

7.2.6 The Effect of Saturation on the Synchronous Machine Model

Saturation effectively modifies the ordinary differential equations describing the behaviour of the voltages used to model the synchronous machine. Equations (6.2.13),

(6.2.14), (6.2.17) and (6.2.18) become respectively

$$pE'_q = (Ef + (X_d - X'_d)I_d - S_d E'_q)/T'_{do} \tag{7.2.18}$$

$$pE'_d = (-(X_q - X'_q)I_q - S_q E'_d)/T'_{qo} \tag{7.2.19}$$

$$pE''_q = (S_d E'_q + (X_d - X''_d)I_d - S_d E''_q)/T''_{do} \tag{7.2.20}$$

$$pE''_d = (S_q E'_d - (X'_q - X''_q)I_q - S_q E''_d)/T''_{qo} \tag{7.2.21}$$

where S_d and S_q are the direct and quadrature axis saturation factors.

Where subtransients are considered then equations (7.2.15) and (7.2.16) are replaced by

$$E''_q - Vt_q = Ra \cdot I_q - X''_{ds} \cdot I_d \tag{7.2.22}$$

$$E''_d - Vt_d = Ra \cdot I_d + X''_{qs} \cdot I_q. \tag{7.2.23}$$

7.2.7 Representation of Saturated Synchronous Machines in the Network

Representation of a salient but unsaturated synchronous machine in the network has been discussed in Section 6.6. When saturation occurs, a double adjustment must be made at each step in the solution process [8].

With the notation developed in Section 6.6, the fixed admittance (Y_0) which is included with the network is made up from unsaturated and nonsalient values of reactance, whereas the correct admittance (Y_{ts}) is made up from saturated and salient values. Using Fig. 6.14, as before the adjusting current to account for this change in admittance is

$$I_{adjs} = (\bar{Y}_{ts} - \bar{Y}_0)(\bar{E}'' - \bar{V}). \tag{7.2.24}$$

That is

$$
\begin{bmatrix} I_{adjs_q} \\ I_{adjs_d} \end{bmatrix} = \left[\frac{1}{(Ra^2 + X''_{ds}X''_{qs})} \begin{bmatrix} Ra & X''_{ds} \\ -X''_{qs} & Ra \end{bmatrix} - \frac{1}{(Ra^2 + X''_d \cdot X''_q)} \begin{bmatrix} Ra & X_{dq} \\ -X_{dq} & Ra \end{bmatrix} \right] \begin{bmatrix} E''_q - V_q \\ E''_d - V_d \end{bmatrix}
$$

$$
\begin{bmatrix} I_{adjs_q} \\ I_{adjs_d} \end{bmatrix} = \begin{bmatrix} I^{(1)}_{adjs_q} \\ I^{(1)}_{adjs_d} \end{bmatrix} + \left[\frac{1}{(Ra^2 + X''_{ds}X''_{qs})} \begin{bmatrix} Ra & X_{dqs} \\ -X_{dqs} & Ra \end{bmatrix} - \frac{1}{(Ra^2 + X''_d X''_q)} \begin{bmatrix} Ra & X_{dq} \\ -X_{dq} & Ra \end{bmatrix} \right] \begin{bmatrix} E''_q - V_q \\ E''_d - V_d \end{bmatrix}. \tag{7.2.25}
$$

The current $\bar{I}^{(1)}_{\text{adjs}}$ is similar to $\bar{I}_{\text{adj}}$ developed in Section 6.6:

$$
\begin{bmatrix} I^{(1)}_{\text{adjs}_q} \\ I^{(1)}_{\text{adjs}_d} \end{bmatrix} = \frac{\frac{1}{2}(X''_{ds} - X''_{qs})}{(Ra^2 + X''_{ds}X''_{qs})} \begin{bmatrix} 0 & 1 \\ 1 & 0 \end{bmatrix} \cdot \begin{bmatrix} E''_q - V_q \\ E''_d - V_d \end{bmatrix}
\tag{7.2.26}
$$

The current injected into the network is given by equation (6.6.4) and as the terms in the brackets contain no saliency then in the real and imaginary axis of the network

$$
\begin{bmatrix} I_{\text{inj}_r} \\ I^{(1)}_{\text{adjs}_d} \end{bmatrix} = \begin{bmatrix} I^{(1)}_{\text{adjs}_r} \\ I^{(1)}_{\text{adjs}_m} \end{bmatrix} + \begin{bmatrix} I^{(2)}_{\text{adjs}_r} \\ I_{\text{adjs}_m} \end{bmatrix} + \frac{1}{(Ra^2 + X''_d X''_q)} \begin{bmatrix} Ra & X_{dq} \\ -X_{dq} & Ra \end{bmatrix} \cdot \begin{bmatrix} V_r \\ V_m \end{bmatrix}
\tag{7.2.27}
$$

where $\bar{I}^{(2)}_{\text{adjs}}$ contains saturated but nonsalient reactance terms and $I^{(1)}_{\text{adjs}}$ contains salient and saturated reactance terms

$$
\begin{bmatrix} I^{(1)}_{\text{adjs}_r} \\ I^{(1)}_{\text{adjs}_m} \end{bmatrix} = \frac{\frac{1}{2}(X''_{ds} - X''_{qs})}{(Ra^2 + X''_{ds}X''_{qs})} \begin{bmatrix} -\sin 2\delta & \cos 2\delta \\ \cos 2\delta & \sin 2\delta \end{bmatrix} \cdot \begin{bmatrix} E''_r - V_r \\ E''_m - V_m \end{bmatrix}
\tag{7.2.28}
$$

$$
\begin{bmatrix} I^{(2)}_{\text{adjs}_r} \\ I^{(2)}_{\text{adjs}_m} \end{bmatrix} = \frac{1}{(Ra^2 + X''_{ds}X''_{qs})} \begin{bmatrix} Ra & X_{dqs} \\ -X_{dqs} & Ra \end{bmatrix} \cdot \begin{bmatrix} E''_r - V_r \\ E''_m - V_m \end{bmatrix}
\tag{7.2.29}
$$

Note that the third part of equation (7.2.27) is $\bar{Y}_0 \bar{V}$ and not $\bar{Y}_0 \bar{E}''$. This part of the injected current is merely the current flowing through $\bar{Y}_0$ and could be eliminated if $\bar{Y}_0$ was not included in the network. The conditioning of the network would be reduced, however, and in certain systems this could lead to numerical problems.

7.2.8 Inclusion of Synchronous Machine Saturation in the Transient Stability Program

Only two subroutines need modification to allow saturation effects in synchronous machines to be modelled. In both cases, an iterative solution is necessary for each saturating machine, although in most instances the number of iterations is small.

Saturation is a function of the voltage behind armature resistance and Potier reactance. Assuming the second method of salient machine saturation is being used, then from equation (7.2.13)

$$
\begin{aligned}
S_d &= 1 + C_{n_d}(Ep_q)^{n-1} \\
S_q &= 1 + C_{n_q}(Ep_d)^{n-1}
\end{aligned}
\tag{7.2.30}
$$

where

$$C_{n_q} = \frac{X_q}{X_d} C_{nd} \tag{7.2.31}$$

and

$$Ep_q = V_q + Ra \cdot I_q - Xp \cdot I_d$$
$$Ep_d = V_d + Ra \cdot I_d + Xp \cdot I_q \tag{7.2.32}$$

and I_d and I_q are given by equations (7.2.22) and (7.2.23).

A Jacobi iterative technique is quite adequate to establish the initial conditions of the synchronous machine and this can be incorporated in the relevant subroutine shown in Section 2 of Fig. 6.20.

During the time solution, however, saturation can vary over a large range of values and a Newton form of iteration is an advantage especially if large integration steps are used.

Redefining equation (7.2.30) as

$$f_1 = 1 - S_d + C_{nd}(Ep_q)^{n-1}$$
$$f_2 = 1 - S_q + C_{n_q}(Ep_d)^{n-1} \tag{7.2.33}$$

the elements of a 2×2 Jacobian matrix can be found. However, elements $\partial f_1/\partial S_d$ and $\partial f_2/\partial S_d$ are small with respect to the other two elements and if Ra is considered to be zero then the four elements reduce to

$$\frac{\partial f_1}{\partial S_d} = \frac{[(n-1)C_{nd}(Ep_q)^{n-2} \cdot Xp(X_d'' - Xp)(E_q'' - Vt_q)]}{[(S_d - 1)Xp + X_d'']} - 1$$

$$\frac{\partial f_2}{\partial S_q} = \frac{[(n-1)C_{n_q}(Ep_d)^{n-2} \cdot Xp(X_q'' - Xp)(E_d'' - Vt_d)]}{[(S_q - 1)Xp + X_q'']} - 1 \tag{7.2.34}$$

$$\frac{\partial f_1}{\partial S_q} = \frac{\partial f_2}{\partial S_d} = 0.$$

This decouples the Newton method and each saturation factor may be solved independently [9]:

$$S_d^{(p+1)} = S_d^{(p)} - f_1^{(p)}/(\partial f_1/\partial S_d)^{(p)}$$
$$S_q^{(r+1)} = S_q^{(r)} - f_2^{(r)}/(\partial f_2/\partial S_q)^{(r)}. \tag{7.2.35}$$

Despite the advantages of a Newton form of solution, it can be found to be divergent if too great an integration step length is used. Analysis of the functions f_1 and f_2 show that they have discontinuities, when $X_d'' = (S_d - 1)Xp$ and $X_q'' = (S_q - 1)Xp$ respectively, although otherwise are almost linear. It is therefore necessary to monitor this iterative procedure and modify the step length if necessary.

The evaluation of S_d and S_q should be performed twice during each iteration. Considering Fig. 6.21, this is during the calculation of the injected currents into the network and the calculation of the nonintegrable variables. Provided the discontinuity is not encountered, convergence is achieved in one or two iterations at each re-evaluation especially if the saturation factors are extrapolated at the beginning of each step.

7.3 DETAILED TURBINE MODEL

More detailed turbine models than the one described in the previous chapter are often required for the following reasons.

(i) A longer-term transient stability study or a dynamic stability study is to be made.

(ii) The turbine is a two-shafted cross-compound machine which has a separate generator on each shaft.

(iii) Generator overspeed is such that an interceptor valve may operate during the study.

A generalised model to accommodate the different types of compound turbine has been developed by the IEEE [10]. As with the generalised AVR model, by setting certain gains to either zero or unity and time constants to either infinity (very large) or zero, the model can be reduced to any desired form. An interceptor valve can easily be incorporated as shown in Fig. 7.6.

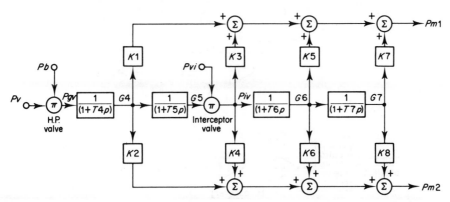

Figure 7.6
Generalised detailed turbine model including H.P. and interceptor valves

Some normal compound turbine configurations are shown in Fig. 7.7 and Table 7.1 gives typical values for these configurations using the generalised model. A hydroturbine can also be represented and the values given in Table 7.1 are justified by the method of representing a lead–lag circuit described in Chapter 6, with the time constant T_5 set at $\frac{1}{2}Tw$ in the case of the simplest model.

The full set of equations for the detailed turbine model is

$$pG_4 = (P_{gv} - G_4)/T_4 \qquad (7.3.1)$$

$$pG_5 = (G_4 - G_5)/T_5 \qquad (7.3.2)$$

$$Piv = G_5 \cdot Pvi \qquad (7.3.3)$$

$$pG_6 = (Piv - G_6)/T_6 \qquad (7.3.4)$$

$$pG_7 = (G_6 - G_7)/T_7 \qquad (7.3.5)$$

208

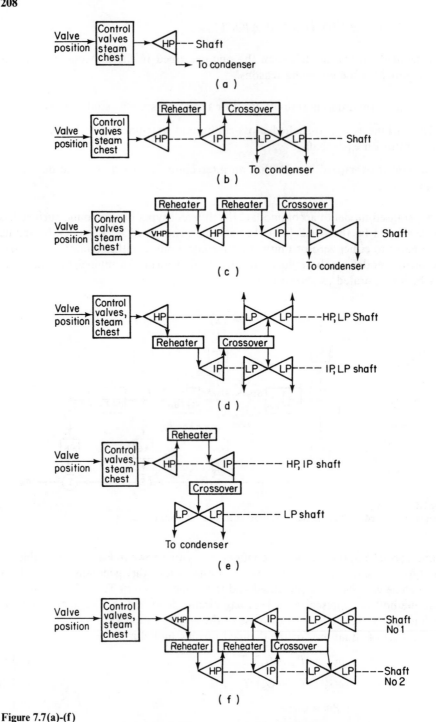

Figure 7.7(a)-(f)
Common steam turbine configurations [10]: (a) nonreheat; (b) tandem compound, single reheat; (c) tandem compound, double reheat; (d) cross compound, single reheat; (e) cross compound, single reheat; (f) cross compound, double reheat (© 1982 IEEE)

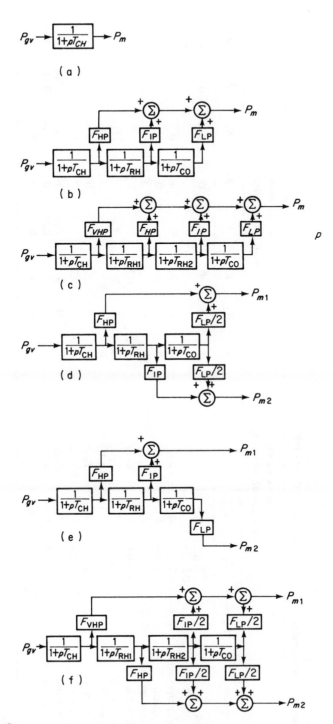

Figure 7.7(a)-(f)
(*continued*). Approximate linear models [10] © 1982 IEEE)

Table 7.1
Parameters used in generalised detailed turbine model [10] (© 1982 IEEE)

Turbine system	Figure	Time constants with typical values (s)				Fractions with typical values (p.u.)							
		T_4	T_5	T_6	T_7	K_1	K_2	K_3	K_4	K_5	K_6	K_7	K_8
Nonreheat	7.7a	T_{CH} 0.2–0.5	—	—	—	1	0	0	0	0	0	0	0
Tandem compound, single reheat	7.7b	T_{CH} 0.1–0.4	T_{RH} 4–11	T_{CO} 0.3–0.5	—	F_{HP} 0.3	0	F_{IP} 0.4	0	F_{LP} 0.3	0	0	0
Tandem compound, double reheat	7.7c	T_{CH} 0.1–0.4	T_{RH1} 4–11	T_{RH2} 4–11	T_{CO} 0.3–0.5	F_{VHP} 0.22	0	F_{HP} 0.22	0	F_{IP} 0.3	0	F_{LP} 0.26	0
Cross compound, single reheat	7.7d	T_{CH} 0.1–0.4	T_{RH} 4–11	T_{CO} 0.3–0.5	—	F_{HP} 0.3	0	0	F_{IP} 0.3	$\frac{1}{2}F_{LP}$ 0.2	$\frac{1}{2}F_{LP}$ 0.2	0	0
Cross compound, single reheat	7.7e	T_{CH} 0.1–0.4	T_{RH} 4–11	T_{CO} 0.3–0.5	—	F_{HP} 0.25	0	F_{IP} 0.25	0	0	F_{LP} 0.5	0	0
Cross compound, double reheat	7.7f	T_{CH} 0.1–0.4	T_{RH1} 4–11	T_{RH2} 4–11	T_{CO} 0.3–0.5	F_{VHP} 0.22	0	0	F_{HP} 0.22	$\frac{1}{2}F_{IP}$ 0.14	$\frac{1}{2}F_{IP}$ 0.14	$\frac{1}{2}F_{LP}$ 0.14	$\frac{1}{2}F_{LP}$ 0.14
Hydro	6.9a	0	$\frac{1}{2}T_W$	—	—	−2	0	3	0	0	0	0	0

$$P_{m1} = K_1 \cdot G_4 + K_3 \cdot Piv + K_5 \cdot G_6 + K_7 \cdot G_7 \tag{7.3.6}$$

$$P_{m2} = K_2 \cdot G_4 + K_4 \cdot Piv + K_6 \cdot G_6 + K_8 \cdot G_7. \tag{7.3.7}$$

Also note that

$$\sum_{n=1}^{8} Kn = 1 \tag{7.3.8}$$

and that, in the initial steady state, the interceptor valve, if present, will be fully open $(Pvi = 1)$ in which case

$$G_4 = G_5 = Piv = G_6 = G_7 = P_{m1} + P_{m2}. \tag{7.3.9}$$

The speed governor controlling the interceptor valve is similar to that controlling the HP turbine except that it is set to operate at some overspeed value of slip (k_ω) and not about synchronous speed. Equation (6.3.12) can be modified in this case to

$$pG_1 = [R(1 + T_2 p)(2\pi f_0 (1 + k_\omega) - \omega) - G_1]/T_1. \tag{7.3.10}$$

7.4 INDUCTION MACHINES

An approach similar to that used to construct the synchronous machine models is required if induction machines are to be explicitly modelled [4, 11]. However, speed cannot be assumed to vary only slightly and this basic difference requires that the equations describing the behaviour of induction machines be somewhat different from those developed for a synchronous machine.

7.4.1 Mechanical Equations

It is necessary to express the equation of motion of an induction machine in terms of torque and not power. Also symmetry of the rotor makes its angular position unimportant, and slip (S) usually replaces angular velocity (ω) as the variable, where

$$S = (\omega_0 - \omega)/\omega_0. \tag{7.4.1}$$

Assuming negligible windage and friction losses and smooth mechanical shaft power, the equation of motion is

$$pS = (Tm - Te)/(2Hm) \tag{7.4.2}$$

where Hm is the inertia constant measured in kW s/kV A established at synchronous speed. The mechanical torque (Tm) and electrical torque (Te) are assumed to be positive when the machine is motoring.

The mechanical torque Tm will normally vary with speed, the relationship depending on the type of load. A commonly used characteristic is

$$Tm \propto \{(\text{speed})^k\}$$

where $k = 1$ for fan-type loads and $k = 2$ for centrifugal pumps. A more elaborate torque/speed characteristic can be used for a composite load, i.e.

$$Tm \propto \{a + b(\text{speed}) + c(\text{speed})^2\} \tag{7.4.3}$$

which can include the effect of friction when start-up is being considered.

In terms of slip the torque is thus

$$Tm = A + BS + CS^2 \qquad (7.4.4)$$

where

$$A \propto \{a + b + c\}$$
$$B \propto \{b + 2c\}$$
$$C \propto c.$$

The values of A, B and C are determined from the initial (steady-state) loading of the motor and hence its initial value of slip.

The electrical torque Te is related to the air gap electrical power by the electrical frequency which is assumed constant and hence

$$Te = Re(\bar{E} \cdot \bar{I}_1^*)/2\pi f_0. \qquad (7.4.5)$$

7.4.2 Electrical Equations

A simplified equivalent circuit for a single-cage induction motor is shown in Fig. 7.8, with R_1 and X_1 referring to the stator and R_2 and X_2 referring to the rotor resistance and reactance respectively. In a similar manner to the transient model of a synchronous machine, an induction motor may be modelled by a Thevenin equivalent circuit of a voltage E' behind the stator resistance R_1 and a transient reactance X'. The transient reactance is the apparent reactance when the rotor is locked stationary and the slip (S) is unity and is given by

$$X' = X_1 + \frac{X_2 \cdot Xm}{(X_2 + Xm)} \qquad (7.4.6)$$

where Xm is the magnetising reactance of the machine. The rate of change of transient voltage is given by

$$p\bar{E}' = -j2\pi f \cdot S\bar{E}' - (\bar{E}' - j(X_0 - X')\bar{I}_1)/T_0' \qquad (7.4.7)$$

where the rotor open-circuit time constant T_0' is

$$T_0' = \frac{(X_2 + Xm)}{2\pi f_0 R_2} \qquad (7.4.8)$$

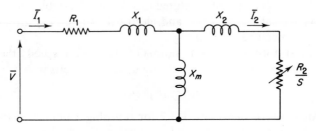

Figure 7.8
Steady-state equivalent circuit of a single cage induction motor

and the open-circuit reactance X_0 is

$$X_0 = X_1 + Xm. \tag{7.4.9}$$

The reactances are unaffected by rotor position and the model is described in the real and imaginary components used for the network, that is, in the synchronously rotating frame of reference. Thus, for a full description of the model, the following equations are used:

$$V_r - E'_r = R_1 \cdot I_{1r} - X' \cdot I_{1m} \tag{7.4.10}$$

$$V_m - E'_m = R_1 \cdot I_{1m} + X' \cdot I_{1r} \tag{7.4.11}$$

$$pE'_r = 2\pi f_0 S E'_m - (E'_r + (X_0 - X')I_{1m})/T'_0 \tag{7.4.12}$$

$$pE'_m = -2\pi f_0 S E'_r - (E'_m - (X_0 - X')I_{1r})/T'_0. \tag{7.4.13}$$

A transient stability program incorporating an induction motor model uses the transient and open-circuit parameters, but it is often convenient to allow the stator, rotor and magnetising parameters to be specified and let the program derive the former parameters.

For completeness, the electrical torque may now be written as

$$Te = (E'_r \cdot I_{1r} + E'_m \cdot I_{1m})/\omega_0. \tag{7.4.14}$$

7.4.3 Electrical Equations when the Slip is Large

Single-cage induction motors have low starting torques and it is often difficult to bring them to speed without either reducing the load or inserting external resistance in the rotor circuit. As a result of the low starting torque, when the slip exceeds the point of maximum torque, the single-cage model is often insufficiently accurate. These problems are overcome by the use of a double-cage or deep-bar rotor model.

7.4.3.1 Cage Factor

When a torque slip characteristic of the motor is available, then a simple solution is to modify the torque-slip characteristic of the single-cage motor model. Double-cage or deep-bar rotors have a resistance and reactance which varies with slip. A cage factor Kg can be included which allows for the variations of rotor resistance:

$$R_2 = R_2(0)(1 + Kg \cdot S) \tag{7.4.15}$$

where $R_2(0)$ is the rotor resistance at zero slip.

It is usually convenient to make the cage factor larger than that necessary to describe the change in rotor resistance. In this way, the torque-slip characteristics of the model can be made similar to that of the motor without the need to vary the rotor reactance with slip. The result of varying the rotor resistance is to modify the open-circuit transient time constant only, and this can be done quite simply at each integration step of the simulation.

Rotor reactance does not vary with slip as greatly as rotor resistance, provided saturation effects are ignored, and its effect on the open-circuit transient time constant is thus small. Transient reactance (X') varies with rotor reactance. However, this variation on the term $(X_0 - X')$ in equations (7.4.12) and (7.4.13) is insignificant. Thus, the only major effect of varying rotor reactance is in equation (7.4.10) and (7.4.11) which requires a technique similar to that adopted in the synchronous machine model to account for saturation and saliency. However, the gains obtained in using two-cage factors are insignificant and a single-cage factor varying rotor resistance is usually adopted.

7.4.3.2   Double-cage Rotor Model

An alternative to the cage factor is the use of a better rotor model, though this is often restricted by the unavailability of suitable data.

Induction motor loads having double-cage or deep-bar rotors can be represented in a similar manner to a single-cage motor [12, 13]. It is assumed that the end-ring resistance and that part of the leakage flux which links the two secondary windings, but not the primary, are neglected. The steady-state equivalent circuit shown in Fig. 7.9 can thus be obtained where R_3 and X_3 are the resistance and reactance of the additional rotor winding. A circuit similar to that of Fig. 7.8 can be obtained by substituting the two parallel rotor circuit branches by a single series circuit, where

$$R_2(S) = \frac{R_2 \cdot R_3(R_2 + R_3) + S^2(R_2 \cdot X_3^2 + R_3 \cdot X_2^2)}{(R_2 + R_3)^2 + S^2(X_2 + X_3)^2} \tag{7.4.16}$$

$$X_2(S) = \frac{R_2^2 X_3 + R_3^2 X_2 + S^2(X_2 + X_3)X_2 \cdot X_3}{(R_2 + R_3)^2 + S^2(X_2 + X_3)^2}. \tag{7.4.17}$$

At any instant during a transient stability study, the rotor impedance may be assumed to be the steady-state value given above.

Analysis similar to that used in developing equations (7.4.10) to (7.4.13) gives

$$V_r - E_r'' = R_1 \cdot I_{1r} - X'' \cdot I_{1m} \tag{7.4.18}$$

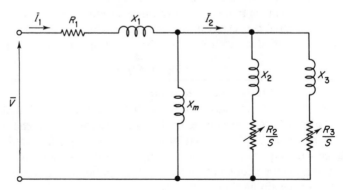

Figure 7.9
Steady-state equivalent circuit of a double-cage induction motor

$$V_m - E''_m = R_1 \cdot I_{1m} + X'' \cdot I_{1r} \tag{7.4.19}$$

$$pE''_r = -2\pi f_0 \cdot S(E'_m - E''_m) + pE'_r + (E'_r - E''_r - (X' - X'')I_{1m})/T''_0 \tag{7.4.20}$$

$$pE''_m = 2\pi f_0 \cdot S(E'_r - E''_r) + pE'_m + (E'_m - E''_m + (X' - X'')I_{1r})/T''_0 \tag{7.4.21}$$

$$Te = E''_r \cdot I_{1r} + E''_m \cdot I_{1m} \tag{7.4.22}$$

with equations (7.4.12) and (7.4.13) applying also.

The parameters for the model, when the motor has a double-cage rotor are given by equations (7.4.6), (7.4.8), (7.4.9) and

$$X'' = X_1 + \frac{X_2 \cdot X_3 \cdot Xm}{(X_2 \cdot X_3 + X_2 \cdot Xm + X_3 \cdot Xm)} \tag{7.4.23}$$

$$T''_0 = \frac{X_3 + (X_2 \cdot Xm)/(X_2 + Xm)}{2\pi f_0 R_3}. \tag{7.4.24}$$

If the rotor is of the deep-bar type, then the parameters of the equivalent double-cage type may be determined using equation (7.4.16) and (7.4.17). The rotor parameters at zero slip are

$$R_2(0) = \frac{R_2 \cdot R_3}{(R_2 + R_3)} \tag{7.4.25}$$

$$X_2(0) = \frac{(R_2^2 \cdot X_3 + R_3^2 \cdot X_2)}{(R_2 + R_3)^2} \tag{7.4.26}$$

and at standstill are

$$R_2(1) = \frac{X_2^2 \cdot R_3 + X_3^2 \cdot R_2 + R_2 \cdot R_3(R_2 + R_3)}{(X_2 + X_3)^2 + (R_2 + R_3)^2} \tag{7.4.27}$$

$$X_2(1) = \frac{X_2 \cdot X_3(X_2 + X_3) + R_2^2 \cdot X_3 + R_3^2 \cdot X_2}{(X_2 + X_3)^2 + (R_2 + R_3)^2}. \tag{7.4.28}$$

This set of nonlinear equations may be solved using Newtonian techniques but by substituting:

$$R_3 = \frac{R_2 \cdot R_2(0)}{(R_2 - R_2(0))} \tag{7.4.29}$$

and

$$X_3 = \frac{X_2 \cdot Xx}{(X_2 - Xx)} \tag{7.4.30}$$

where

$$Xx = X_2(1) - \frac{(R_2(1) - R_2(0))^2}{(X_2(0) - X_2(1))} \tag{7.4.31}$$

the number of variables reduces to two and a simple iterative procedure yields a result in only a few iterations [14]. A reasonable starting value is $X_2 \simeq \frac{3}{2}X_0$ derived from assuming $R_2 \simeq \frac{1}{5}R_3$ and $X_2 \simeq \frac{5}{2}X_3$.

7.4.4   Representation of Induction Machines in the Network

This is quite simple compared to the representation of a synchronous machine as neither saliency nor saturation are normally considered in the induction machine models. They may, therefore, be considered as injected currents in parallel with fixed admittance.

Modifying equations (7.4.18) and (7.4.19) gives a machine current of

$$\bar{I}_1 = \bar{Y}(\bar{V} - \bar{E}'') \tag{7.4.32}$$

or

$$\begin{bmatrix} I_{1r} \\ I_{1m} \end{bmatrix} = \frac{1}{(R_1^2 + X''^2)} \begin{bmatrix} R_1 & X'' \\ -X'' & R_1 \end{bmatrix} \cdot \begin{bmatrix} V_r - E''_r \\ V_m - E''_m \end{bmatrix} \tag{7.4.33}$$

The injected current into the network which includes $\bar{Y}$ is thus

$$\begin{bmatrix} I_{\text{inj}_r} \\ I_{\text{inj}_m} \end{bmatrix} = \frac{-1}{(R_1^2 + X''^2)} \begin{bmatrix} R_1 & X'' \\ -X'' & R_1 \end{bmatrix} \cdot \begin{bmatrix} E''_r \\ E''_m \end{bmatrix} \tag{7.4.34}$$

where the minus sign confirms the induction machine is assumed to be motoring.

7.4.5   Inclusion of Induction Machines in the Transient Stability Program

This is relatively straightforward using the same format as developed for synchronous machines. Most induction machines are equipped with contactors which respond to terminal conditions such as undervoltage and it is sometimes necessary to model this equipment. The characteristics and logic associated with contactors are included in Section 7.7 (Relays).

7.5   A.C.–D.C. CONVERSION

The use of high-voltage and/or high-current d.c. systems is now sufficiently wide-spread to require the inclusion of d.c. converter models as a standard part of a comprehensive transient stability program. Further, rectification equipment is also required in many industrial processes, notably smelters and chlorine producers, and these are sufficiently large-load items to warrant good modelling.

The dynamic behaviour of h.v.d.c. links immediately after a large disturbance either on the d.c. side or close to the converter a.c. terminals requires much more elaborate models [1, 2]. When analysing small perturbations and dynamic stability, it is often assumed that the converter equipment operates in a controlled manner almost instantaneously when compared with the relatively slow a.c. system dynamics. In these cases, it is quite acceptable to use a modified steady-state (or quasi-steady-state) model, the modifications being due to the different constraints imposed by the load-flow and stability studies. Such a model is also suitable for representing large rectifier loads during a.c. system disturbances, with further modifications necessary to represent abnormal rectifier operating modes.

Further to the basic assumptions listed in Appendix II, the following need to be made here.

- The implementation of delay angle control is instantaneous.
- The transformer tap position remains unchanged throughout the stability study unless otherwise specified.
- The direct current is smooth, though its actual value may change during the study.

7.5.1 Rectifier Loads

Large rectifier loads generally consist of a number of bridges connected in series and/or parallel, each bridge being phase-shifted relative to the others. With these configurations, high pulse numbers can be achieved resulting in minimal distortion of the supply voltage without filtering. Rectifier loads can therefore be modelled as a single equivalent bridge with a sinusoidal supply voltage at the terminals but without representation of passive filters. This model is shown in Fig. 7.10.

Rectifier loads can utilise a number of control methods. They can use diode and thyristor elements in full- or half-bridge configurations. In some cases, diode bridges are used with tap changer and saturable reactor control. The effect of the saturable reactors on diode conduction is identical to delay angle control of a thyristor over a limited range of delay angles. All these different control methods can be modelled using a controlled rectifier with suitable limits imposed on the delay angle (α) [15].

7.5.1.1 Static Loads

Operating under constant current control, the d.c. equations are

$$V_d = I_d R_d + V_{\text{load}} \tag{7.5.1}$$

$$I_d = \frac{(A \cdot I_{d_s} - V_{\text{load}})}{(A + R_d)} \tag{7.5.2}$$

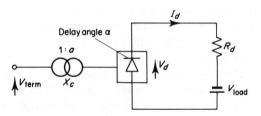

Figure 7.10
Static rectifier load equivalent circuit

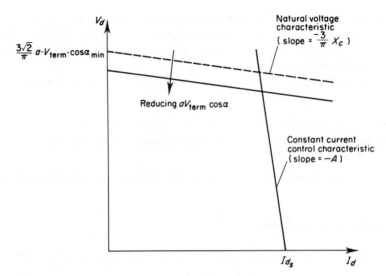

Figure 7.11
Simple rectifier control characteristic

where A is the constant current controller gain and I_{d_s} is the nominal d.c. current setting as shown in Fig. 7.11.

Constant current cannot be maintained during a large disturbance as a limit of delay angle will be reached. In this event, the rectifier control specification will become one of constant delay angle and equation (7.5.2) becomes

$$I_d = \frac{[(3\sqrt{2}/\pi)aV_{\text{term}}\cos\alpha_{\min} - V_{\text{load}}]}{[R_d + (3X_c/\pi)]}. \tag{7.5.3}$$

Protection limits and disturbance severity determines the rectifier operating characteristics during the disturbance. Shutdown occurs if I_d reaches a set minimum or zero and the voltage V_{load} will cause shutdown before the a.c. terminal voltage reaches zero. The action of the rectifier load system is thus described in Appendix II by equations (II.2.5), (II.2.7), (II.2.10), (II.2.12), (II.2.13), (II.5.1) and either (7.5.2) or (7.5.3).

7.5.1.2 Dynamic Loads

The basic rectifier load model assumes that current on the d.c. side of the bridge can change instantaneously. For some types of rectifier loads, this may be a valid assumption, but the d.c. load may well have an overall time constant which is significant with respect to the fault clearing time. In order to realistically examine the effects which rectifiers have on the transient stability of the system, this time constant must be taken into account. This requires a more complex model to account for extended overlap angles, when low commutating voltages are associated with large d.c. currents.

When the delay angle (α) reaches a limiting value, the dynamic response of the d.c.

current (I_d) is given by

$$V_d = I_d \cdot R_d + V_{\text{load}} + L_d \cdot pI_d \tag{7.5.4}$$

where L_d represents the equivalent inductance in the load circuit. Substituting for V_d using equation (II.2.5) gives

$$pI_d = \frac{1}{T_{dc}} \left\{ \frac{3\sqrt{2}}{\pi R_d} a V_{\text{term}} \cos \alpha - \left(\frac{3X_c}{R_d \pi} + 1 \right) I_d - \frac{V_{\text{load}}}{R_d} \right\} \tag{7.5.5}$$

where $T_{dc} = L_d/R_d$.

The controller time constant may also be large enough to be considered. However, in transient stability studies where large disturbances are usually being investigated, faults close to the rectifier load force the delay angle (α) to minimum very quickly. Provided the rectifier load continues to operate, the delay angle will remain at its minimum setting throughout the fault period and well into the post-fault period until the terminal voltage recovers. The controller will, therefore, not exert any significant control over the d.c. load current. Ignoring the controller time constant can therefore be justified in most studies.

7.5.1.3 Abnormal Modes of Operation

The slow response of the d.c. current when a large disturbance has been applied to the a.c. system can cause the rectifier to operate in an abnormal mode.

After a fault application near the rectifier, the near normal value of d.c. current (I_d) needs to be commutated by a reduced a.c. voltage. This causes the commutation angle (μ) to increase and it is possible for it to exceed 60°. This mode of operation is beyond the validity of the equations and to model the dynamic load effects accurately it is necessary to extend the model.

The full range of rectifier operation can be classified into four modes [16].

Mode 1—Normal operation. Only two valves in the bridge are involved in simultaneous commutation at any one time. This mode extends up to a commutation angle of 60°.

Mode 2—Enforced delay. Although a commutation angle greater than 60° is desired, the forward voltage across the incoming thyristor is negative until either the previous commutation is complete or until the firing angle exceeds 30°. In this mode, μ remains at 60° and α ranges up to 30°.

Mode 3—Abnormal operation. In this mode, periods of three-phase short circuit and d.c. short circuit exist when two commutations overlap. During this period there is a controlled safe short circuit which is cleared when one of the commutations is complete. During the short-circuit periods, four valves are conducting. Commutation cannot commence until 30° after the voltage crossover.

Mode 4—Continuous three-phase and d.c. short circuit caused by two commutations taking place continuously. In this mode, the commutation angle is 120° and the a.c. and d.c. current paths are independent.

The waveforms for these modes are shown in Fig. 7.12 and Table 7.2 summarises the conditions for the different modes of operation. Equations (II.2.5) and (II.2.7)

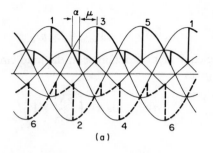

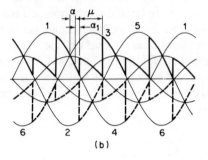

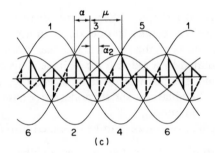

Figure 7.12
Rectifier voltage waveforms showing different modes of operation: (a) mode 1, $\mu < 60°$; (b) mode 2, $\mu = 60°$ with enforced delay α_1; (c) mode 3, $\mu > 60°$ with short-circuit period α_2

Table 7.2
Rectifier modes of operation

Mode	Firing angle	Overlap angle
1	$0° \leqslant \alpha \leqslant 90°$	$0° \leqslant \mu < 60°$
2	$0° \leqslant \alpha \leqslant 30°$	$60°$
3	$30° \leqslant \alpha \leqslant 90°$	$60° \leqslant \mu < 120°$
4	$30° \leqslant \alpha \leqslant 90°$	$120°$

do not apply for a rectifier operating in mode 3 and they must be replaced by

$$V_d = \frac{3\sqrt{6}}{\pi} a V_{\text{term}} \cos \alpha' - \frac{9}{\pi} X_c \cdot I_d \qquad (7.5.6)$$

and

$$I_d = \frac{a}{\sqrt{6}X_c} V_{\text{term}}(\cos \alpha' - \cos \gamma'). \qquad (7.5.7)$$

Fourier analysis of the waveform leads to the relationship between a.c. and d.c. current given by equation (II.2.9) where the factor k is now

$$k = \frac{3(\underline{/-2\alpha'} - \underline{/-2\gamma'} - j2\mu)}{4(\cos \alpha' - \cos \gamma')} \qquad (7.5.8)$$

where

$$\alpha' = \alpha - 30°$$

and

$$\gamma' = \gamma + 30°. \qquad (7.5.9)$$

A graph showing the value of k for various delay angles (α) and commutation angles (μ) is shown in Fig. 7.13.

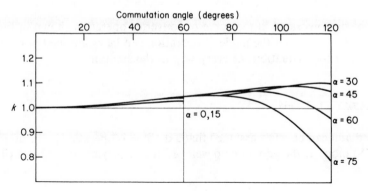

Figure 7.13
Variation of k in expression $I_p = k(3\sqrt{2}/\pi)I_d$

7.5.1.4   Identification of Operating Mode

The mode in which the rectifier is operating can be determined simply by use of a current factor K_I. The current factor is defined as

$$K_I = \frac{\sqrt{2}X_c \cdot I_d}{a V_{\text{term}}}. \qquad (7.5.10)$$

Substitution in this, using the relevant equations, yields limits for the modes.

Mode 1:

$$K_I \leqslant \cos(60° - \alpha) \tag{7.5.11}$$

and

$$K_I \leqslant 2\cos(\alpha) \quad \text{for rectifier operation.}$$

Mode 2:

$$K_I \leqslant \frac{\sqrt{3}}{2}. \tag{7.5.12}$$

Mode 3:

$$K_I < \frac{2}{\sqrt{3}} \qquad\qquad \text{when } \alpha \leqslant 30°$$

$$K_I < \frac{2}{\sqrt{3}}\cos(\alpha - 30°) \quad \text{when } \alpha > 30°. \tag{7.5.13}$$

Mode 4:

$$K_I = \frac{2}{\sqrt{3}} \qquad\qquad \text{when } \alpha \leqslant 30°$$

$$K_I = \frac{2}{\sqrt{3}}\cos(\alpha - 30°) \quad \text{when } \alpha > 30°. \tag{7.5.14}$$

This can be demonstrated in the curve of converter operation shown in Fig. 7.14.

It can thus be seen that the mode of operation can be established prior to solving for the rectifier load equations at every step in the solution.

7.5.2   D.C. Link

Provided that it can be safely assumed that a d.c. link is operating in the quasi-steady-state (QSS) mode 1, the equations developed for converters in Appendix II can be

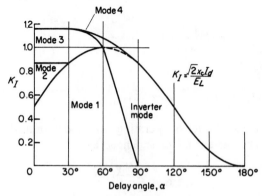

Figure 7.14
Converter operation [16]

used. That is, the converters are considered to be controllable and fast acting so that the normal steady-state type of model can be used at each step in the transient stability study.

The initial steady-state operating conditions of the d.c. link will have been determined by a load-flow and in this, the control type, setting and margin will have been established.

7.5.2.1 Constant Current Control

During the solution process at each iteration the control mode must be established. This can be done by assuming mode 1 (i.e. with the rectifier on c.c. control) and by combining equations (II.3.2), (II.5.1) and (II.5.2) a d.c. current can be determined as

$$I_{d_{\text{mode 1}}} = \frac{I_{d_{s_r}} - [(3\sqrt{2}/\pi)a_i \cdot V_{\text{term}_i} \cdot \cos \gamma_{i_c}]/A_r}{[1 + (R_d - (3/\pi)X_{c_i})/A_r]}. \tag{7.5.15}$$

Assuming this current to be valid, then d.c. voltages at each end of the link can be calculated using equations (II.2.5) and (II.3.1). The d.c. link is operating in mode 2 (i.e. with the inverter on c.c. control) if

$$V_{d_r \text{ mode 1}} - V_{d_i \text{ mode 1}} < 0. \tag{7.5.16}$$

The d.c. current for mode 2 operation is given by

$$I_{d_{\text{mode 2}}} = \frac{I_{d_{s_i}} + \left(\dfrac{3\sqrt{2}}{\pi}\alpha_r \cdot V_{\text{term}_r} \cdot \cos \alpha_{r\,\text{min}}\right)\Big/ A_i}{[(1 + (3/\pi)X_{c_i})/A_i]}. \tag{7.5.17}$$

7.5.2.2 Constant Power Control

For constant power control, under control mode 1, the d.c. current may be determined from the quadratic equation

$$\left(\frac{R_d}{2} - \frac{3}{\pi}X_{c_i}\right)I_{d_{\text{mode 1}}}^2 + \left(\frac{3\sqrt{2}}{\pi}\alpha_i \cdot V_{\text{term}_i} \cdot \cos \gamma_{i_c}\right)I_{d_{\text{mode 1}}} - P_{d_s} = 0 \tag{7.5.18}$$

where P_{d_s} is the setting at the electrical mid-point of the d.c. system, that is

$$P_{d_s} = (P_{d_{s_r}} + P_{d_{s_i}})/2. \tag{7.5.19}$$

The correct value for $I_{d_{\text{mode 1}}}$ can then be found from Table 7.3. Control mode 2 is determined using equation (7.5.16) and in this case the following quadratic equation must be solved.

$$k_r \cdot I_{d_{\text{mode 2}}}^2 - k_v \cdot I_{d_{\text{mode 2}}} - P_{d_{\text{marg}}} - P_{d_s} = 0 \tag{7.5.20}$$

where

$$k_r = \frac{R_d}{2} + \frac{3}{\pi}X_{c_r} \tag{7.5.21}$$

Table 7.3

Current setting for constant power control from quadratic equation

I_{d_1}	I_{d_2}	I_d
Within	Outside	I_{d_1}
Outside	Within	I_{d_2}
Within	Within	Greater of I_{d_1} and I_{d_2}
Greater	Greater	$I_{d_{max}}$
Greater	Less	$I_{d_{max}}$
Less	Greater	$I_{d_{max}}$
Less	Less	0

Within $\equiv$ within the range $I_{d_{min}}$ to $I_{d_{max}}$;
Outside $\equiv$ outside the range $I_{d_{min}}$ to $I_{d_{max}}$;
Greater $\equiv$ greater than $I_{d_{max}}$
Less $\equiv$ less than $I_{d_{min}}$

$$k_v = \frac{3\sqrt{2}}{\pi} a_r \cdot V_{term_r} \cdot \cos \alpha_{r_{min}}. \tag{7.5.22}$$

If the link is operating under constant power control but with a current margin then for control mode 2

$$-k_r I_{d_{mode\,2}}^2 + (k_v - k_r I_{d_{marg}}) I_{d_{mode\,2}} + k_v I_{d_{marg}} - P_{d_{s_r}} = 0. \tag{7.5.23}$$

It is possible for the d.c. link to be operating in control mode 2 despite satisfying the inequality of equation (7.5.16). This occurs when the solution indicates that the rectifier firing angle (α_r) is less than the minimum value $(\alpha_{r_{min}})$. In this case the delay angle should be set to its minimum and a solution in mode 2 is obtained.

It is also possible, that when the link is operating close to the changeover between modes, convergence problems will occur in which the control mode changes at each iteration. This can easily be overcome by retaining mode 2 operation whenever detected for the remaining iterations in that particular time step.

7.5.2.3 D.C. Power Modulation

It has been shown in the previous section that under the constant power control mode, the d.c. link is not responsive to a.c. system terminal conditions, i.e. the d.c. power transfer can be controlled disregarding the actual a.c. voltage angles. Since, generally, the stability limit of an a.c. line is lower than its thermal limit, the former can be increased in system involving d.c. links by proper use of the fast converter controllability.

The d.c. power can be modulated in response to a.c. system variables to increase system damping. Optimum performance can be achieved by controlling the d.c. system so as to maximise the responses of the a.c. system and d.c. line simultaneously following the variation of terminal conditions.

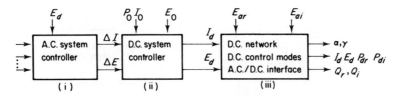

Figure 7.15
A.C.–d.c. dynamic control structure: (i) a.c. system controller; (ii) d.c. system controller; (iii) a.c.–d.c. network

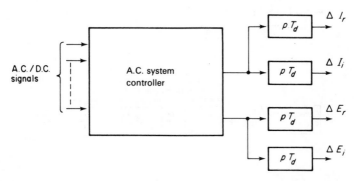

Figure 7.16
A.C. system controller

The dynamic performance under d.c. power modulation is best modelled in three separate levels [17]. These levels, illustrated in Fig. 7.15, are the a.c. system controller (i), the d.c. system controller (ii) and the a.c.–d.c. network (iii).

(i) The a.c. system controller uses a.c. and/or d.c. system information to derive the current and voltage modulation signals. A block diagram of the controller and a.c.–d.c. signal conditioner is shown in Fig. 7.16.

(ii) The d.c. system controller receives the modulation signals ΔI and ΔE and the steady-state specifications for power P_0 current I_0 and voltage E_0. Fig. 7.17(a) illustrates the power controller model, which develops the scheduled current setting; it is also shown that the current order undergoes a gradual increase during restart, after a temporary blocking of the d.c. link.

The rectifier current controller, Fig. 7.17(b), includes signal limits and rate limits, transducer time constant, bandpass filtering and a voltage dependent current order limit (VDCOL).

The inverter current controller, Fig. 7.17(c), includes similar components plus a communications delay and the system margin current (Im).

Finally the d.c. voltage controller, including voltage restart dynamics, is illustrated in Fig. 7.17(d).

(iii) The d.c. current I_d and voltage E_d derived in the d.c. system controller constitute the input signals for the a.c.–d.c. network model which involves the steady-state solution of the d.c. system (neglecting the d.c. line dynamics which are included in the d.c. system controller). Here the actual a.c. and d.c. system quantities are

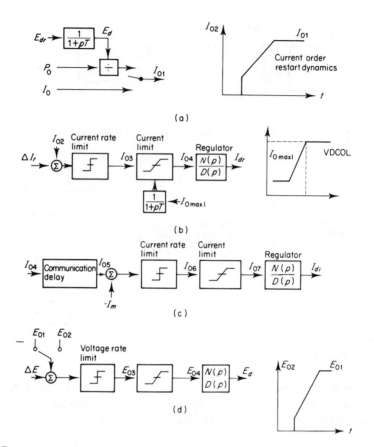

Figure 7.17
D.C. system controller: (a) power controller; (b) rectifier current controller; (c) inverter current
controller: (d) d.c. voltage controller

calculated, i.e. control angles, d.c. current, voltage, active and reactive power. The converter a.c. system constraints are the open-circuit secondary voltages E_{ar} and E_{ai}.

7.5.3 Representation of Converters in the Network

7.5.3.1 Rectifiers

The static-load rectifier model can be included in the overall solution of the transient stability program in a similar manner to the basic loads described in Chapter 6.

From the initial load flow, nominal bus shunt admittance ($\bar{y}_0$) can be calculated for the rectifier. This is included directly into the network admittance matrix $[Y]$. The injected current into the network in the initial steady state is therefore zero. In general

$$I_{\text{inj}} = (\bar{y}_0 - \bar{y})\bar{V}_{\text{term}} \qquad (7.5.24)$$

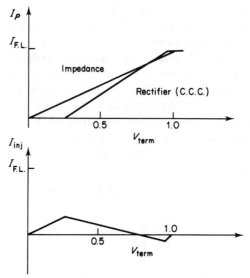

Figure 7.18
Difference between impedance and static load rectifier characteristic (for $V_{\text{load}} \neq 0$)

where

$$\bar{y}_0 = \frac{\bar{S}_0^*}{|V_{\text{term}_0}|^2} \qquad (7.5.25)$$

and

$$\bar{y} = \frac{\bar{S}^*}{|V_{\text{term}}|^2}. \qquad (7.5.26)$$

The static-load rectifier model does not depart greatly from an impedance characteristic and is well behaved for low terminal voltages, the injected current tending to zero as the voltage approaches zero. Fig. 7.18 compares the current due to a rectifier with that due to a constant impedance load. As the injected current is never large, the iterative solution for all a.c. conditions is stable.

When the rectifier model is modified to account for the dynamic behaviour of the d.c. load its characteristic departs widely from that of an impedance. Immediately after a fault application, the voltage drops to a low value but the injected current magnitude does not change significantly. Similarly on fault clearing, the voltage recovers instantaneously to some higher value while the current remains low.

When the load characteristic differs greatly from that of an impedance, the sequential solution technique can exhibit convergence problems [8], especially when the voltage is low. With small terminal voltages, the a.c. current magnitude of the rectifier load is related to the d.c. current but the current phase is greatly affected by the terminal voltage. Small voltage changes in the complex plane can result in large variations of the voltage and current phase angles.

To avoid the convergence problems of the sequential solution, an alternative algorithm has been developed [15]. This combines the rectifier and network solutions into a unified process. It, however, does not affect the sequential solution of the other components of the power system with the network.

The basis of this approach is to reduce the a.c. network, excluding the rectifier, to an equivalent Thevenin source voltage and impedance as viewed from the primary side of the rectifier transformer terminals. This equivalent of the system, along with the rectifier, can be described by a set of nonlinear simultaneous equations which can be solved by a standard Newton–Raphson algorithm. The solution of the reduced system yields the fundamental a.c. current at the rectifier terminals.

To obtain the network equivalent impedance, it is only necessary to inject 1 p.u. current into the network at the rectifier terminals while all other nodal injected currents are zero. With an injected current vector of this form, a solution of the network equation (6.5.1) gives the driving point and transfer impedances in the resulting voltage vector

$$[\bar{Z}] \equiv [\bar{V}'] = [\bar{Y}]^{-1}[\bar{I}'_{inj}] \tag{7.5.27}$$

where

$$[\bar{I}'_{inj}] = \begin{bmatrix} 0 \\ \vdots \\ 0 \\ \bar{I}_r \\ 0 \\ \vdots \\ 0 \end{bmatrix} \quad \text{and} \quad \bar{I}_r = 1 + j0. \tag{7.5.28}$$

The equivalent circuit shown in Fig. 7.19 can now be applied to find the rectifier current $(\bar{I}_p)$ by using the Newton–Raphson technique.

The effect of the rectifier on the rest of the system can be determined by superposition:

$$[\bar{V}] = [\bar{V}^\sigma] + [\bar{Z}]I_p \tag{7.5.29}$$

where

$$[\bar{V}^\sigma] = [\bar{Y}]^{-1}[\bar{I}^\sigma_{inj}] \tag{7.5.30}$$

and $[\bar{I}^\sigma_{inj}]$ are the injected currents due to all other generation and loads in the system.

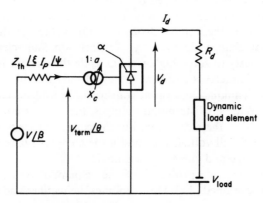

Figure 7.19
Equivalent system for Newton–Raphson solution

If the network remains constant, vector $[\bar{Z}]$ is also constant and thus only needs re-evaluation on the occurrence of a discontinuity.

Thus the advantages of the unified and sequential methods are combined. That is, good convergence for a difficult element in the system is achieved while the programming for the rest of the system remains simple and storage requirements are kept low.

The equivalent system of Fig. 7.19 contains seven variables ($V_{\text{term}}, I_p, \theta, \psi, \alpha, V_d$ and I_d). With these variables four independent equations can be formed. They are equation (II.2.5) and

$$V \underline{/\beta} - V_{\text{term}} \underline{/\theta} - Z_{\text{th}} \underline{/\xi} \cdot I_p \underline{/\psi} = 0 \qquad (7.5.31)$$

$$V_d \cdot I_d - \sqrt{3}\alpha V_{\text{term}} \cdot I_p \cdot \cos(\theta - \psi) = 0. \qquad (7.5.32)$$

Equation (7.5.31) is complex and represents two equations. Substituting for V_d and I_p using equations (II.2.11) and (7.5.1) reduces the number of variables to five. A fifth equation is necessary and with constant current control, that is with the delay angle (α) within its limits, this can be written as

$$I_d - I_{d_{\text{sp}}} = 0. \qquad (7.5.33)$$

Equation (II.2.5), suitably reorganised, and equations (7.5.31) to (7.5.33) represent $[F(X)] = 0$ of the Newton–Raphson process and

$$[X]^{\text{T}} = [E_r, \theta, \psi, \alpha, I_d]. \qquad (7.5.34)$$

When the delay angle reaches a specified lower limit (α_{min}), the control specification, given by equation (7.5.33), changes to

$$\alpha - \alpha_{\text{min}} = 0 \qquad (7.5.35)$$

and equation (II.2.5) is no longer valid. The d.c. current (I_d) is now governed by the differential equation (7.5.5). If the trapezoidal method is being used, this equation can be transformed into an algebraic form similar to that described in Chapter 6. Equation (II.2.5) is replaced by

$$I_d = ka \cdot E_r \cdot \cos\alpha - kb = 0. \qquad (7.5.36)$$

The variables ka and kb contain information from the beginning of the integration step only and are thus constant during the iterative procedure.

$$ka = h/(2 + kc \cdot h) \qquad (7.5.37)$$

$$kb = (1 - 2kc \cdot ka)I_d(t) + \frac{3\sqrt{2}}{\pi T_{dc} \cdot R_d} a \cdot ka \cdot V_{\text{term}}(t) \cos\alpha(t) + \frac{2 \cdot ka \cdot V_{\text{load}}}{T_{dc} \cdot R_d} \qquad (7.5.38)$$

where

$$kc = \left(\frac{3X_c}{\pi R_d}\right) + 1/T_{dc} \qquad (7.5.39)$$

and t represents the time at the beginning of the integration step and h is the step length.

Commutation angle μ is not explicitly included in the formulation, and since these equations are for normal operation, the value of k in equation (II.2.11) is close to unity and may be considered constant at each step without loss of accuracy. On convergence, μ may be calculated and a new k evaluated suitable for the next step.

In mode 3 operation, the value of k becomes more significant and for this reason the number of variables is increased to six to include the commutation angle μ. The equations $[F(X)] = 0$ for the Newton–Raphson method in this case are

$$V\underline{/\beta} - V_{\text{term}}\underline{/\theta} - Z_{\text{th}}\underline{/\xi} \cdot \frac{\sqrt{6}}{\pi} f(\mu) \cdot I_d\underline{/\psi} = 0 \qquad (7.5.40)$$

$$\frac{\sqrt{2}}{\pi} a \cdot V_{\text{term}} \cdot \cos(\theta - \psi) \cdot f(\mu) - \frac{\sqrt{6}}{\pi} a \cdot V_{\text{term}} \cos\alpha' + \frac{3X_c}{\pi} I_d = 0 \qquad (7.5.41)$$

$$I_d - ka \cdot a \cdot V_{\text{term}} \cos\alpha' - kb = 0 \qquad (7.5.42)$$

$$\cos(\alpha + \mu + 30) - \cos\alpha' + \frac{\sqrt{6}X_c}{a \cdot V_{\text{term}}} \cdot I_d = 0 \qquad (7.5.43)$$

$$\alpha - \alpha_{\min} = 0. \qquad (7.5.44)$$

Although k can be calculated explicitly, a linearised form of equation (7.5.8) obtained for $\alpha = 30°$ can be used to simplify the expression. In the range $60° < \mu < 120°$, the value of k can be obtained from

$$f(\mu) = 1.01 - 0.0573\mu \qquad (7.5.45)$$

where μ is measured is radians.

In mode 4, the a.c. and d.c. systems are both short circuited at the rectifier and operate independently. In this case the system equivalent of Fig. 7.19 reduces to that shown in Fig. 7.20. The network equivalent can be solved directly and the d.c. current is obtained from the algebraic form of the differential equation (7.5.5).

7.5.3.2 D.C. Links

The problems associated with dynamic rectifier loads do not occur when the d.c. link is represented by a quasi-steady-state model. Each converter behaves in a manner similar to that of a converter for a static rectifier load. A nominal bus shunt admittance

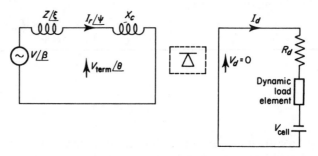

Figure 7.20
Rectifier load equivalent in mode 4 operation

($\bar{y}_0$) is calculated from the initial load flow for both the rectifier and inverter ends and injected currents are used at each step in the solution to account for the change from steady state calculated from equation (7.5.24). Note that the steady-state shunt admittance at the inverter ($\bar{y}_{0_i}$) will have a negative conductance value as power is being supplied to the network. This is not so for a synchronous or induction generator as the shunt admittance serves a different purpose in these cases.

7.5.4 Inclusion of Converters in the Transient Stability Program

A flow diagram of the unified algorithm is given in Fig. 7.21 [18]. It is important to note that the hyperplanes of the functions used in the Newtonian iterative solution

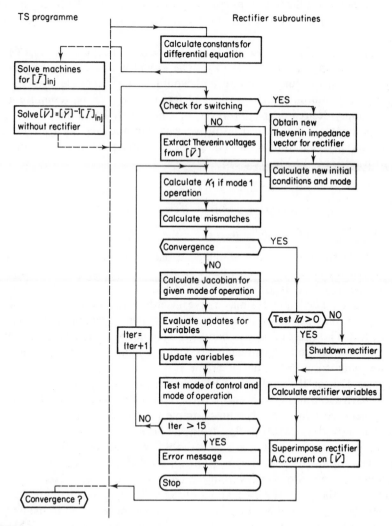

Figure 7.21
Unified algorithm flow diagram

process are not linear and good initial estimates are essential at every step in the procedure. A common problem in converter modelling is that the solution converges to the unrealistic result of converter reactive power generation. It is therefore necessary to check against this condition at every iteration. With integration step lengths of up to 25 mS, however, convergence is rapid.

7.6 STATIC VAR COMPENSATION SYSTEMS

The use of static VAR compensation systems (SVS) to maintain an even voltage profile at load centres remote from generation has become common. An SVS can have a large VAR rating and therefore to consider it as a fixed shunt element can produce erroneous results in a transient stability study. Also an SVS may be installed to improve stability in which case good modelling is essential for both planning and operation.

The model of the SVS shown in Fig. 7.22 is based on representations developed by CIGRE Working Group 31–01 [19]. The model is not overly complex as this would make data difficult to obtain and would be incompatible with the overall philosophy of a multimachine transient program. The SVS representation can be simplified to any desired degree, however, by suitable choice of data.

The basic control circuit consists of two lead–lag and one lag transfer function connected serially. The differential equations describing the action of the control circuit with reference to Fig. 7.22 are

$$pB_1 = [K(1 + T_2 p)(V_{SV_{set}} - V'_{SV}) - B_1]/T_1 \qquad (7.6.1)$$

$$pB_2 = [(1 + T_4 p)B_1 - B_2]/T_3 \qquad (7.6.2)$$

$$pB_3 = [B_2 - B_3]/T_5. \qquad (7.6.3)$$

Although electronically produced, the dead band may be considered as a physical linkage problem as shown in Fig. 7.23(a). In this example, the input (x) and output (y) move vertically. The diagram shows the initial steady-state condition in which x and y are equal. The input (x) may move in either direction by an amount $D_b/2$ before y moves. Beyond this amount of travel, y follows x, lagging by $D_b/2$ as depicted in Fig. 7.23(b). The effect of a dead band can be ignored by setting D_b to zero.

Stepped output permits the modelling of SVS when discrete capacitor (or inductor) blocks are switched in or out of the circuit. It is usual to assume that all blocks are of equal size. During the study, the SVS operates on the step nearest to the control setting. Iterative chattering can occur if the control system output (B_3) is on the boundary between two steps. The most simple remedy is to prevent a step change until B_3 has moved at least 0.55 B_{step} from the mean setting of the step.

The initial MVAR loading of the SVS should be included in the busbar loading schedule data input. However, it is possible for an SVS to contain both controllable and uncontrollable sections (e.g. variable reactor in parallel with fixed capacitors or vice versa). It is the total MVAR loading of the SVS which is, therefore, included in the busbar loading. Only the controllable part should be specified in the SVS model input and this is removed from the busbar loading leaving an uncontrollable MVAR load which is converted into a fixed susceptance associated with the network.

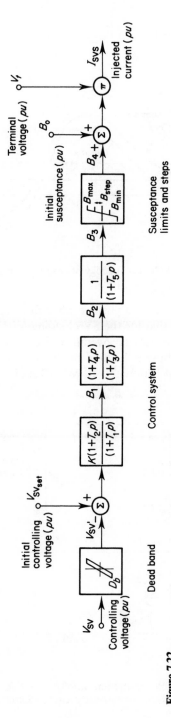

Figure 7.22
Composite static VAR compensation system (SVS) model

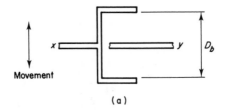

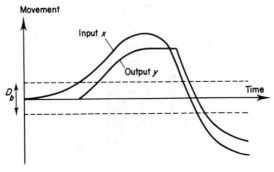

Figure 7.23
Dead band analogy and effect: (a) physical analogy of dead band; (b) the effect of a dead band on output

Note that busbar load is assumed positive when flowing out of the network. The sign is therefore opposite to that for the SVS loading.

In order to clarify this, consider an overall SVS operating in the steady state as shown in Fig. 7.24(a). The busbar loading in this case must be specified as -50 MVAR and it may be varied between -10 MVAR and -80 MVAR provided the voltage remains constant.

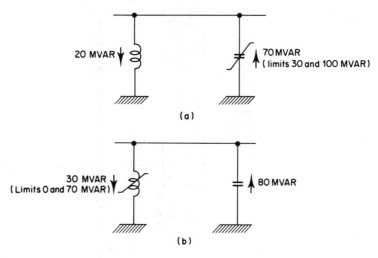

Figure 7.24
Example of an overall SVS controllable and uncontrollable sections. (a) Example of overall SVS using controllable capacitors; (b) alternative to overall SVS in (a) using a controllable reactor

Table 7.4
Examples of MVAR loading specification for SVS shown in Fig. 7.24(a)

Example	Initial loading	Maximum limit	Minimum limit
1	70	100	30
2	50	80	10
3	−30	0	−70

The SVS may be specified in a variety of ways, some more obvious than others, the response of the system being identical. Three possible specifications are given in Table 7.4. In the first example, the network static load will be +20 MVAR while in the second case the static load will be zero. The third example may be represented by an overall SVS as shown in Fig. 7.24(b).

It is convenient, when specifying the initial steady-state operation, to use MVAR. However, this is a function of the voltage and hence all MVAR settings must be converted to their equivalent per unit susceptance values prior to the start of the stability study.

7.6.1 Representation of SVS in the Overall System

The initial MVAR loading of the SVS is converted into a shunt susceptance (B_0) and added to the total susceptance at the SVS terminal busbar. During the system study, the deviation from a fixed susceptance device is calculated (B_4) and a current equivalent to this deviation is injected into the network.

A reduction in controlling voltage V_{sv} will cause the desired susceptance B_4 to increase. That is the capacitance of the SVS will rise and the MVAR output will increase.

The injected current $(\bar{I}_{inj})$ into the network is given by

$$\bar{I}_{inj} = -\bar{V}\bar{Y} \tag{7.6.4}$$

where

$$Y = 0 + jB_4.$$

Although not necessary for the solution process, the MVA output from the SVS into the system is given by

$$\bar{S} = \bar{V}\bar{I}_{svs}^*$$

and hence

$$Q = |V|^2(B_4 + B_0). \tag{7.6.5}$$

7.7 RELAYS

Relay characteristics may be applied to a transient stability program and the effect of relay operation automatically included in system studies. This permits checking of relay settings and gives more realistic information as to system behaviour after a disturbance, assuming 100% reliability of protective equipment. Reconstruction of the events after fault occurrences may also be carried out.

Unit protection only responds to faults within a well-defined section of a power system and as the faults are prespecified, the operation of unit protection schemes can equally be specified in the switching data input. Thus, only nonunit protection needs to be modelled and of these overcurrent, undervoltage and distance schemes are the most common.

7.7.1 Instantaneous Overcurrent Relays

Instantaneous or fixed time delay overcurrent relays are readily modelled. The operating point of the relay should be specified in terms of p.u. primary current thus avoiding the need to specifically model the current transformer. However, the location of the current transformer must be specified, e.g. at busbar A on branch to B, so that the correct signals are used by the relay model. The only other piece of information required is delay time (t_{del}) between the relay operation time and the circuit-breaker arc extinction time (t_{cb}).

Initially, the circuit-breaker operating time (t_{cb}) is set to some large value as it must be assumed that the steady-state current is less than the relay setting. At the end of each time step (e.g. at time t) the current at the current transformer location is evaluated and if it exceeds the relay setting, the effective circuit-breaker time is set to

$$t_{cb} = t + t_{del}. \qquad (7.7.1)$$

The integration then proceeds until the time step nearest to t_{cb} when circuit-breaker opening is simulated by reducing the relevant branch admittance to a very small value. Alternatively, the integration step length can be adjusted to open the circuit at time t_{cb}.

During the period between relay operation and t_{cb} the simulation of relay drop-off may be desired. In this case, if current falls below a prespecified percentage of relay setting current then t_{cb} is reset to a large value.

7.7.2 Inverse Definite Minimum Time Lag Overcurrent Relays

The inverse time characteristics of induction disc and similar relays may easily be included in an overcurrent relay. This may be accomplished by defining several points on the characteristic and interpolating, but curve fitting is better if a simple function can be found.

For example, an overcurrent relay conforming to British Standard BS 142 would appear to be accurately modelled by defining seven points on the curve as shown in Fig. 7.25(a). However, when plotted on a log–log graph as in Fig. 7.25(b), the errors are more obvious and can exceed the accuracy limits laid down in the standard if care is not taken. However, acceptable accuracy can be obtained by using the approximation

$$t_{op} = 3.0/[\log(I)] \qquad \text{for } 1.1 \leqslant I \leqslant 20 \qquad (7.7.2)$$

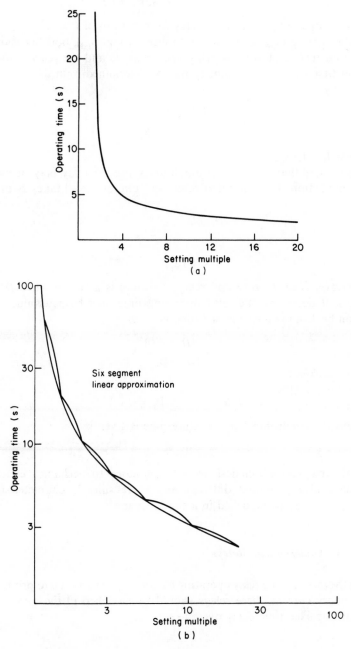

Figure 7.25
Inverse definite minimum time lag overcurrent relay (from BS142 (1966): (a) linear scale;
(b) logarithmic scale

and

$$t_{op} = \infty \qquad \text{for } i < 1.1$$

where t_{op} is the operating time of the relay for a current of I.

Plug bridge setting (S_{pb}) and time multiplier setting (S_{tm}), both measured in per unit, can be incorporated into the relay characteristic and the relay induction disc travel (D_t) at time t due to a current I_t may be determined from the previous travel at time $t - h$ by

$$D_t = D_{t-h} + \frac{h \cdot \log \{\frac{1}{2}(I_t + I_{t-h})/S_{pb}]}{3S_{tm}} \qquad (7.7.3)$$

provided that $I_t \geqslant 1.1 S_{pb}$.

For currents less than the definite minimum value the relay may be assumed to reset by spring action. Assuming that resetting from full travel takes 2s then

$$D_t = D_{t-h} - \frac{h}{2S_{tm}} \qquad (7.7.4)$$

when

$$I_t < 1.1 S_{pb}.$$

Initially, travel is set to zero and relay operation is assumed when D equals or exceeds 1 p.u. If necessary, the relay operating time may be determined by linear interpolation backwards over the last time interval:

$$t_{op} = t - \frac{(D_t - 1.0)}{(D_t - D_{t-h})} h \qquad (7.7.5)$$

when

$$D_t \geqslant 1.0 \qquad D_{t-h} < 1.0$$

and from this the circuit-breaker operating time is given by

$$t_{cb} = t_{op} + t_{del}. \qquad (7.7.6)$$

Many static relays have been designed which conform to mechanical characteristics but they have also permitted different and more suitable characteristics to be developed. These may be modelled in a similar manner.

7.7.3 Undervoltage Relays

Apart from the fact that the relay operating current is proportional to primary voltage and not primary current, these relays should be modelled in the same manner as instantaneous or fixed time delay overcurrent relays.

7.7.4 Induction Machine Contactors

The transient analysis of industrial power systems usually require that many induction machines are modelled. During a disturbance the voltage levels throughout the system will fluctuate and may result in the machine being disconnected from the system, albeit temporarily. Other machines may be on automatic stand-by to maintain essential services.

It is, therefore, necessary to include models of induction machine contactors in a transient stability program. Undervoltage protection is usually associated with the contactors and can be modelled in the normal manner.

7.7.5 Directional Overcurrent Relay

A directional overcurrent relay requires a voltage signal as well as current. The relay may operate only when the phase difference of the two signals is within prescribed limits and all other constraints are satisfied.

7.7.6 Distance Relays

As in practice, both busbar voltage and branch current signals are required, from which an apparent impedance $Z_s\underline{/\theta_s}$ of the system at the relaying point can be calculated. This is then compared with the relay characteristic to determine operation.

A typical three-zone distance protection relay is shown in Fig. 7.26. Assuming circular characteristics, then the settings of the relay may be identified by forward reach $Z_{rf}\underline{/\theta_{rf}}$ measured in impedance (complex) coupled with backward reach R_b expressed as a per unit of forward reach. From this information, the centre $(p + jq)$

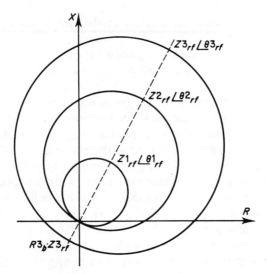

Figure 7.26
Three-zone distance relay characteristic

and radius (a) of each of the three circles in the impedance plane can be established:

$$a = \tfrac{1}{2}Z_{rf}(1 + R_b)$$
$$p = \tfrac{1}{2}Z_{rf}(1 - R_b)\cos\theta_{rf} \qquad (7.7.7)$$
$$q = \tfrac{1}{2}Z_{rf}(1 - R_b)\sin\theta_{rf}.$$

In the example in Fig. 7.26, R_b for zones 1 and 2 is zero.

The equation of the boundary of an operating zone is

$$(Z\cdot\cos\theta - p)^2 + (Z\cdot\sin\theta - q)^2 - a^2 = 0 \qquad (7.7.8)$$

and hence operation is defined when

$$Z_s^2 - 2Z_s(p\cdot\cos\theta_s + q\cdot\sin\theta_s) + (p^2 + q^2 - a^2) \leqslant 0. \qquad (7.7.9)$$

Tomato, lens, quadrilateral or other complex characteristics may be constructed by combining several simple characteristics of this type.

Each zone has a fixed time delay associated with it so that the timing for circuit-breaker action is the same as that described previously.

7.7.7 Incorporating Relays in the Transient Stability Program

Nonunit protection equipment usually only trips the local circuit breaker. Therefore, it is necessary to create dummy busbars so that a faulted branch can be switched out correctly. This can be done automatically during the data input stage in the same manner as described in Chapter 6 for faults located on branches. Thus, a faulted branch may have several dummy busbars associated with it and care should be taken to ensure all are adequately identified. Protected branches which are not directly faulted need not be modelled as accurately and the whole branch may be removed if the circuit breaker at either end is opened.

Relay characteristics should be checked at the end of each time solution and reconvergence after a discontinuity. It is not necessary to perform the check at each iteration however. This reduces the computational effort associated with relays and permits more complex relay characteristics to be modelled at critical points in the system.

Induction machine switching should not be simulated by creating dummy branches which can be removed from the network whenever necessary. While this is a feasible solution, it is extremely wasteful to computational storage and effort. A more satisfactory method is to identify the state of the machine, i.e. either switched in or out, by a simple flag and when switched out to solve for the machine with zero stator current and likewise remove its injected current from the network.

The network, however, usually includes a shunt admittance representing the machine in the initial steady state. This problem may be overcome by injecting another current to compensate for this admittance whenever a machine is switched out. Alternatively, a machine liable to switching need not have its equivalent shunt admittance included in the network at any time during the study. This simplifies periods when the machine is switched out, but requires a different injected current to the usual when switched in.

A minor problem occurs when induction machines, which are initially switched out of service, are included in the input data. An estimate of the full-load active power of the machine must be specified so that the load characteristics of the machine can be adequately defined. Also induction motors on stand-by for automatic start-up must be modelled accurately if sensible run-up simulation is to be achieved.

7.8 UNABLANCED FAULTS

The models developed so far for transient stability analysis have assumed balanced three-phase operation even during the fault period. Although three-phase faults are the most onerous, there are occasions when unsymmetrical fault conditions need to be analysed. It is possible to develop three-phase models of all power system equipment but the development effort plus the extra computational costs restrict this type of program to very simple systems. Unbalanced fault studies are relatively rare and the unbalance only occurs for a short period of the study thus the need for a three-phase model is limited, and makes full scale development unattractive.

A more practical approach is the use of symmetrical components. The negative- and zero-sequence component system models can be added to the existing single-phase (positive-sequence) model without major disruption and can be easily removed when not required.

7.8.1 Negative-sequence System

Of the two additional symmetrical component systems the negative sequence is the easier. It is very similar to the positive-sequence system.

The negative-sequence impedances of the components of the transmission network and static loads are usually the same as for the positive-sequence impedances and hence no additional storage is required. Phase displacement in transformer banks is of the opposite sign to that for the positive sequence. While phase displacement can be ignored during balanced operation, it must be established if phase quantities are to be calculated during unbalanced operation. A simple clock notation with each hour representing $30°$ shift is suitable for this purpose.

The negative-sequence impedance of synchronous machines is different from the positive-sequence impedance. The flux produced by negative-sequence armature current rotates in the opposite direction to the rotor, unlike that produced by positive-sequence current, which is stationary with respect to the rotor. Rotor currents induced by this flux prevent it from penetrating deeply into the rotor. The flux path oscillates rapidly between the positive-sequence d- and q-axis subtransient flux paths and the negative-sequence reactance X_2 may conveniently be defined as

$$X_2 = (X_d'' + X_q'')/2. \tag{7.8.1}$$

This reactance is the same as the reactance which represents the machine in the positive-sequence network. The negative-sequence resistance is given by [20]

$$R_2 \simeq Ra + \tfrac{1}{2}Rr \tag{7.8.2}$$

where Rr is the rotor resistance. While R_2 and Ra will differ, the overall difference between the negative-sequence impedance (Z_2) and positive-sequence impedance representing the machine is so small as to be neglected in most cases. Further, rotating machinery does not generate negative-sequence e.m.f.s and hence there is no negative-sequence Norton injected current.

Thus, ignoring d.c. equipment, the overall negative-sequence network is identical to the positive-sequence network with all injected currents set to zero.

Negative-sequence currents have a braking effect on the dynamic behaviour of rotating machinery. For a synchronous machine, where torque and power may be assumed to be equivalent, the mechanical breaking power (Pb) is [20]

$$Pb = I_2^2(R_2 - Ra) \tag{7.8.3}$$

which may be added directly into the mechanical equation of motion given by equation (6.2.4).

A similar expression can be found for negative-sequence-breaking torque in an induction machine where the speed of the rotor and the negative-sequence currents are taken into account.

7.8.2 Zero-sequence System

The zero-sequence system differs greatly from the other two sequence systems.

The zero-sequence impedance of transmission lines is higher and for a transformer

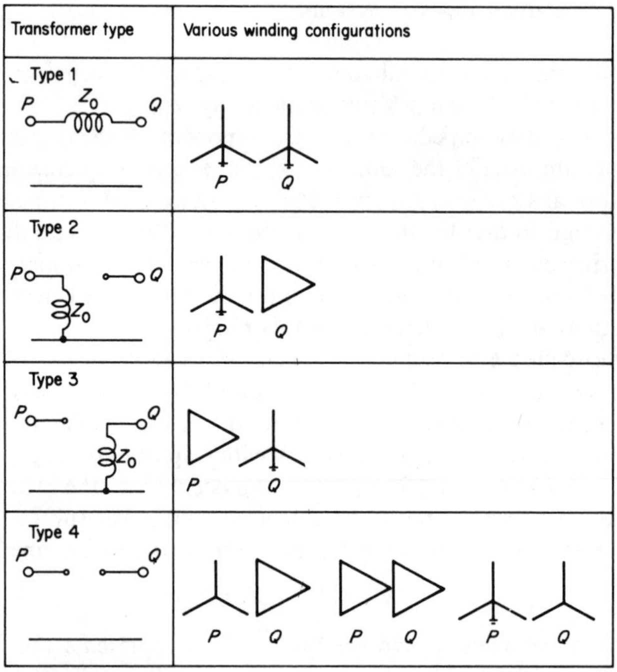

Figure 7.27
Modelling of zero-sequence equivalent networks of transformers

its value and location depends on the phase connection and neutral arrangements. Figure 7.27 shows the zero-sequence models for various typical transformer connections. By replacing the open circuit of transformer types 2, 3 and 4 with a very low admittance, the topology of the zero-sequence network can be made the same as for the other sequence networks.

The zero-sequence impedance of rotating machinery must be specified in the data input so that its inverse can be included in the zero-sequence system admittance matrix. As with the negative-sequence system model, there is no zero-sequence e.m.f. generated and hence there is no Norton injected current into this system.

7.8.3 Inclusion of Negative- and Zero-sequence Systems for Unsymmetrical Faults

The major effect of unsymmetrical faults is to increase the apparent fault impedance. On fault application, the negative- and zero-sequence impedances of the system at the point of fault are calculated. These are simply the inverse of the self-admittances at the point of fault and are determined in the same manner as described by equations (7.5.27) and (7.5.28). Depending on the type of fault, the fault impedance is modified to include the negative- and zero-sequence impedance. The fault impedance then remains constant until changed by either branch switching or fault removal.

If negative-sequence-breaking effects are to be included, it is necessary to evaluate the negative-sequence current in the relevant machines at each iteration. This is done by injecting the negative-sequence current, determined at the point of fault, into the negative-sequence system admittance matrix $[\bar{Y}_2]$:

$$[\bar{Y}_2][\bar{V}_2] = [\bar{I}f_2] \qquad (7.8.4)$$

where $[\bar{I}f_2]$ is a zero vector except at the point of fault. The vector $[\bar{V}_2]$ contains the negative-sequence voltages at all busbars from which the machine negative-sequence currents are readily obtained.

If phase information is required, then the zero-sequence voltages at all busbars also need to be determined, depending on the type of fault. This is done in an identical manner to that used for the negative-sequence system.

7.9 GENERAL CONCLUSIONS

This chapter has extended the capabilities of the transient stability analysis program developed in Chapter 6. This has been achieved by producing more advanced models of some basic power system components and also by introducing models of less frequently simulated equipment.

A transient stability program need not necessarily contain all the models described in order to completely describe a power system. Conversely, a program containing all these refinements is not necessarily adequate for a particular system. It must be anticipated that transient stability programs will be continuously refined as tighter operating constraints coupled with new control strategies are introduced.

7.10   REFERENCES

1. J. Arrillaga, C. P. Arnold and B. J. Harker, 1983. *Computer Modelling of Electrical Power Systems* Wiley & Sons, London.
2. D. A. Woodford, A. M. Gole and R. W. Menzies, 1983. Digital Simulation of DC links and AC machines, *IEEE Trans.* **PAS-102** 1616–1623.
3. S. B. Crary, 1945. *Power System Stability—Steady-State Stability* (vol. 1) Wiley & Sons Ltd., New York.
4. D. W. Olive, 1966. New techniques for the calculation of dynamic response, *IEEE Trans.* **PAS-85** 767–777.
5. T. J. Hammons, and D. J. Winning, 1971. Comparisons of synchronous-machine models in the study of the transient behaviour of electrical power systems, *Proc. IEEE* **118** 1442–1458.
6. S. Beckwith, 1937. Approximating Potier reactance, *AIEE Trans.* 813.
7. A. H. Knable, 1956. *Electrical Power Systems Engineering—Problems and Solutions* McGraw-Hill, New York.
8. H. W. Dommell, and N. Sato, 1972. Fast transient stability solutions, *IEEE Trans.* **PAS-91** 1643–1650.
9. C. P. Arnold, 1976. Solutions of the multi-machine power-system stability problem. *PhD Thesis* Victoria University of Manchester, UK.
10. IEEE Committee Report, 1973. Dynamic models for steam and hydro turbines in power-system studies, *IEEE Trans.* **PAS-92** 1904–1915.
11. D. S. Brereton, D. G. Lewis, and C. C. Young, 1957. Representation of induction motor loads during power-system stability studies, *AIEE Trans.* **PAS-76** 451–461.
12. H. E. Jordan, 1979. Synthesis of double-cage induction motor design, *AIEE Trans.* **PAS-78** 691–695.
13. C. P. Arnold, and E. J. P. Pacheco, 1979. Modelling induction motor start-up in a multi-machine transient stability program. *IEEE PES Summer Meeting* Vancouver, B.C., Canada.
14. E. J. P. Pacheco, 1975. Induction motor starting in an electrical power-system transient-stability programme. *MSc Dissertation*, Victoria University of Manchester, UK.
15. C. P. Arnold, K. S. Turner, and J. Arrillaga, 1980. Modelling rectifier loads for a multi-machine transient-stability programme, *IEEE Trans.* **PAS-99** 78–85.
16. D. B. Giesner, and J. Arrillaga, 1970. Operating modes of the three-phase bridge converter, *Int. J. Elect. Eng. Educ.* **8** 373–388.
17. IEEE Working Group on Dynamic Performance and Modeling of DC Systems, 1980. Hierarchical structure.
18. K. S. Turner, 1980. Transient stability analysis of integrated a.c. and d.c. power systems, *PhD Thesis* University of Canterbury, New Zealand.
19. CIGRE Working Group 31-01, 1977. Modelling of static shunt VAR systems for system analysis, *Electra* (51) 45–74.
20. E. W. Kimbark, 1956. *Power System Stability: Synchronous Machines* Wiley & Sons, New York.

8. ANALYSIS OF ELECTRO-MAGNETIC TRANSIENTS

8.1 INTRODUCTION

The previous two chapters have described the computation of electromechanical transients in power systems, where the main concern is the oscillatory behaviour of the generators with respect to each other following transmission faults and switching operations.

Such disturbances also cause temporary overvoltages and overcurrents in the power system, which need to be accurately predicted for the design of protective systems and insulation co-ordination. These studies come under the general umbrella of electromagnetic transient analysis, and the degree of representation of the plant components depends on the type of study, e.g. lightning surges and transient recovery voltages (in micro-seconds), switching surge distribution (in milliseconds) or in-rush currents (up to seconds).

Fourier and Laplace transformation techniques are of limited value for general purpose transient simulation programs because such programs may need to handle multiple switching operations which cannot be specified in advance (e.g. the voltage-dependent closing of surge-divertor gaps).

Most existing general purpose programs perform transient simulation in the time domain based on Bergeron's method [1]. This method uses linear relationships (characteristics) between current and voltage which are invariant from the point of view of an observer travelling with the wave.

The discrete steps (or time intervals) of the digital solution cause truncation errors that often lead to numerical instability. The use of the trapezoidal rule for the integration of the ordinary differential equations has proved invaluable in this respect.

In the 1960s Professor Dommel [2] combined the method of characteristics and the trapezoidal rule into a generalised algorithm which permits the accurate simulation of transients in networks involving distributed as well as lumped parameters. This algorithm has gained universal acceptance under the name EMTP (electromagnetic transients program) and has become a general tool in power system transient simulation. This chapter describes the formulation and computer implementation of the basic EMTP.

8.2   TRANSMISSION LINE EQUIVALENT

Let us consider the differential length of line shown in Fig. 8.1. The voltage and current wave propagation along the lossless line (at a point x) are related to the line's distributed inductance L' and capacitance C', by the equations

$$-\frac{\partial v}{\partial x} = L'\frac{\partial i}{\partial t} \tag{8.2.1a}$$

$$-\frac{\partial i}{\partial x} = C'\frac{\partial v}{\partial t}. \tag{8.2.1b}$$

The general solutions of equations (8.2.1a) and (8.2.1b) are

$$i(x, t) = f_1(x - at) + f_2(x + at) \tag{8.2.2a}$$

$$v(x, t) = Zf_1(x - at) - Zf_2(x + at) \tag{8.2.2b}$$

where f_1 and f_2 are arbitrary functions of the variables $(x - at)$ and $(x + at)$ to be determined from problem boundary and initial conditions. The physical interpretation of $f_1(x - at)$ is a wave travelling at velocity a in the forward direction and of $f_2(x + at)$ is a wave travelling at velocity a in the backward direction.

Z and a are the surge impedance and velocity of propagation respectively and for the lossless line their values are

$$Z = \sqrt{L'/C'} \tag{8.2.3}$$

$$a = 1/\sqrt{L'C'}. \tag{8.2.4}$$

The required branch equation is obtained by multiplying equation (8.2.2a) by Z and adding it to (8.2.2b):

$$v(x, t) + Zi(x, t) = 2Zf_1(x - at). \tag{8.2.5}$$

In equation (8.2.5) the left-hand side $(v + Zi)$ is constant when $(x - at)$ is constant.

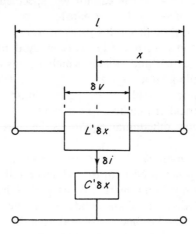

Figure 8.1
Differential length of line

This can be interpreted by becoming a fictitious observer travelling along the line with the wave. Then $(x - at)$ and $(v + Zi)$ will appear constant all along the line.

If the travel time to get from terminal k to terminal m of a line of length d is

$$\tau = d/a = d\sqrt{L'C'} \tag{8.2.6}$$

then the expression $v + Zi$ seen by the observer when leaving terminal m at time $t - \tau$ must be the same when he arrives at terminal k at time t, i.e.

$$v_m(t - \tau) + Zi_{m,k}(t - \tau) = v_k(t) + Z(-i_{k,m}(t)). \tag{8.2.7}$$

From this equation the following two-part equation results:

$$i_{k,m}(t) = (1/Z)v_k(t) + I_k(t - \tau) \tag{8.2.8a}$$

and by analogy

$$i_{m,k}(t) = (1/Z)v_m(t) + I_m(t - \tau) \tag{8.2.8b}$$

where the current sources I_k and I_m are known from previous computed values:

$$I_k(t - \tau) = -(1/Z)v_m(t - \tau) - i_{m,k}(t - \tau) \tag{8.2.9a}$$

$$I_m(t - \tau) = -(1/Z)v_k(t - \tau) - i_{k,m}(t - \tau). \tag{8.2.9b}$$

By way of example, if we assume that the computer solution uses a fixed step $\Delta t = 0.1$ ms and the line travel time is $\tau = 1$ ms then $\tau = 10\Delta t$, and therefore the value of τ to be used is 10 time steps back.

The corresponding equivalent of the lossless line is illustrated in Fig. 8.2, which shows that the two line terminals are not directly connected.

Equations (8.2.8) provide an exact solution for the lossless line at its terminals and is the basis of Bergeron's method [1].

The effect of line attenuation can be approximated with sufficient accuracy by adding half of the line resistance R at each end or, even better, by adding $R/4$ at the terminals and $R/2$ in the middle of the line.

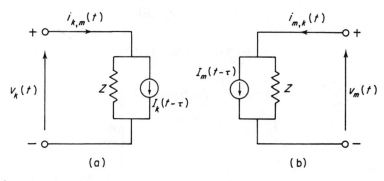

Figure 8.2
Equivalent circuit of lossless line between terminals k and m

8.3 LINEAR EQUIVALENTS DERIVED FROM THE TRAPEZOIDAL RULE

Power system plant components, other than transmission lines, are normally simulated by equivalent circuits consisting of combinations of voltage or current sources, resistances, inductances and capacitances. The type of equivalents used to represent generators and transformers depends on the type of transient disturbance under consideration.

For instance, during short-duration (or fast) transients, such as switching surges, there is no need to represent the power sources in great detail. The generators can then be modelled as voltage sources behind subtransient reactances. However, during long-duration (or slow) transient analysis, such as the overvoltages caused by load rejection, the generators need to be represented in much greater detail [3]. In these cases appropriate matrix admittances can be derived from the steady-state equivalents [4] for the transformers, taking into consideration winding connections, leakage inductances and even magnetising admittances. Nonlinear saturation effects need special consideration (see Section 8.5.2).

The representation of composite loads in transient studies is an important subject which so far has been given very little coverage.

The lumped components representing generators, transformers and loads can be replaced at each time step by a current source in parallel with a resistance. These two components are derived using the trapezoidal rule (see Fig. 8.3) as follows.

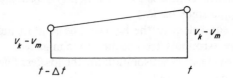

Figure 8.3
Trapezoidal rule

8.3.1 Resistance

This case, shown in Fig. 8.4, is straightforward, i.e.

$$v_k(t) - v_m(t) = Ri_{k,m}(t) \tag{8.3.1}$$

or

$$i_{k,m}(t) = (1/R)(v_k(t) - v_m(t)). \tag{8.3.2}$$

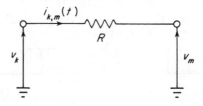

Figure 8.4
Resistance

8.3.2 Inductance

The differential equation for the inductance L of Fig. 8.5(a) is

$$v_k(t) - v_m(t) = L\frac{di_{k,m}}{dt} \tag{8.3.3}$$

which must be integrated from a known state at $t - \Delta t$ to the unknown one at t, i.e.

$$i_{k,m}(t) = i_{k,m}(t - \Delta t) + \frac{1}{L}\int_{t-\Delta t}^{t} (v_k - v_m)dt. \tag{8.3.4}$$

Using the trapezoidal rule equation (8.3.4) can be replaced by

$$i_{k,m}(t) = i_{k,m}(t - \Delta t) + \frac{1}{L}\frac{\Delta t}{2}[(v_k(t) - v_m(t)) + (v_k(t - \Delta t) - v_m(t - \Delta t))]$$

$$= I_{k,m}(t - \Delta t) + \frac{\Delta t}{2L}(v_k(t) - v_m(t)) \tag{8.3.5}$$

where

$$I_{k,m}(t - \Delta t) = i_{k,m}(t - \Delta t) + \frac{\Delta t}{2L}(v_k(t - \Delta t) - v_m(t - \Delta t)). \tag{8.3.6}$$

This is illustrated in Fig. 8.5(b).

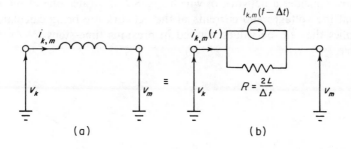

$$\equiv$$

(a)   (b)

Figure 8.5
Inductance

8.3.3 Capacitance

The capacitance C of a branch k, m (Fig. 8.6(a)) is represented by the equation

$$v_k(t) - v_m(t) = \frac{1}{C}\int_{t-\Delta t}^{t} i_{k,m}(t)dt + v_k(t - \Delta t) - v_m(t - \Delta t). \tag{8.3.7}$$

Integration by the trapezoidal rule gives

$$i_{k,m}(t) = (2C/\Delta t)(v_k(t) - v_m(t)) + I_{k,m}(t - \Delta t) \tag{8.3.8}$$

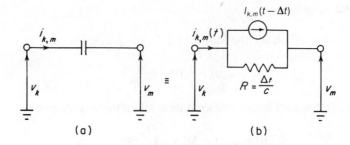

Figure 8.6
Capacitance

where

$$I_{k,m}(t - \Delta t) = - i_{k,m}(t - \Delta t) - (2C/\Delta t)(v_k(t - \Delta t) - v_m(t - \Delta t)). \qquad (8.3.9)$$

The resulting equivalent is illustrated in Fig. 8.6(b).

8.4 NODAL SOLUTION

In the expressions developed in the previous section for the branch currents the transmission lines, resistances, inductances and capacitances are considered as linear elements. These currents have been expressed as functions of the node voltages.

Let us now consider a network of which Fig. 8.7 illustrates one of the nodes and suppose that the voltages and currents of the network are being calculated at time t. This implies that all the values derived in previous time steps $t - \Delta t$, $t - 2\Delta t$, ... are available.

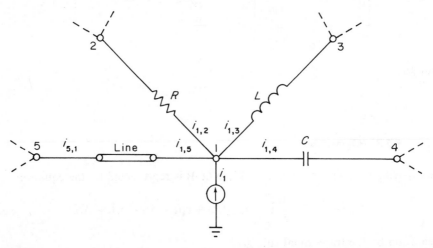

Figure 8.7
Connections involving terminal 1

Using nodal analysis in Fig. 8.7 we can write

$$i_{1,2}(t) + i_{1,3}(t) + i_{1,4}(t) + i_{1,5}(t) = i_1(t) \tag{8.4.1}$$

or, in the absence of current sources, $i_1(t) = 0$.

In terms of the nodal voltages the elements of equation (8.4.1) are

$$i_{1,2}(t) = \frac{1}{R}[v_1(t) - v_2(t)] \tag{8.4.2}$$

$$i_{1,3}(t) = \frac{\Delta t}{2L}[v_1(t) - v_3(t)] + I_{1,3}(t - \Delta t) \tag{8.4.3}$$

$$i_{1,4}(t) = \frac{2C}{\Delta t}[v_1(t) - v_4(t)] + I_{1,4}(t - \Delta t) \tag{8.4.4}$$

$$i_{1,5}(t) = \frac{1}{Z}v_1(t) + I_{1,5}(t - \tau) \tag{8.4.5}$$

and substituting in equation (8.4.1) gives

$$\left(\frac{1}{R} + \frac{\Delta t}{2L} + \frac{2C}{\Delta t} + \frac{1}{Z}\right)v_1(t) - \frac{1}{R}v_2(t) - \frac{\Delta t}{2L}v_3(t) - \frac{2C}{\Delta t}v_4(t)$$

$$= i_1(t) - I_{1,3}(t - \Delta t) - I_{1,4}(t - \Delta t) - I_{1,5}(t - \tau). \tag{8.4.6}$$

Note that the lossless line current $I_{1,5}$ is selected at time $(t - \tau)$ and not at $(t - \Delta t)$.

The whole network can be represented by the following system of linear algebraic equations:

$$[G][v(t)] = [i(t)] - [I] \tag{8.4.7}$$

where

[G] is the nodal conductance matrix

[v(t)] is the column vector of the n node voltages

[i(t)] is the column vector of current sources

[I] is the column vector of past history current sources.

Since the elements of [G] involve the time step Δt such a conductance matrix can only be constant for as long as Δt remains unchanged. It is thus preferable to work with a fixed step length Δt. However, this may create difficulties in cases where the beginning of a step must be placed at unspecified instants (such as is the case with the commutation switching instants of h.v.d.c. convertors). While the use of a variable Δt is a straightforward computation task, its implementation would make the algorithm less efficient.

The choice of Δt is not critical as long as the oscillations of highest frequency are still represented by an appropriate number of points. Changing Δt influences mainly the phase position of the high-frequency oscillations while their amplitude remain practically unaffected.

Since the network normally contains some known voltage sources, matrix equation (8.4.7) is subdivided into two subjects of nodes, A (consisting of unknown voltages) and B (with known voltages). Thus equation (8.4.7) becomes

$$\begin{bmatrix} [G_{AA}] & [G_{AB}] \\ [G_{BA}] & [G_{BB}] \end{bmatrix} \cdot \begin{bmatrix} [v_A(t)] \\ [v_B(t)] \end{bmatrix} = \begin{bmatrix} [i_A(t)] \\ [i_B(t)] \end{bmatrix} - \begin{bmatrix} [I_A] \\ [I_B] \end{bmatrix} \tag{8.4.8}$$

and the unknown voltage vector $[v_A(t)]$ is obtained from

$$[G_{AA}][v_A(t)] = [I_{total}] - [G_{AB}][v_B(t)] \tag{8.4.9}$$

with

$$[I_{total}] = [i_A(t)] - [I_A]. \tag{8.4.10}$$

If the step length Δt is constant $[v_A(t)]$ results from the solution of a system of linear equations, where only the right-hand side of equation (8.4.9) needs to be recalculated at each time step.

8.5 COMPUTATION ASPECTS

The computer implementation of the basic transient simulation algorithm is as follows. First the matrices $[G_{AA}]$ and $[G_{AB}]$ of equation (8.4.9) are built following the standard rules for the formation of the nodal admittance matrix in steady-state analysis. Then $[G_{AA}]$ is triangularised outside the time-step loop and also at every subsequent switching event and when some of the elements are altered due to the piecewise linear representation of nonlinear components.

Next the vector $[I_{total}]$ is computed at every step (forward solution) and this is followed by back substitution to solve for $[v_A(t)]$ using the existing triangularised conductance matrix. This process is illustrated in Fig. 8.8.

Most of the elements of the conductance matrices $[G_{AA}]$ and $[G_{AB}]$ are zero and this sparsity is exploited, as in the steady-state solution, by storing only the nonzero elements and/or using an optimal ordering elimination scheme [5].

In systems involving only lossless lines and lumped parameters connected between nodes and ground, or from nodes of subset A to source nodes of subset B, matrix $[G_{AA}]$ is purely diagonal and the equations can be solved independently node by node. However, sparsity-oriented solutions automatically exploit the diagonal matrix structure and thus accept off-diagonal elements without any restrictions.

The formation of $[I_{total}]$ at each step requires information of the specified currents $[I_A(t)]$ and the past history currents $[I_A]$ before going into the forward solution.

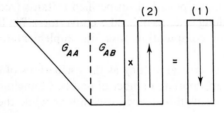

Figure 8.8
Repeated solution of linear equations involving triangular factorisation: (1) forward solution; (2) back substitution

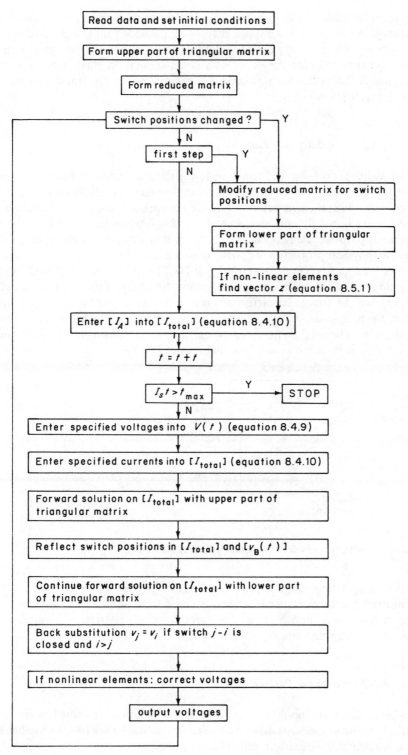

Figure 8.9
Flow diagram of the transient solution

Any specified voltage and/or current excitations (there may be none, as in the case of discharge of capacitor banks) are read or calculated from prespecified standard wave functions. Once $[v_A(t)]$ has been derived, the past history records are updated for the next time step (or preset to zero if the simulation starts from zero initial conditions). A flow diagram including the main steps of the transient solution is shown in Fig. 8.9.

8.5.1 Switching and Time-varying Conditions

Network switches may change their state during the analysis according to prespecified rules. The switching, whether caused by circuit-breaker operations or commutations between valves in static converters, alters the network topology. The actual switches themselves are normally considered ideal (i.e. $R = 0$ when closed and $R = \infty$ when open), but appropriate elements may be connected in series or parallel with the switch to simulate physical properties (e.g. stray capacitances, time-varying resistances, etc.). To reflect the topology, matrices $[G_{AA}]$ and $[G_{AB}]$ will have to be altered. However there is no need to repeat the triangular factorisation fully. If the nodes with switches are placed last the triangular factorisation is only carried out for the nodes without switches, the remaining nodes producing a reduced matrix which needs to be altered following the switching event. A graphical display is shown in Fig. 8.10 and the program logic is included in the flow diagram of Fig. 8.9.

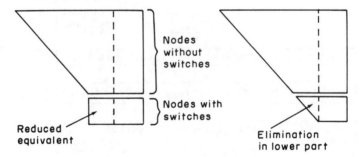

Figure 8.10
Reduction for network equivalent: (a) initially; (b) after each change

In the case of time-varying parameters although the topology of the network remains fixed, the conductance matrices also need to be altered to follow the changes of the time-varying impedances. The procedure of restricted triangularisation described above applies equally to this case.

8.5.2 Nonlinear Parameters

For a given network topology, the basic nodal solution described so far relates exclusively to linear elements, whereas a practical network may involve nonlinearities such as transformer saturation, arc behaviour, etc.

The representation of nonlinearities is a difficult subject and will in general require iterative procedures, often of difficult convergence. Practical solution techniques are available in specific cases, such as that of a single nonlinearity with a sufficiently regular branch characteristic.

A compensation technique can then be used with the nonlinear branch omitted from the matrix and being replaced by a current injection as illustrated in Fig. 8.11(a). In this case the network solution $[v(t)]$ is found by superposition as the value $[v^\circ(t)]$ obtained without the nonlinear branch (k, m), plus the contribution from the injected current $i_{k,m}$, i.e.

$$[v(t)] = [v^\circ(t)] - [z]i_{k,m} \qquad (8.5.1)$$

where vector $[z]$ is the precalculated difference of the m and k columns of $[G_{AA}]^{-1}$.

From matrix equation (8.5.1) a straight line is derived for $v_{k,m}$ as a function of $i_{k,m}$ (with slope $z_{th} = z_k - z_m$), which is plotted in Fig. 8.11(b), and the unknown value of the current injection is thus found from the intersection with the nonlinear characteristic.

The compensation technique, explained above for the case of a nonlinear resistance, can also be used for inductive nonlinearities by transforming the flux–current characteristics into voltage–current characteristics. This is done by first expressing

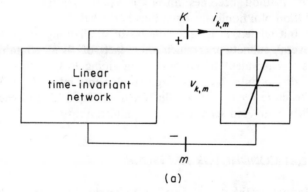

(a)

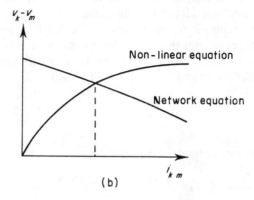

(b)

Figure 8.11
Compensation method for single nonlinearity (a) circuit; (b) characteristics

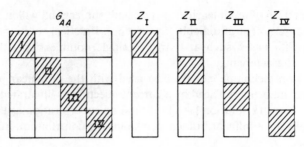

Figure 8.12
Subdivision of the network topology

the flux as the integral of the voltage and then using the trapezoidal rule on this integral.

Saturation nonlinearities can normally be represented by piecewise linear approximations, often with only two or three slopes [6]. In such cases there is no need for iterations and the reduced conductance matrix only needs to be retriangularised at points of transition between slopes.

In the presence of multiple (n) nonlinearities the scalar Thevenin impedance z_{th} of the compensation method becomes an $n \times n$ square matrix and an iterative, simultaneous solution of n nonlinear equations is needed.

An alternative, but related, technique consists of subdividing the topology of the network into several subnetworks, such that each of them contains only one nonlinearity; this is equivalent to networks containing lossless lines and, as was explained earlier, do not introduce off-diagonal elements into $[G_{AA}]$. This yields a block-diagonal structure for $[G_{AA}]$, as illustrated in Fig. 8.12, and each nonlinear parameter can then be treated separately as explained above.

8.6   MULTICONDUCTOR NETWORKS

The linear transformation theory described in Appendix I has already been applied in Chapters 3 and 4 directly to three-phase network components with coupled lumped inductances. In multiconductor load-flow studies the transmission lines are modelled by equivalent-π circuits. In transient analysis the matrix equivalents replace the scalar quantities of the series impedance and shunt capacitance. As shown in Chapter 3 the multiconductor π-circuits can be used to model lines with any number of conductors, phases or parallel lines using the same right of way.

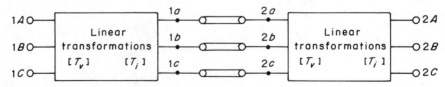

Figure 8.13
Three-phase to nodal transformation

However the computer solution permits the use of more accurate models of transmission lines using distributed parameters. This requires the use of a double transformation, with first a modal transformation to convert the coupled equations of the phase domain into decoupled equations in the modal domain. Each mode can then be described by the same nodal equations derived for single-phase lines and these modal nodal equations are then transformed back into the phase domain.

With reference to Fig. 8.13 for a three-phase line the following modal equations apply

$$i_{1a,2a}(t) = \frac{1}{Z_a} v_{1a}(t) + I_{1a,2a}(t - \tau_a) \tag{8.6.1a}$$

$$i_{1b,2b}(t) = \frac{1}{Z_b} v_{1b}(t) + I_{1b,2b}(t - \tau_b) \tag{8.6.1b}$$

$$i_{1c,2c}(t) = \frac{1}{Z_c} v_{1c}(t) + I_{1c,2c}(t - \tau_c) \tag{8.6.1c}$$

where

$$I_{1a,2a}(t - \tau_a) = -\frac{1}{Z_a} v_{2a}(t - \tau_a) - i_{2a,1a}(t - \tau_a). \tag{8.6.2}$$

Then equations (8.6.1) are transformed back to the phase domain:

$$[i_{1,2}{}^{\text{phase}}] = [Z^{\text{phase}}]^{-1} [v_1{}^{\text{phase}}] + [I_{1,2}{}^{\text{phase}}] \tag{8.6.3}$$

where

$$[Z^{\text{phase}}]^{-1} = [T_i][Z^{\text{mode}}]^{-1}[T_v]^{-1} \tag{8.6.4}$$

$$[I_{1,2}{}^{\text{phase}}] = [T_i][T_{1,2}{}^{\text{mode}}]. \tag{8.6.5}$$

The transformations $[T_i]$ and $[T_v]$, assumed real, are defined in Appendix III.

Matrix equation (8.6.3) is incorporated in the nodal analysis (equation (8.4.7)) similarly to the case of the single-phase line (equation (8.2.8)). However the 3×3 conductance matrix $[Z^{\text{phase}}]^{-1}$ contains nine elements instead of one. Also, on the right-hand side, vector $[I_{1,2}{}^{\text{phase}}]$ has three elements. Although the nodal equations are in phase quantities the past history must be recorded in modal quantities.

When the three-phase line is perfectly balanced, simple transformation matrices can be used to decouple the line equations; a commonly used transformation is the $\alpha, \beta, 0$ components. The assumption of perfect symmetry is realistic in cases of two conductors, such as h.v.d.c. lines. It is also often justified for three-phase lines when line transpositions are used. However transpositions are designed to balance the fundamental (or power) frequency, whereas during transient conditions many other frequencies are generated and in such cases the presence of transpositions may even increased the asymmetry of the line [7]. Moreover, parallel three-phase transmission lines cannot be assumed balanced either. In general, therefore, appropriate eigenvector matrices as described in Appendix III must be found for each particular case.

8.7 FREQUENCY DEPENDENCE

To a lesser or greater extent the equivalent models of all power plant components are affected by frequency dependence. In practice, the need for frequency dependence models is restricted to the transmission lines, particularly for the ground-return mode because the earth impedance is highly dependent on frequency.

A solution in the time domain is needed in transient studies involving switchings and nonlinearities, as explained earlier. The steady-state behaviour of a multiconductor transmission line at a discrete frequency has been described (Appendix III) by the phasor equations

$$-\left[\frac{dV}{dx}\right] = [Z'] \cdot [I]$$

$$-\left[\frac{dI}{dx}\right] = [Y'] \cdot [V].$$

These equations apply to any frequency; it is possible to use superposition and Fourier transformation to derive the time response from the individual responses at each frequency.

In Fourier transformations, the frequency spectrum of the output function is obtained by multiplying the frequency spectrum of the input function by the transfer function. This multiplication can be converted to the time domain by means of the convolution theorem, which makes it possible to analyse the problem in the frequency domain.

The following efficient convolution formulation [8], compatible with the electromagnetic transient program, has been designed by Snelson [9]:

$$b_k = v_k - Z_1 i_k \qquad (8.7.1a)$$

$$b_m = v_m - Z_1 i_m \qquad (8.7.1b)$$

$$f_k = v_k + Z_1 i_k \qquad (8.7.1c)$$

$$f_m = v_m + Z_1 i_m \qquad (8.7.1d)$$

where

$$Z_1 = \lim_{\omega \to \infty} Z(j\omega). \qquad (8.7.2)$$

For a transmission line between nodes k and m the following integrals are needed at each time step:

$$b_k(t) = \int_0^\infty \{a_1(u) f_m(t-u) + a_2(u) f_k(t-u)\} du \qquad (8.7.3a)$$

$$b_m(t) = \int_0^\infty \{a_1(u) f_k(t-u) + a_2(u) f_m(t-u)\} du \qquad (8.7.3b)$$

where $a_1(u)$ and $a_2(u)$ are weighting functions which are precalculated by inverse Fourier transformation. The simple nonrepetitive form of these weighting functions [10] makes the numerical integration of equation (8.7.3) easy.

With b_k and b_m known at each time step, equation (8.7.2) provides two linear algebraic equations.

For the transmission line branch between nodes k and m equation (8.2.8a) would be replaced by

$$v_k(t) - Z_1 i_{k,m}(t) = b_1(t) \tag{8.7.4}$$

which maintains the conventional form.

However, the model described applies only to single-phase lines or balanced multiphase lines for which a modal transformation derives the ground-return mode to which the frequency dependence adjustment must be applied.

When line symmetry cannot be assumed, the solution requires first the derivation of eigenvalues and eigenvectors at different frequencies. This information is then used to change the scalar multiplications $a(u)f(t-u)$ to matrix vector products [9]. This process is far more complicated and approximate frequency-independent matrices should be used if acceptable results can be achieved.

8.8   ILLUSTRATIVE STUDIES

8.8.1   Line Energisation

This study and the field test comparisons were carried out on the Brazilian Jaguara–Taquaril transmission system [11], a simplified sketch of which is illustrated in Fig. 8.14.

The transmission line, of nominal voltage 345 kV, is 398 km long, transposed and uses twin bundle conductors (2 × 954 MCM-ACSR) protected with two shield wires of transposed galvanised steel (EHS-$\frac{3}{8}$″). The configuration of the conductors on the tower is shown in Fig. 8.15 and an equivalent circuit of the test system in Fig. 8.16. The distributed parameters of the line sequence components (calculated to 60 Hz)

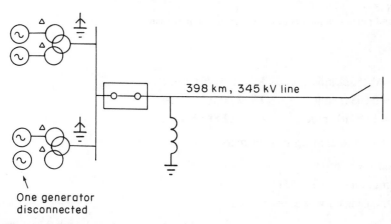

One generator
disconnected

Figure 8.14
Simplified sketch of the test system [11]

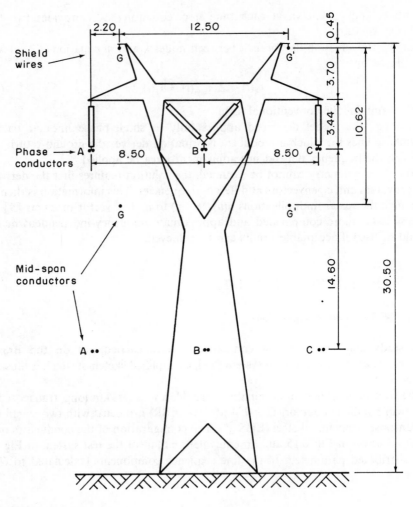

Figure 8.15
Shield and phase conductors at the tower and in mid-span

were

$$R^+ = 0.255 \,\Omega/\text{mile} \qquad R^0 = 0.5178 \,\Omega/\text{mile}$$
$$X^+ = 0.603 \,\Omega/\text{mile} \qquad X^0 = 2.0385 \,\Omega/\text{mile}$$
$$C^+ = 18.99 \,\text{nF/mile} \qquad C^0 = 12.88 \,\text{nF/mile}.$$

Generators and step-up transformers:

$$X_s(\text{self}) = 77.65 \,\Omega$$
$$X_m(\text{mutual}) = -21.95 \,\Omega.$$

Magnetising impedance neglected.

Prior to energisation the voltage was 328 kV (or 0.95 p.u.) and this voltage was used as the internal e.m.f. behind subtransient reactance. A three-phase reactor was

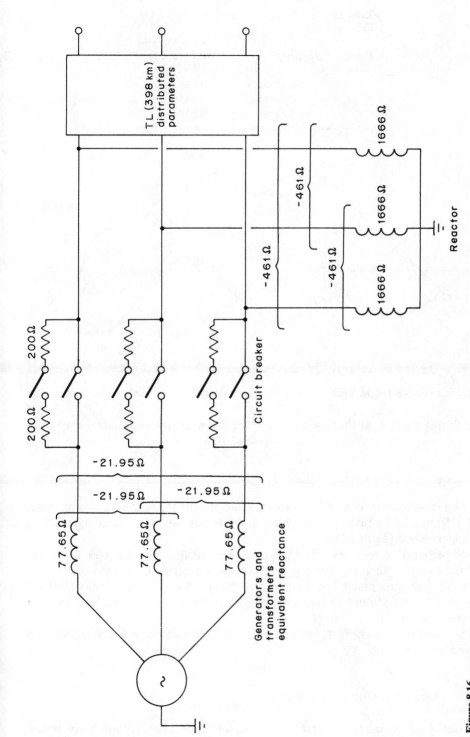

Figure 8.16
Equivalent test circuit

Table 8.1
Closing times

Phase	Auxiliary contacts	Main contacts
A	8.45 ms	15.85 ms
B	7.15 ms	14.45 ms
C	8.10 ms	15.10 ms

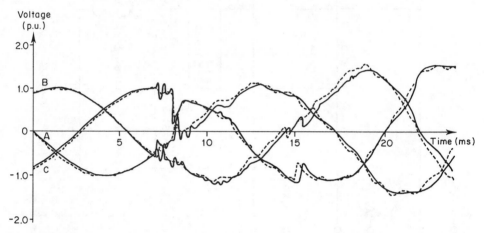

Figure 8.17
Energisation transient of the Jaguara–Taquaril line: recorded (full curve); calculated (broken curve)

used during the tests at the sending end of the line with self- and mutual reactances of

$$X_s = 1666\,\Omega$$
$$X_m = -461\,\Omega.$$

From close examination of the oscillographic records the following closing times, with reference to the instant when phase A passes through zero and going negative, were determined (Table 8.1).

The preinsertion resistors, $400\,\Omega$ per pole, were divided in two halves, to the left and right of the switch respectively, to avoid the connection of two switches to one node, which the program did not permit. An average value of soil resistivity of $100\,\Omega\,\text{m}$ was used for the ground return, a relatively low value because the line does not traverse arid regions.

The comparison of calculated results with actual tests, shown in Fig. 8.17, indicates very close agreement.

8.8.2 Transient Recovery Voltage

Calculated versus test results were carried out for the recovery voltage in the system of Fig. 8.16 following a short-circuit fault set on phase A at 0.75 miles from the

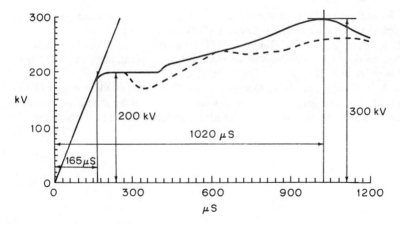

Figure 8.18
Transient recovery voltage: recorded (full curve); calculated (broken curve)

sending end. The circuit breaker cleared the fault in four cycles and when the breaker opened, the effective r.m.s. value of the short-circuit current was 6450 A, with an offset of 7% due to the d.c. component. The A-phase pole was the last to open because of the large fault current (the other phases only carried line charging currents).

The reactances of the generators, transformers and shunt reactors were as described in the previous section. For the present study the capacitance of the transformer banks were also required (which for the Jaguara plant was $0.02\,\mu F$ per phase).

The calculated recovery voltage is illustrated in Fig. 8.18 by the broken curve, and the recorded wave by the full curve. The two curves were practically the same for the first $300\,\mu s$ and beyond that time the largest observed difference between them is 16%.

8.9   REFERENCES

[1] L. Bergeron, 1949. *Du Coup de Belier en Hydraulique au Coup de Foudre en Electricite* Dunod, Paris (Translated version: Wiley, New York 1961).

[2] H. W. Dommel, 1969. Digital Computer Solution of Electromagnetic Transients in Single and Multiple Networks, *IEEE Trans.* **PAS-88** 388–99.

[3] R. J. Byerly and T. M. McCauley, 1969. Mathematical models and computing techniques for generator and transmission line voltages following load rejection, *Proc. 6th IEEE PICA Conf.* pp. 410–425.

[4] J. Arrillaga, C. P. Arnold and B. J. Harker, 1983. *Computer Modelling of Electrical Power Systems* Wiley & Sons, London.

[5] B. Stott and E. Hobson, 1971. Solution of large power-system networks by ordered elimination: a comparison of ordering schemes *Proc. IEE* **118** 125–134.

[6] E. J. Dolan, D. A. Gillies and E. W. Kimbark, 1972. Ferroresonance in a transformer switched with an EHV line, *IEEE Trans.* **PAS-91** 1273–80.

[7] J. Arrillaga, E. Acha, T. J. Densem and P. S. Bodger, 1986. Ineffectiveness of transmission line transpositions at harmonic frequencies, *Proc. IEE* **133** 99–104.

[8] A. Budner, 1970. Introduction of frequency-dependent line parameters into an Electromagnetic Transients Program, *IEEE Trans.* **PAS-89** 88–97.

[9] J. K. Snelson, 1972. Propagation of travelling waves on transmission lines. Frequency dependent parameters *IEEE Trans.* **PAS-91** 85–91.

[10] W. S. Meyer and H. W. Dommel, 1974. Numerical modelling of frequency-dependent transmission line parameters in an Electromagnetic Transients Program, *IEEE Winter Power Meeting, New York* Paper T74080-8.

[11] C. A. F. Cunha and H. W. Dommel, 1973. Computer simulation of field tests on the 345 kV Jaguara-Taquaril line, *II Seminario Nacional de Producao e Transmissao de Energia Eletrica, Belo Horizonte, Brazil* Paper BH/GSP/12 (in Portuguese).

9. ANALYSIS OF HARMONIC PROPAGATION

9.1 INTRODUCTION

This chapter describes the parameters and techniques involved in the derivation of the network harmonic impedances. This information constitutes the basis of harmonic penetration studies, which involve the computation of harmonic currents and voltages throughout the a.c. system in the presence of one or more current harmonic sources.

Although most of the existing computer algorithms still use a single-phase model, for accurate harmonic frequency analysis three-phase modelling of the power system is necessary. The harmonic injections may in general be unbalanced and the transmission system will always include impedance imbalance and circuit coupling.

In common with the load-flow algorithms discussed in previous chapters the analysis of harmonic penetration uses the nodal admittance matrix and linear transformation techniques to interconnect the various plant components of a network represented by their equivalent circuits.

9.2 TRANSMISSION LINE MODELS

The derivation of series and shunt impedances of a three-phase transmission line has been described in Chapter 3 (Section 3.3) with reference to load-flow analysis. The following relationships were arrived at:

$$[\Delta V_{abc}] = [Z_{abc}] \cdot [I_{abc}] \tag{9.2.1}$$

$$[V_{abc}] = [P'_{abc}] \cdot [Q_{abc}]. \tag{9.2.2}$$

Although the form of the equations remains the same, some modifications are required for the series impedances in the harmonic models.

The self-impedance per kilometre of conductor 'a' with earth return (Z_{aa}), and the mutual impedance per kilometre between conductors 'a' and 'b' (Z_{ab}) are expressed as [1]

$$Z_{aa} = R_a + R_g + j(X_{aa} + X_g) \tag{9.2.3}$$

$$Z_{ab} = R_g + j(X_{ab} + X_g) \tag{9.2.4}$$

where R_a is the a.c. resistance of conductor 'a', X_{aa} is the self-reactance of conductor 'a', X_{ab} is the mutual reactance between conductors 'a' and 'b', and R_g, X_g are Carson's earth-return corrections [2]. The effect of earth resistivity (ρ) on the self-reactance X_{aa} can be assessed from the approximate expression

$$Z_{aa} = 0.00289 f \log\left(\frac{660\sqrt{\rho/f}}{\text{GMR}}\right) \quad \Omega/\text{km}. \tag{9.2.5}$$

For long lines skin effect can have considerable influence on the resonant voltage level. A practical method of calculating the skin effect resistance ratios has been suggested by Lewis and Tuttle [3] by approximating ACSR conductors to uniform tubes having the same inside and outside diameters as the aluminium conductors. That method has been used to calculate the ratios plotted in Fig. 9.1.

The skin effect is demonstrated in Fig. 9.2 for the case of the 220 kV, 230 km line illustrated in Fig. 9.3. The vertical broken line in Fig. 9.1 indicates the skin effect ratio for that line at the half-wavelength resonant frequency.

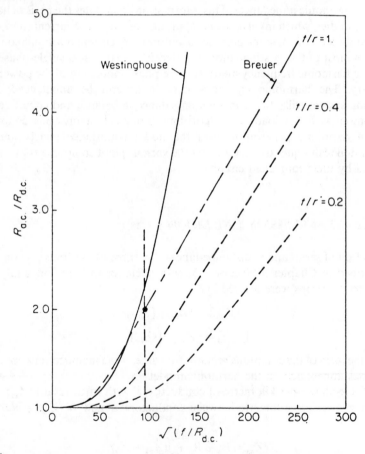

Figure 9.1
Skin effect resistance ratios for different models. The full circle indicates the skin effect ratio for the Islington to Kikiwa line at the half-wavelength resonant frequency. Broken curves indicate ACSR conductors with various tube ratios

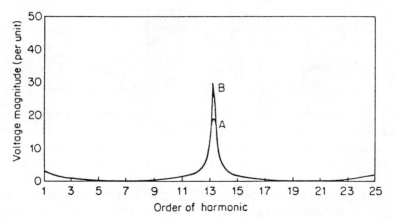

Figure 9.2
The effect of skin effect modelling: curve A, skin effect included; curve B, no skin effect

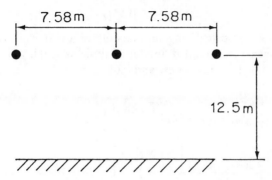

Figure 9.3
Conductor information for the Islington to Kikiwa line: conductor type, Zebra (54/3.18 + 7/3.18); length, 230 km; resistivity, 100 Ω m

9.2.1 The Equivalent-π Model

For the purpose of harmonic penetration studies the series-shunt nominal π representation of the line is inaccurate and an equivalent-π model is used instead [4].

The equivalent-π model, illustrated in Fig. 9.4 for the case of a single-phase line, is obtained from the nominal π model by applying correction factors to the series impedance and shunt admittance, i.e.

$$\frac{\sinh(x\sqrt{Z'Y'})}{x\sqrt{Z'Y'}} \qquad \text{for the series impedance} \qquad (9.2.6)$$

$$\frac{\tanh(x\sqrt{Z'Y'}/2)}{x\sqrt{Z'Y'}/2} \qquad \text{for the shunt admittance.} \qquad (9.2.7)$$

In the case of a multiconductor transmission line, the nominal π series impedance

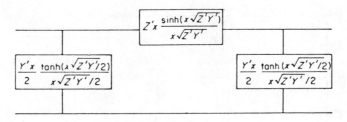

Figure 9.4
The equivalent-π model of a long transmission line

and shunt admittance matrices per unit distance $[Z']$ and $[Y']$ are square and their size is fixed by the number of mutually coupled conductors.

The derivation of the equivalent-π model for harmonic penetration studies is similar to that of the single-phase lines, except that it involves the evaluation of hyperbolic functions of the propagation constant which is now a matrix:

$$[\gamma] = ([Z'][Y'])^{1/2}. \qquad (9.2.8)$$

There is no direct way of calculating sinh or tanh of a matrix, thus a method using eigenvalues and eigenvectors, called 'modal analysis' is employed [5] to derive the following expressions for the series and shunt impedance components of the equivalent-π model:

$$[Z]_{\text{EPM}} = l[Z'] \cdot [M] \cdot \left[\frac{\sinh \gamma l}{\gamma l} \right] \cdot [M]^{-1}$$

$$[Y]_{\text{EPM}} = l[M] \cdot \left[\frac{\tanh(\gamma l/2)}{\gamma l/2} \right] \cdot (M)^{-1} \cdot [Y^1]$$

where
    l is the transmission line length
    $[M]$ is the matrix of normalised eigenvectors
    γ are the eigenvalues of the mutually coupled circuits.

A detailed description of the modal analysis method is given in Appendix III.

9.3 TRANSFORMER MODELS

The representation of transformer impedances at fundamental frequency has been discussed in Chapters 2 and 3. Section 2.2 described the single-phase model suitable for symmetrical load-flows and Section 3.4 the three-phase models needed for unbalanced studies. The parameters of these models need to be modified to take into account frequency dependence.

As the internal resonant frequencies of high-voltage power transformers occur well above the range of interest for harmonic penetration studies, the interwinding capacitances and capacitances to ground of transformers have very little effect on the accuracy of the results.

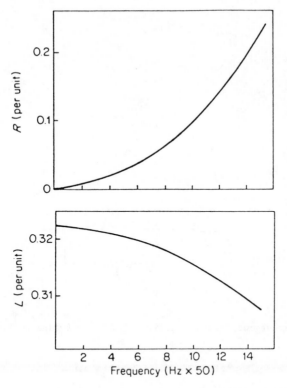

Figure 9.5
Frequency dependence of transformer model: (a) per unit resistance versus frequency; (b) per unit inductance versus frequency (note supressed zero of scale)

The frequency dependence of the resistance accounts for the increased transformer core losses with frequency due to skin effect. Fig. 9.5 shows the change in resistance and inductance with frequency for a practical transformer.

Assuming that transformers are not operated in saturation, various representations have been suggested to replace the leakage inductance. These are shown in Fig. 9.6.

In Fig. 9.6(a), X_{50} is the leakage reactance at 50 Hz [6]. In Figure 9.6(b), $R = 0.1026\,khX_{50}\,(J + h)$ where J is the ratio of hysteresis to eddy current losses, taken as 3 for silicon steels, and $k = 1/(J + 1)$. As an alternative model the values of R and X are scaled to 80% of the values at 50 Hz [7]. In Fig. 9.6(c), $90 < V^2/SR_s < 110$ and $13 < SR_p/V^2 < 30$, with S being the rated power of the transformer. Typical values (per unit) of R_s and R_p are 0.04 and 60 for a 30 MVA transformer and 0.01 and 20 for the case of a 100 MVA transformer.

Considering the wide range of models, further work is clearly needed in this area to provide more specific information related to particular transformer ratings and characteristics.

Whenever the effect of transformer magnetic nonlinearity is considered relevant, the magnetising current harmonics must be calculated and represented as current-injecting sources.

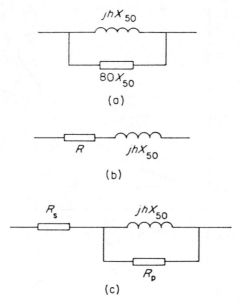

Figure 9.6
Transformer models suggested for harmonic penetration (where h is the harmonic order)

9.4   REPRESENTATION OF SYNCHRONOUS MACHINES

In the presence of rotor saliency a generator becomes a frequency convertor [8]. However in practice the harmonic levels produced by frequency conversion are not significant and generally it can be assumed that synchronous generators produce no harmonic voltages. They can therefore be modelled by a shunt impedance at the generator terminals.

A linear reactance derived from either the subtransient or negative sequence inductances is often used [9], both having similar values.

In the absence of a more generally accepted model, an empirical linear model is suggested which consists of the full subtransient reactance with a power factor of 0.2.

9.5   LOAD MODELLING

When carrying out harmonic penetration studies in transmission systems, it is not usual to represent the system from generators right through to individual consumer loads. At some point down the network the elements are aggregated into an equivalent circuit. Typically, equivalent circuits are used at the points of supply (POS) to distribution authorities, who reticulate power to individual consumers within the load centres.

The methods available for determining the equivalent harmonic impedance of supply authority networks are as follows.

(i) Direct measurement, performed at a sufficient number of frequencies to enable satisfactory interpolation. Limitations in measurement techniques make this method very time consuming and difficult, especially for a number of points of supply.

(ii) Derivation of component characteristics, i.e. motors and industrial plant, by using statistical diversity data. This approach, while difficult, is under consideration for system stability studies [10] and could be extended to harmonic studies.

(iii) Use of the known fundamental frequency real and reactive power flow at the point of supply.

There is considerable variation in impedance with frequency and load level for industrial and domestic customers. Moreover, industrial loads often have capacitors installed for power factor compensation which can cause series and parallel resonances. Various models [6, 11, 12] have been proposed for consumer loads, some of them relating to individual components and others as component aggregate models.

Various suggested combinations of the real and reactive power demand at fundamental frequency are shown in Fig. 9.7. Converting these models into a suitable form for inclusion into the system admittance matrices is straightforward.

- In model A, suggested by Pesonen et al [6], h is the harmonic order, V the nominal voltage and $k = 0.1h + 0.9$.
- In model B the reactance is assumed to be frequency dependent while the parallel resistance is kept constant.

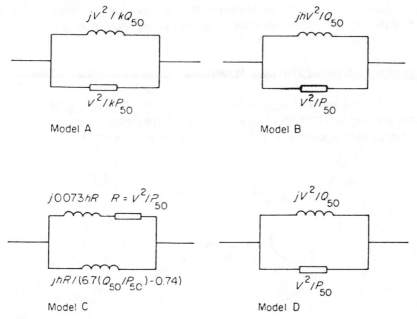

Model A   Model B

Model C   Model D

Figure 9.7
Load models for harmonic penetration studies

- Model C was derived by measurements on medium voltage loads using audiofrequency ripple control generators [6].
- Finally in model D the load impedance, calculated at 50 Hz, remains constant for all frequencies [13].

9.6 ALGORITHM DEVELOPMENT

The requirements to be met for accurate harmonic modelling are as follows.

- Transmission lines must be represented with provision for skin effect and standing wave phenomena.
- Load, transformer, generator, shunt capacitor and filter models should be included.
- Nodal admittance matrices should be formed for any range of frequencies and not restricted to harmonic multiples of the fundamental.
- It should be possible to calculate system impedances at any busbar.
- The possibility of current injections at multiple locations in the system needs to be considered.
- The network (assumed linear and passive) must be solved to obtain system voltages at all nodes for all frequencies.
- Line current flows should be calculated at each frequency.
- Output data need to be plotted to make interpretation easier.

These requirements use standard power system techniques involving the solution of simultaneous linear equations. However, the nature of the problem will determine which of the above features will need to be used in any particular study.

9.6.1 Balanced Harmonic Penetration

In Figure 9.8 two sets of balanced harmonic currents, I_{1h} and I_{2h}, of order h, are injected into any two busbars of an a.c. system; a large power system is likely to have a number of such injections. It is assumed that the a.c. system is linear and passive

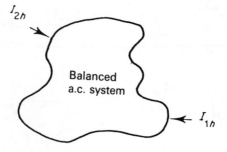

Figure 9.8
Balanced current injection into a balanced a.c. system

and therefore the principle of superposition may be applied to enable each harmonic to be considered independently.

The resultant system harmonic voltages are calculated by direct solution of the linear equation:

$$[I_h] = [Y_h][V_h] \tag{9.6.1}$$

where $[Y_h]$ is the system admittance matrix.

On the assumption of a balanced a.c. system, the model will only include the positive-sequence component impedances.

The above algorithm can model the steady-state behaviour of a power system but unfortunately the harmonic behaviour of a physical system changes as loads, generators and line configuration alter.

9.6.2 Unbalanced Harmonic Penetration

The three-phase nature of the power system always results in some load or transmission line asymmetry as well as circuit coupling. These effects give rise to unbalanced self- and mutual admittances of the network elements.

A more accurate representation of the unbalanced conditions is illustrated in Fig. 9.9. The current injections, i.e. $I_{1h}-I_{3h}$ and $I_{4h}-I_{6h}$, can be unbalanced in magnitude and phase angle. In a similar manner to the balanced system, the current injections for each frequency are presumed constant and known, and the linear equation (9.6.1) is solved directly to obtain the three-phase harmonic voltages.

For the three-phase system, the elements of the admittance matrix are themselves 3×3 matrices consisting of self- and transfer admittances. Fig. 9.10 indicates the nature of the analysis where h sets of linear equations are solved.

The injected currents at most a.c. busbars will be zero, since the sources of the harmonics considered are generally from static convertors. To calculate an admittance matrix for the reduced portion of a system comprising just the injection busbars, it is necessary to form the admittance matrix with those buses at which harmonic injection occurs, ordered last. Advantage is taken of the symmetry and sparsity of the admittance matrix [14], using a row-ordering technique to reduce the amount

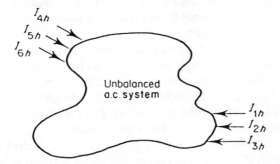

Figure 9.9
Unbalanced current injection into an unbalanced a.c. system

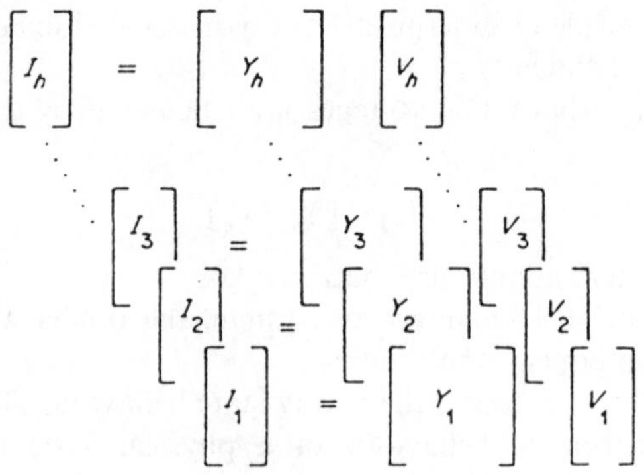

$$\begin{bmatrix} I_h \end{bmatrix} = \begin{bmatrix} Y_h \end{bmatrix} \begin{bmatrix} V_h \end{bmatrix}$$

$$\begin{bmatrix} I_3 \end{bmatrix} = \begin{bmatrix} Y_3 \end{bmatrix} \begin{bmatrix} V_3 \end{bmatrix}$$

$$\begin{bmatrix} I_2 \end{bmatrix} = \begin{bmatrix} Y_2 \end{bmatrix} \begin{bmatrix} V_2 \end{bmatrix}$$

$$\begin{bmatrix} I_1 \end{bmatrix} = \begin{bmatrix} Y_1 \end{bmatrix} \begin{bmatrix} V_1 \end{bmatrix}$$

Figure 9.10
Solution of h sets of linear simultaneous equations

of off-diagonal element build-up. The matrix is triangulated using Gaussian elimination, down to but excluding the rows of the specified buses.

The resulting matrix equation for an n-node system with $n - j + 1$ injection points is

$$\begin{bmatrix} 0 \\ \vdots \\ 0 \\ \hline I_j \\ \vdots \\ I_n \end{bmatrix} = \begin{bmatrix} & & \\ & & \\ & & \\ \hline & & Y_{jj} \cdots Y_{jn} \\ 0 & & \vdots \ddots \vdots \\ & & Y_{nj} \cdots Y_{nn} \end{bmatrix} \cdot \begin{bmatrix} V_1 \\ \vdots \\ V_{j-1} \\ \hline V_j \\ \vdots \\ V_n \end{bmatrix} \qquad (9.6.2)$$

As a consequence, $I_j \cdots I_n$ remain unchanged since the currents above these in the current vector are zero. The reduced matrix equation is

$$\begin{bmatrix} I_j \\ \vdots \\ I_n \end{bmatrix} = \begin{bmatrix} Y_{jj} & \cdots & Y_{jn} \\ \vdots & \ddots & \vdots \\ Y_{nj} & \cdots & Y_{nn} \end{bmatrix} \cdot \begin{bmatrix} V_j \\ \vdots \\ V_n \end{bmatrix} \qquad (9.6.3)$$

and the order of the admittance matrix is three times the number of injection busbars. The elements are the self- and transfer admittances of the reduced system as viewed from the injection busbars. Whenever required, the impedance matrix may be obtained for the reduced system by matrix inversion.

Reducing a system to provide an equivalent admittance matrix is an essential part of filter design where the system, as viewed from a specific bus, is required; it is also useful where a number of converters are connected to the a.c. system at different points, as in Fig. 9.11. In this example the reduced admittance matrix is of order 9.

In harmonic penetration studies the currents from the converters are assumed to be known. In general, however [15] any voltage distortion present at the convertor terminals affects the firing angles and hence the harmonic current injection into the system. The solution of this problem is iterative and not suited to the large matrices associated with the a.c. system. However, during each iteration only the converter terminal voltages are required. These can be obtained in the example above by

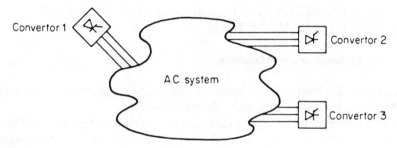

Figure 9.11
Three converters attached to different busbars on the a.c. system

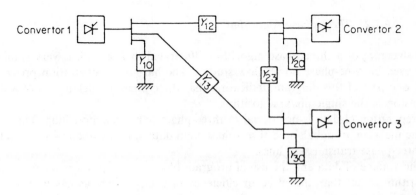

Figure 9.12
Reduced three-converter system

reducing the a.c. system to a three-bus equivalent system for the three converters, as indicated in Fig. 9.12 where each of the admittances represents a 3×3 matrix.

Restricted measurements on the physical network limit the ability to compare a three-phase model with test results. The data obtained from live three-phase systems only includes the phase voltages and currents of the coupled phases; to compare measured and simulated impedances at a current injection busbar it is thus necessary to derive equivalent phase impedances from the 3×3 admittance matrix.

By making $I_1 = 1 \angle 0°$ p.u., $I_2 = 1 \angle -120°$ p.u., $I_3 = 1 \angle 120°$ p.u., the matrix equation

$$\begin{bmatrix} I_1 \\ I_2 \\ I_3 \end{bmatrix} = \begin{bmatrix} Y_{11} & Y_{12} & Y_{13} \\ Y_{21} & Y_{22} & Y_{23} \\ Y_{31} & Y_{32} & Y_{33} \end{bmatrix} \cdot \begin{bmatrix} V_1 \\ V_2 \\ V_3 \end{bmatrix} \tag{9.6.4}$$

can be solved for V_1, V_2 and V_3, yielding the following equivalent phase impedances:

$$Z_1 = \frac{V_1}{I_1} \qquad Z_2 = \frac{V_2}{I_2} \qquad Z_3 = \frac{V_3}{I_3}. \tag{9.6.5}$$

9.7 COMPUTATIONAL REQUIREMENTS OF HARMONIC PENETRATION ALGORITHMS

9.7.1 Single-phase Modelling

The structure of a single-phase harmonic penetration algorithm is illustrated in Fig. 9.13, which involves simple and efficient software.

Only component data from the fundamental frequency system are used to derive information at harmonic frequencies. This information is held in the system data base and its processing requires very little extra work. Storage is only required for the nonzero elements of a single admittance matrix and this information is reformed for each frequency.

9.7.2 Three-phase Algorithm

The structure of a three-phase algorithm, illustrated in Fig. 9.14, is very similar to that used in three-phase power-flow studies. The harmonic penetration program is only one part of this diagram, indicating that three-phase modelling is not a direct extension of the single-phase algorithm.

Preparation of data is not trivial in three-phase harmonic modelling. This is due to the inclusion of unbalanced three-phase load data, and the frequency dependence of three-phase transmission lines.

The volume of data and the use of program blocks are of primary concern to this algorithm rather than the speed or efficiency of computation, as has been the case in the development of algorithms for power flow or transient stability simulation. The number of separate program blocks is a function of the multiple use of software, having regard for practical program debugging and maintenance.

The first block of Fig. 9.14 calculates the transmission line parameters for each frequency over a required range, using the equivalent-π model. The second program block completes the database by reading line data from the first and adding it to the balanced load and other component data required.

Data formation for both the harmonic penetration and three-phase power-flow studies is performed by the same software.

The three-phase a.c./d.c. power flow provides sufficient information of the converter operating state [16] to derive the harmonic current injections. The current injections from a number of convertors connected at any busbars in the a.c. system are then available for the analysis of the penetration of these harmonic currents into the a.c. system.

9.7.3 Three-phase Harmonic Penetration

A three-phase harmonic penetration program, illustrated in the structure diagram of Fig. 9.15, should include the following features:

● provision for the representation of three-phase mutually coupled transmission line data

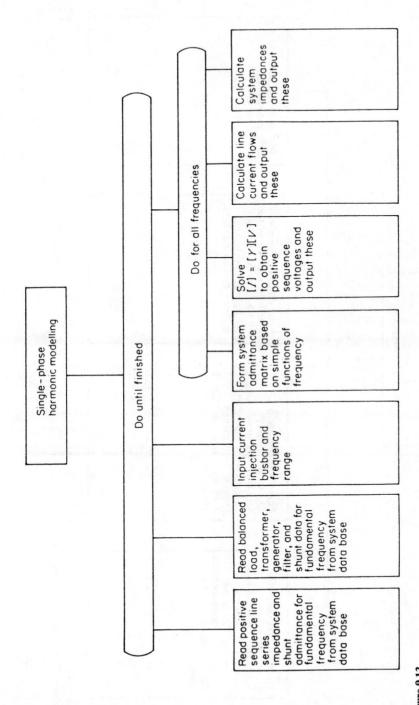

Figure 9.13
Structure diagram of single-phase modelling

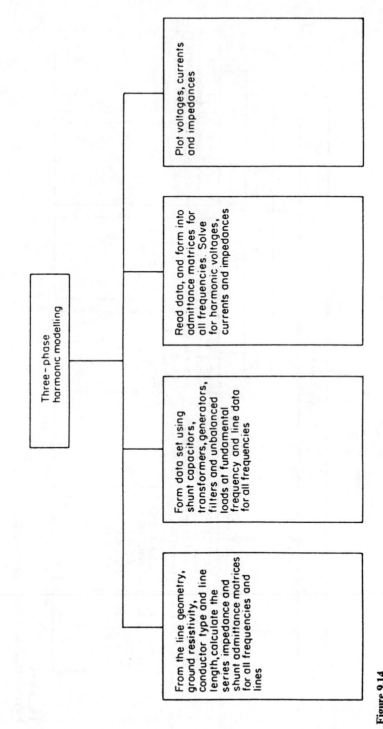

Figure 9.14
Structure diagram of three-phase modelling

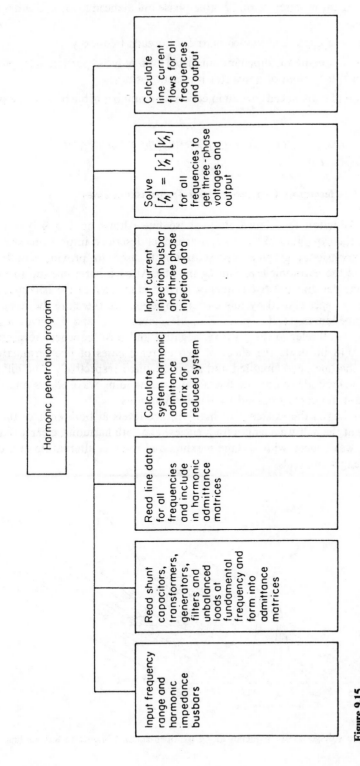

Figure 9.15
Structure diagram of three-phase harmonic penetration program

- provision for the representation of three-phase unbalanced loads and other system components
- formation of separate admittance matrices for each frequency
- derivation of three-phase impedance matrices for a reduced portion of the network, suitable for filter design or converter interaction studies.
- specification of unbalanced current injections at a number of busbars on the system.

9.8 APPLICATION OF THE HARMONIC PENETRATION ALGORITHM

9.8.1 Harmonics Generated along Transmission Lines

The 220 kV Islington to Kikiwa three-phase line (shown in Fig. 9.3) is used to demonstrate the capability of the computer model described in previous sections.

A three-dimensional graphic representation is used to provide simultaneous information of the harmonic levels along the line. At each harmonic (up to the 25th harmonic), one per unit positive-sequence current is injected at the Islington end of the line. The voltages caused by this current injection are therefore the same as the calculated impedances, i.e. V_+ gives Z_{++}, V_- gives Z_{+-} and V_0 gives Z_{+0} (the subscripts $+$, $-$, 0 refer to the positive, negative and zero sequences respectively).

Figs 9.16–9.18 illustrate the effect of two extreme cases of line termination (at Kikiwa), i.e. the line open-circuited and short-circuited respectively. The differences in harmonic magnitudes along the line are due to standing wave effects and shifting of the resonant frequencies caused by line terminations.

Fig. 9.16 indicates the existence of high voltage levels at both ends of the open-circuited line at the half-wavelength frequencies. The 25th harmonic clearly illustrates the standing wave effect, with voltage maxima and minima alternating at a quarter of the wavelength intervals.

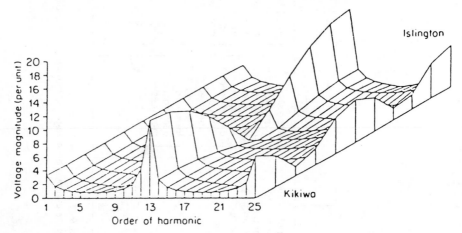

Figure 9.16
Positive-sequence voltage versus frequency along the open-ended Islington to Kikiwa line

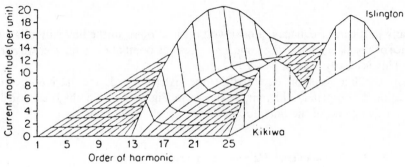

Figure 9.17
Positive-sequence current along the open-ended line for a 1 per unit positive-sequence current. injection at Islington

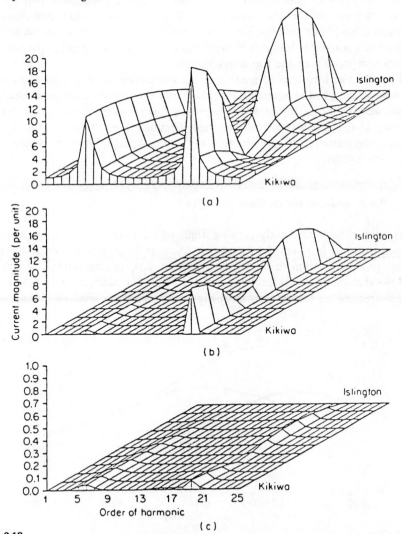

Figure 9.18
Sequence currents along the short-circuited line for a 1 per unit positive-sequence current injection at Islington: (a) positive-sequence current; (b) negative-sequence current; (c) zero-sequence current

At any particular frequency, a peak voltage at a point in the line will indicate the presence of a peak current of the same frequency at a point about a quarter wavelength away. This is clearly seen in Fig. 9.17.

When the line is short-circuited at the extreme end, the harmonic current penetration is completely different, as shown in Fig. 9.18(a). The high current levels at the receiving end of the line are due to the short-circuit condition.

9.8.2 Zero-sequence Harmonics in Transmission Lines connected to Static Converters

It is the zero-sequence penetration, rather than the positive sequence, that provides relevant information for the assessment of possible harmonic interference in neighbouring telephone systems. The presence of zero sequence in a transmission line connected to a convertor bridge is entirely due to asymmetries in either the convertor a.c. plant components or the transmission line itself.

In Fig. 9.18 the locations of maximum zero-sequence current (plot (c)) coincide with those of the positive sequence (plot (a)), and the highest level produced in the test line, about 10% of the injected positive sequence current, occurs at the 19th harmonic, at the Kikiwa end of the short-circuited line. However, the levels of zero-sequence current are low (notice the scale change between positive- and zero-sequence plots).

9.8.3 Differences in Phase Voltages

In conventional harmonic analysis using single-phase positive-sequence models [17], a transmission line is assumed to have one resonant frequency. However the use of the three-phase algorithm to model the Islington–Kikiwa unbalanced transmission line shows that the resonant frequencies are different for each phase. In this case the spread of frequencies can be seen from Fig. 9.19 to be approximately 6 Hz.

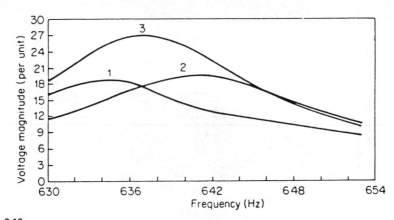

Figure 9.19
Three-phase resonant frequencies of the Islington to Kikiwa line with a 1 per unit positive-sequence current injection (skin effect included)

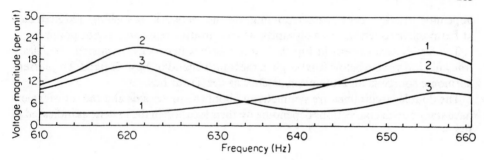

Figure 9.20
Three-phase resonant frequencies for the transposed line

The different magnitudes of the resonant frequencies (up to 30%) of the three-phases partly explains the problems encountered with correlating single-phase modelling and measurement on the physical network. The results clearly indicate that harmonics in the transmission system are unbalanced and three-phase in nature.

Normal transposition of a transmission line into three equal length sections, to balance the line at fundamental frequency, can have a detrimental effect at harmonic frequencies. For instance the modelling of transpositions in the Islington to Kikiwa line produces the results illustrated in Fig. 9.20, which shows the existence of two resonant peaks separated by almost 40 Hz for the half wavelength, (i.e. at 620 and 656 Hz for phases 3 and 1 respectively).

9.8.4   Harmonic Impedances of an Interconnected System

This section considers the progressive formation of the harmonic impedances of an interconnected system from the individual component characteristics. This will

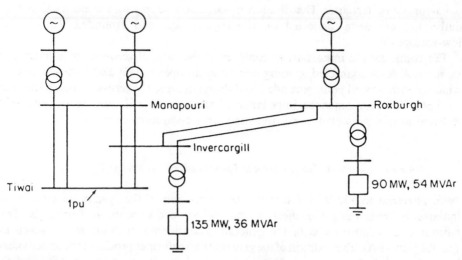

Figure 9.21
Test system including load and generation

hopefully provide some understanding of the network modelling requirements at harmonic frequencies, in a situation where intuitive reasoning is not possible.

The test system, shown in Fig. 9.21, is a nine-bus network comprising the 220 kV transmission system below Roxburgh in the South Island of New Zealand. The current harmonic source is an aluminium smelter at the Tiwai bus.

The double circuit lines are symmetrical about the tower axis and the transformers have star or delta connections depending on their location in the system, as indicated in Fig. 9.22.

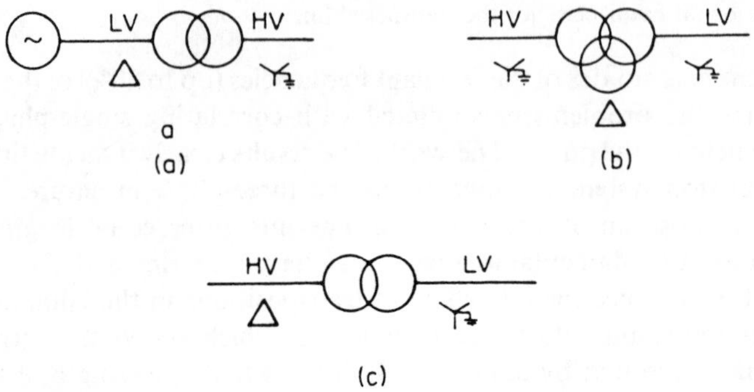

Figure 9.22
Transformer connections: (a) generating station; (b) transmission substation; (c) distribution substation

Generator transformers have deltas on the generator or low-voltage, side and grounded star connection on the high-voltage side. Transmission substation transformers have grounded star on the high-voltage and low-voltage windings with delta-connected tertiaries. Distribution transformers supplying the electrical supply authorities are delta-connected on the high-voltage and grounded star on the low-voltage side.

The connection is important in considering the flow of zero-sequence harmonic currents. A delta-connected winding will act as an open circuit and a star-connected winding, with neutral point grounded, as a short circuit to the zero-sequence harmonic currents. The zero-sequence impedance of the system will thus be considerably different to that presented to positive- or negative-sequence currents.

9.8.4.1   Generator, Transformer and Load Impedances at Roxburgh

With reference to Fig. 9.21 a step by step formation of the system impedances is initiated by examining the effect of the various components at Roxburgh. The harmonic impedance locus of the generator, considered in isolation, is shown in Fig. 9.23 (curve A). The addition of the generator transformer produces the impedances locus of curve B. Finally, curve C illustrates the damping effect of a 90 MW and 54 MVAR load connected through a transformer to the Roxburgh bus.

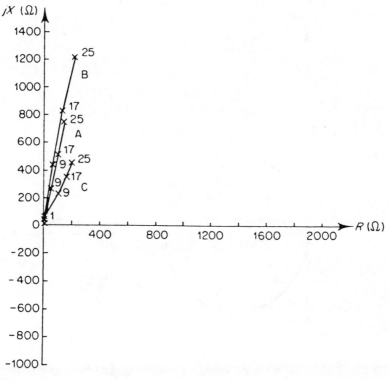

Figure 9.23
Polar plot of the generator, transformer and load impedances at Roxburgh: curve A, generator only; curve B, generator and generator transformer; curve C, generator, generator transformer, load (100%) and load transformer

9.8.4.2 Interconnection between Invercargill and Roxburgh

The double 220 kV transmission line between Invercargill and Roxburgh in isolation (i.e. open-circuited at Roxburgh) has the impedance locus of Fig. 9.24. At fundamental frequency the line is capacitive, although this is difficult to observe. As the frequency increases the line approaches a series resonance, at which point the impedance is very small and purely resistive, the phase angle becoming inductive. From this point the impedance increases in magnitude in a clockwise direction with increasing frequency. Somewhere between the 11th and 12th harmonics a parallel resonance occurs, manifested by a large and purely resistive impedance. As frequency increases further the line again becomes capacitive.

The effect of line termination is shown in Fig. 9.25 for a 1 p.u. harmonic injection at Invercargill. When the line is isolated (i.e. corresponding to the locus of Fig. 9.24) the per unit voltages of the various harmonics are illustrated in Fig. 9.25(a) which gives the voltage magnitudes. Figure 9.25(b) gives the voltage phase angles.

The same graphs show corresponding voltage magnitudes and phases when the line is short circuited (Fig. 9.25, curves B) loaded by the generator transformer group (Fig. 9.25, curves C) and by the complete system at Roxburgh (Fig. 9.25, curves D).

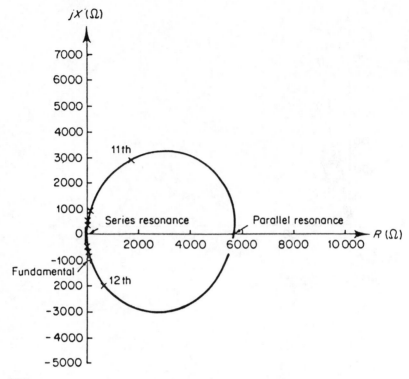

Figure 9.24
Polar plot of the impedance of the open-circuited Invercargill to Roxburgh lines with 50 Hz intervals
marked

9.8.4.3 Left-hand Side of the System

Referring now to the left-hand side of the system, with the lines between Invercargill
and Roxburgh open, Fig. 9.26 illustrates the effect of 1 p.u. harmonic current injections
at Tiwai. Curves A and B show the voltage spectra at Tiwai with the rest of the
system open- and short-circuited respectively.

When generation (Fig. 9.26, curve C) and the load (Fig. 9.26, curve D) are added
with the associated transformers, similar effects to the previous section are observed.
The resonant points lie between those of the open- and short-circuit cases, with
reduced magnitudes as compared with the extreme cases of termination.

9.8.4.4 Complete Test System

By combining the two individual systems considered in the two preceding subsections
the progressive formation of the test network (Fig. 9.21) is completed.

Fig. 9.27 compares the voltage magnitudes at the Tiwai bus for different loading
conditions. Curve A shows the effect of the transmission system in isolation (i.e. with
all the generators and loads disconnected); the resulting resonance frequencies of the
interconnected system do not correspond to those of the two individual parts (ii) and

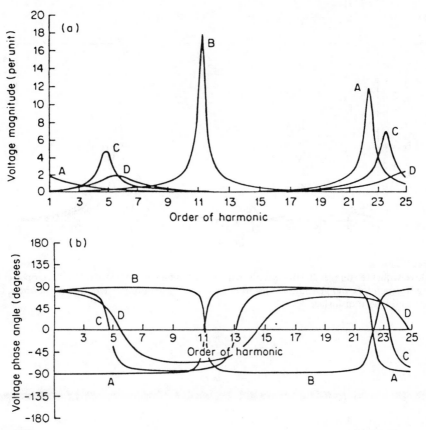

Figure 9.25
Positive-sequence voltages at Invercargill versus frequency for different terminations of the Roxburgh to Invercargill lines: (a) voltage magnitudes; (b) voltage phase angles. Curves A, open circuit; curves B, short circuit; curves C, generator and generator transformer; curves D, generator, generator transformer, load and load transformer

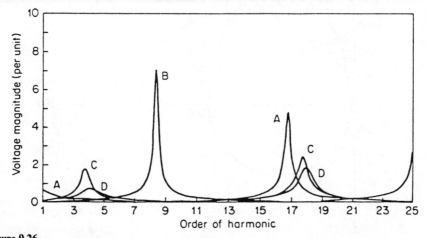

Figure 9.26
Positive-sequence voltage magnitude at Tiwai versus frequency for different terminations: curve A, open circuit; curve B, short circuit; curve C, generator and generator transformer; curve D, generator, generator transformer, load and load transformer

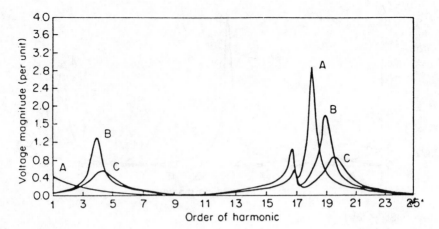

Figure 9.27
Positive-sequence voltages at Tiwai versus frequency for different terminations: curve A, open circuit; curve B, generators and generator transformers; curve C, generators, generators transformers, loads and load transformers

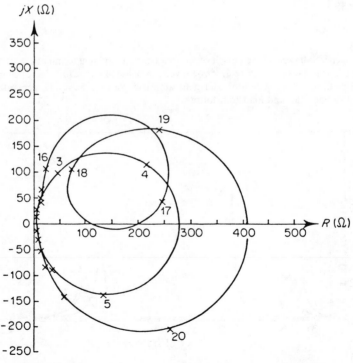

Figure 9.28
Positive-sequence impedances of the test system from Tiwai with harmonic intervals indicated

(iii) of the system. There are now two resonances around the 18th harmonic, one smaller in magnitude. The effect of this latter resonance is to create an extra loop in the impedance locus (as shown in Fig. 9.28).

Fig. 9.28 illustrates the progressive complexity of the impedance locus as the a.c. system increases.

9.8.4.5 Three-phase Impedances of the Test system at Tiwai

The unbalanced nature of the transmission network can be illustrated by plotting the three individual equivalent phase impedances. Fig. 9.29 shows that the imbalance is low at fundamental frequency, but increases towards the first parallel resonance which occurs between the 4th and 5th harmonics, where the magnitude differences are of the order of 30%. This effect is mainly caused by differences in the mutual impedances between phases, resulting from the asymmetry in transmission line conductor geometries.

The series resonance at the 11th harmonic exhibits low levels of imbalance and the second parallel resonance between the 19th and the 20th harmonics again shows considerable differences in the impedances between phases. High levels of imbalance at parallel resonant frequencies assist in explaining the difficulties being experienced with correlating single-phase simulation results with measured tests [12, 17].

While most system loads are nearly balanced, this is not the case with single-phase traction supplies [18]. This effect has been simulated by reducing phase 1 load by 10% and increasing phase 3 load by 10%. The results, also plotted in Fig. 9.29, indicate that the level of impedance imbalance at the parallel points increases with load imbalance.

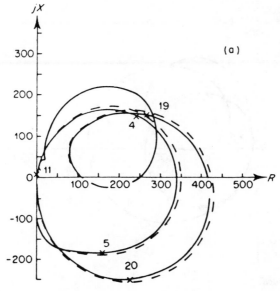

Figure 9.29(a)–(c)
Equivalent phase impedances for the test system: (a) red phase; (b) yellow phase; (c) blue phase.——, balanced load; ----, unbalanced load

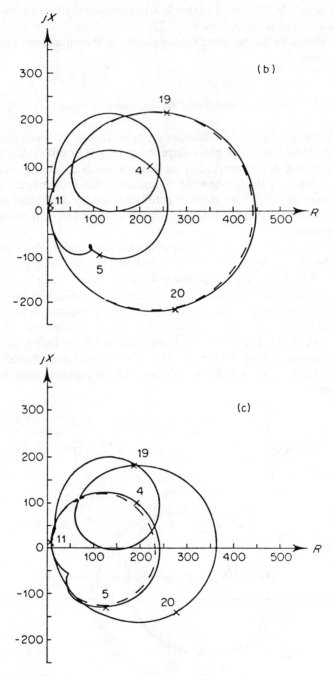

Figure 9.29 (*continued*)

9.9 REFERENCES

[1] E. Clarke, 1943. *Circuit Analysis of A.C. Power Systems* vol. I, John Wiley, New York.

[2] J. R. Carson, 1926. Wave propagation in overhead wires with ground return. *Bell Systems Tech. J.* **5** 539–554.

[3] V. A. Lewis and P. D. Tuttle, 1958. The resistance and reactance of aluminium conductors steel reinforced. *Trans. AIEE* **PAS-77** 1189–1215.

[4] E. W. Kimbark, 1950. '*Electrical Transmission of Power and Signals* Wiley & Sons, New York.

[5] W. I. Bowman and J. M. McNamee, 1964. Development of equivalent PI and T matrix circuits for long untransposed transmission lines. *IEEE Trans.* **PAS-84** 625–632.

[6] CIGRE Working Group 36–05 (Disturbing Loads) 1981. Harmonics, characteristic parameters, methods of study, estimating of existing values in the network. *Electra* **77** 35–54.

[7] J. Baird 1981. *ELAFANT-audio frequency analysis of a power system* New Zealand Electricity, Wellington.

[8] A. Semlyen, J. F. Eggleston and J. Arrillaga, 1987. Admittance Matrix model of a synchronous machine for harmonic analysis. *IEEE Trans.* **PWRS-2**, 833–40.

[9] T. W. Ross and R. M. A. Smith 1948. Centralised ripple control on high voltage networks. *Proc. IEE* **95** 470–479.

[10] C. Concordia and S. Ihara, (1982). Load representation in power system stability studies. *IEEE Trans.* **PAS-101** 969–975.

[11] D. J. Pileggi, N. Harish Chandra and A. E. Emanuel, 1981. Prediction of harmonic voltage in distribution systems. *IEEE Trans.* **PAS-100** 1307–1315.

[12] M. F. McGranaghan, R. C. Dugan and W. L. Sponsler, 1981. Digital simulation of distribution system frequency-response characteristics. *IEEE Trans.* **PAS-100** 1362–1369.

[13] N. W. Ross, 1972. Modelling a ripple control system. *The New Zealand Electrical Journal* **1972** 116–123.

[14] K. Zollenkopf, 1960. Bi-factorization—basic computational algorithm and programming techniques. Paper presented at a conference on Large Sets of Sparse Linear Equations, Oxford.

[15] R. Yacamini and J. C. de Oliveira, 1980. Harmonics in multiple convertor systems: a generalised approach. *Proc. IEE,* **127** B 96–106.

[16] B. J. Harker, 1980. Steady state analysis of integrated a.c. and d.c. systems, *PhD Thesis* University of Canterbury, New Zealand.

[17] G. D. Breuer, J. H. Chow, T. J. Gentile, *et al* 1982. HVDC–AC harmonic interaction. I. Development of a harmonic measurement system hardware and software. *IEEE Trans.* **PAS-101** 701–708.

[18] D. A. Winthrop, 1983. Planning for a railway traction load on the New Zealand power system. Paper 62 presented at the IPENZ Conference.

10. ANALYSIS OF SYSTEM OPTIMISATION AND SECURITY

10.1 INTRODUCTION

To provide a secure energy supply at minimum operating cost is a very complex process that relies heavily upon on-line computer control.

Optimisation and security are often conflicting requirements and should be considered together. The present computational tools used in the unified solution are contingency analysis to identify potential emergencies and optimal power flow (OPF) to perform dispatch calculations of active and reactive power subject to static security limits.

The more recent versions of OPF interface with contingency analysis and the computation requirements are enormous. A comprehensive survey of the subject, recently made by Stott [1] as part of a special issue of the IEEE on computers in power systems operation, concludes the 'barring unforeseeable major breakthroughs contingency-constrained OPF for large power systems can only be run at satisfactory intervals with much faster processing power than is typical of present Energy Management Systems'.

The basic aspects of OPF are discussed in this chapter using as a basis the fast decoupled power-flow algorithm described in Chapter 2. It must be understood, however, that OPF can take many forms and that the technology will continue to develop in many different ways.

10.2 OBJECTIVES

The aim of optimal power system operation is to try and make the best use of resources subject to a number of requirements over any specified time period. Here are some examples of power system optimization studies, with time scales given in brackets.

- Long-term scheduling for plant maintenance and availability of resources (months/years).
- Short-term scheduling for unit commitment (days).
- Economic allocation of generation base points (minutes).
- Tie-line interchange for frequency control (seconds).
- Plant and unit control (continuous).

Of particular interest to optimal system operation is the solution of the economic dispatch problem required to meet a predicted load.

As explained in the introduction the optimisation problem has to be considered in terms of economy and security. The economic criterion which appears to have universal acceptance is that of minimising production costs of which only those of fuel and maintenance vary significantly with generation output.

The security objective determines local plant loading limits. It also imposes limitations on network structures and loading patterns on a system scale which often conflicts with the economic objective. It is thus important to provide adequate representation of security constraints at the scheduling stage, prior to the use of optimisation techniques.

10.3   FORMULATION OF THE OPTIMISATION PROBLEM [2][3]

With reference to power system operation the optimisation problem consists of minimising a scalar objective function (normally a cost criterion) through the optimal control of a vector $[u]$ of control parameters, i.e.

$$\min f([x], [u]) \tag{10.3.1}$$

subject to

- equality constraints of the power-flow equations

$$[g([x], [u])] = 0 \tag{10.3.2}$$

- inequality constraints on the control parameters (parameter constraints)

$$U_{i,\min} \leqslant U_i \leqslant U_{i,\max} \tag{10.3.3}$$

- dependent variables and dependent functions (functional constraints)

$$X_{i,\min} \leqslant X_i \leqslant X_{i,\max} \tag{10.3.4}$$

$$h_i([x], [u]) \leqslant 0. \tag{10.3.5}$$

Examples of functional constraints are the limits on voltage magnitudes at P, Q nodes and the limits on reactive power at P, V nodes.

The optimal dispatch of real and reactive powers can be assessed simultaneously using the following control parameters:

- voltage magnitude at slack node
- voltage magnitudes at controllable P, V nodes
- taps at controllable transformers
- controllable power P_{Gi}
- phase shift at controllable phase-shifting transformers
- other control parameters.

Let us assume that only part (P_{Gi}) of the total net power (P_{Ni}) is controllable for the purpose of optimisation.

The objective function can then be defined as the sum of instantaneous operating costs over all controllable power generation:

$$f([x],[u]) = \sum_i c_i(P_{Gi}) \qquad (10.3.6)$$

where c_i is the cost of producing P_{Gi}.

The slack node must be included among the nodes with controllable power. If no costs were associated with power generated at the slack node, then the minimisation process would try and assign all power to the slack node.

The minimisation of f provides an optimal dispatch of real and reactive powers with the lowest possible operating costs and the best possible reactive flow. If the dispatch of real power has been decided from other considerations (e.g. stream flow in an all-hydro system), the only remaining problem is that of reactive power dispatch and its optimisation. In this case fewer control parameters are used:

- voltage magnitude at slack node
- voltage magnitudes at controllable P, V nodes
- taps at controllable transformers
- other control parameters.

An appropriate objective function for optimal reactive flow is the total system losses, or

$$f([x],[u]) = \sum_{i=1}^{N} P_{Ni}. \qquad (10.3.7)$$

Since all P_{Ni}, except at the slack node, are already scheduled, equation (10.3.7) can be rewritten as

$$f([x],[u]) = P_1([x],[u]) + \sum_{i=2}^{N} P_{Ni} \qquad (10.3.8)$$

where

$$\sum_{i=2}^{N} P_{Ni}$$

is a constant term.

Therefore, the minimisation of system losses is achieved by minimising the power injected at the slack node. If equation (10.3.6) is used, with the only controllable power at the slack node, then the cost $C_1(P_{G1})$ is minimised and therefore, optimisation of the reactive power flow is a special case of the complete optimisation.

10.4 CONDITIONS FOR MINIMISATION

10.4.1 Strategy for a Two-generator System [4][5]

The objective function of equation (10.3.6) can be expressed as

$$f = f_1 + f_2 = c_1(P_{G1}) + c_2(P_{G2}) \qquad (10.4.1)$$

and the equality constraints

$$g(P_{G1}, P_{G2}) = P_{G1} + P_{G2} - P_D - P_L = 0 \tag{10.4.2}$$

where P_D is the total load demand and P_L the total losses.

If the losses are neglected for the time being, the equality constraint becomes

$$g(P_{G1}, P_{G2}) = P_{G1} + P_{G2} - P_D = 0. \tag{10.4.3}$$

Equations (10.4.1) and (10.4.3) can be plotted in a three-dimensional co-ordinate system as shown in Fig. 10.1. For minimum cost the system must operate as far down as possible on the cost surface while remaining on the constraint plain.

By slicing the cost and constraint surfaces horizontally the minimum point lies where the constraint line $g(P_{G1}, P_{G2}) = 0$ is tangential to the equicost contours $C(P_{G1}, P_{G2})$ as shown in Fig. 10.2.

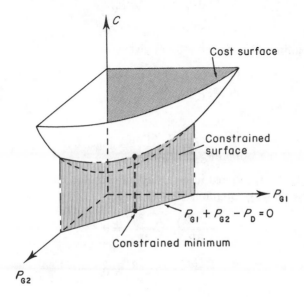

Figure 10.1
Cost surface

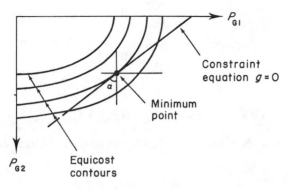

Figure 10.2
Equicost contours

Differentiating the equicost curves gives

$$dC = \frac{\partial C}{\partial P_{G1}} dP_{G1} + \frac{\partial C}{\partial P_{G2}} dP_{G2} = 0 \tag{10.4.4}$$

and the expression for the tangent is

$$\frac{dP_{G2}}{dP_{G1}} = -\frac{\partial C/\partial P_{G1}}{\partial C/\partial P_{G2}}. \tag{10.4.5}$$

Similarly differentiation of the constraint equation

$$g(P_{G1}, P_{G2}) = 0$$

provides the following expression for the tangent:

$$\frac{dP_{G2}}{dP_{G1}} = -\frac{\partial g/P_{G1}}{\partial g/\partial P_{G2}}. \tag{10.4.6}$$

Combining equations (10.4.5) and (10.4.6) gives

$$\frac{\partial C/\partial P_{G2}}{\partial C/\partial P_{G1}} = \frac{\partial g/\partial P_{G2}}{\partial g/\partial P_{G1}} \tag{10.4.7}$$

or

$$\frac{\partial C/\partial P_{G2}}{\partial g/\partial P_{G2}} = \frac{\partial C/\partial P_{G1}}{\partial g/\partial P_{G1}} = \text{const} \overset{\Delta}{=} \lambda \tag{10.4.8}$$

where the constant λ is referred to as a Lagrange multiplier.

Also from rearranging equations (10.4.8) we get

$$\frac{\partial C}{\partial P_{G1}} - \lambda \frac{\partial g}{\partial P_{G1}} = 0$$

$$\frac{\partial C}{\partial P_{G2}} - \lambda \frac{\partial g}{\partial P_{G2}} = 0. \tag{10.4.9}$$

Thus the constrained minimum is characterised by

$$\frac{\partial C^*}{\partial P_{G1}} = 0$$

$$\frac{\partial C^*}{\partial P_{G2}} = 0 \tag{10.4.10}$$

where

$$C^* \overset{\Delta}{=} C - \lambda g = c_1 + c_2 - \lambda(P_{G1} + P_{G2} - P_D). \tag{10.4.11}$$

The partial derivatives of equations (10.4.9) can be obtained from equations (10.4.1) and (10.4.3), i.e.

$$\frac{\partial g}{\partial P_{G1}} = \frac{\partial g}{\partial P_{G2}} = 1 \tag{10.4.12}$$

$$\frac{\partial C}{\partial P_{G1}} = \frac{\partial c_1}{\partial P_{G1}}$$

$$\frac{\partial C}{\partial P_{G2}} = \frac{\partial c_2}{\partial P_{G2}} \tag{10.4.13}$$

and substitution into equations (10.4.9) leads to the optimum dispatch equations

$$\frac{\partial c_1}{\partial P_{G1}} = \frac{\partial c_2}{\partial P_{G2}} = \lambda \tag{10.4.14}$$

which indicate that for optimum dispatch the individual generators must operate at equal incremental production costs.

10.4.2   Generalised Strategy

In general the minimisation of the objective function $f([x], [u])$ can be achieved with reference to the following expanded expression (referred to as the Lagrange function):

$$\mathcal{L} = f([x], [u]) - [\lambda^T] \cdot [g]. \tag{10.4.15}$$

For minimisation, the partial derivatives of $\mathcal{L}$ with respect to all the variables must be equal to zero, i.e. setting them equal to zero will then give the necessary conditions for a minimum:

$$\left[\frac{\partial \mathcal{L}}{\partial \lambda}\right] \equiv [g] = 0 \tag{10.4.16}$$

which is simply the system of power-flow equations (10.4.3)

$$\left[\frac{\partial \mathcal{L}}{\partial x}\right] \equiv \left[\frac{\partial f}{\partial x}\right] - \left[\frac{\partial g}{\partial x}\right]^T \cdot [\lambda] = 0 \tag{10.4.17}$$

$$\left[\frac{\partial \mathcal{L}}{\partial u}\right] \equiv \left[\frac{\partial f}{\partial u}\right] - \left[\frac{\partial g}{\partial u}\right]^T \cdot [\lambda] = 0. \tag{10.4.18}$$

Newton's power-flow solution already produces the matrix of equation (10.4.17) in triangularised form as a by-product and can therefore be used to solve (10.4.17) for $[\lambda]$ with only one repeat solution. Having found $[\lambda]$ from equation (10.4.17) and since $(\partial g / \partial u) = 1$ as shown in equation (10.4.12) $[\nabla f]$, the gradient of the objective function f with respect to $[u]$ can now be calculated with the advantage that, unlike $[\partial f / \partial u]$, this gradient takes the power-flow equality constraints into account.

To take into consideration the inequality constraints, when an improved vector $[u]$ is computed, its components are checked to see whether they lie within the permissible range. If the improvement is made by adding Δu_i to the old value, then the new value will be set to

$$
\begin{aligned}
u_i^{new} &= u_i^{old} + \Delta u_i & &\text{if } u_{imin} < u_i^{new} < u_{imax} \\
u_i^{new} &= u_{imin} & &\text{if } u_i^{old} + \Delta u_i \leqslant u_{imin} \\
u_i^{new} &= u_{imax} & &\text{if } u_i^{old} + \Delta u_i \geqslant u_{imax}.
\end{aligned}
\tag{10.4.19}
$$

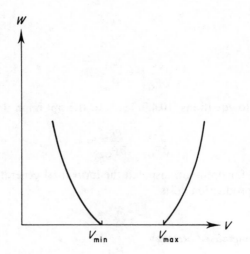

Figure 10.3
Penalty term

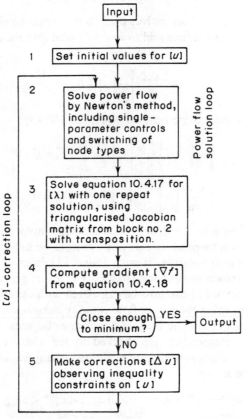

Figure 10.4
Flow diagram of the optimal power flow

When the minimum has been found, the gradient components will be

$$\frac{\partial f}{\partial u_i} \begin{cases} = 0 & \text{if } u_{imin} \leqslant u_{imax} \\ > 0 & \text{if } u_i = u_{imax} \\ < 0 & \text{if } u_i = u_{imin}. \end{cases} \tag{10.4.20}$$

These are the necessary conditions [5] for a minimum, provided the objective function f, all equations in (10.3.2) and all h_i in (10.3.5) are convex functions.

The penalty method is used to handle functional constraints of type (10.3.4) and (10.3.5). If the voltage stays within its permissible range, no penalty term is added but when a limit is exceeded, a penalty term is added of the form shown in Fig. 10.3 and equation (10.4.21):

$$W = k(V - V_{\text{limit}})^2 \tag{10.4.21}$$

with

$$\begin{cases} V_{\text{limit}} = V_{\text{max}} & \text{if } V > V_{\text{max}} \\ V_{\text{limit}} = V_{\text{min}} & \text{if } V < V_{\text{min}} \end{cases}.$$

Thus the objective function becomes

$$f = f([x],[u]) + W \tag{10.4.22}$$

and the modified f is minimised. The penalty term will force the voltage closer to the permissible range. The limit is treated as 'soft' rather than 'rigid' and the lower the factor k is in (10.4.21), the softer the limit will be. Experience has shown that soft constraints are well suited for handling voltage limits on P, Q nodes.

A simplified flow diagram of an optimal power-flow program [3] is shown in Fig. 10.4.

10.4.3 Effect of Transmission Losses

When transmitting power over large distances the energy loss must be taken into account. In this case the following augmented cost function must be used instead:

$$C^* \triangleq \sum_{i=1}^{n} c_i - \lambda \left(\sum_{i=1}^{n} P_{Gi} - P_D - P_L \right). \tag{10.4.23}$$

As in the previous section the effect on P_L by the reactive power flows is ignored, and partial differentiation of equation (10.4.23) yields the following equations for optimum real power dispatch:

$$\frac{\partial C^*}{\partial P_{Gi}} = (IC)_i - \lambda + \lambda \frac{\partial P_L}{\partial P_{Gi}} = 0 \qquad \text{for } i = 1, 2, \ldots, n. \tag{10.4.24}$$

Equation (10.4.24) includes the extra term $\partial P_L / \partial P_{Gi}$, referred to as the incremental transmission loss.

The n optimum dispatch equations (10.4.24), together with the power balance equation (10.3.2), permit the solution of the $n + 1$ unknowns $P_{G1}, \ldots, P_{Gn}$ plus λ.

10.5 SENSITIVITY OF THE OBJECTIVE FUNCTION

A by-product of the optimisation algorithm is the sensitivity of the objective function contained in $[\nabla f]$ and $[\lambda]$. This sensitivity information is valid for any feasible power-flow solution, whether optimal or not.

The gradient components $\delta f / \delta u_i$ are the first-order sensitivities of the objective function with respect to control parameters. At the approximate minimum, they should be close to zero for those parameters that lie in the interior of the permissible range. Gradient components for control parameters that reached a limit give a measure of the costs associated with imposing the limit. As an example, $\delta(\text{losses})/\delta V_k = -105\,\text{MW/p.u.}$ volts with $V_k = V_{k\text{max}} = 1.05\,\text{p.u.}$ indicates expected savings in losses of $1.05\,\text{MW}$ if the upper limit were raised to $1.06\,\text{p.u.}$

The vector $[\lambda]$ can be interpreted as sensitivities of the objective function with respect to all P_{Ni} and Q_{Ni} for which the power- flow was solved in block 2 of the flow diagram.

$$\frac{\partial f}{\partial P_{Ni}} = \lambda_{Pi}$$

$$\frac{\partial f}{\partial Q_{Ni}} = \lambda_{Qi}$$

where λ_{Pi} and λ_{Qi} are the components of $[\lambda]$ associated with the equations for P_{Ni} and Q_{Ni} respectively. As an example, $\lambda_{Pi} = 1.26\,\text{MW/MW}$ in optimal reactive power flow $(f = P_1)$ indicates that an increase of $1\,\text{MW}$ in P_{Ni} would cause a decrease in the power at the slack node by $1.26\,\text{MW}$, which amounts to expected savings in losses of $0.26\,\text{MW}$ by shifting $1\,\text{MW}$ generation to node i.

10.5.1 Input–Output Sensitivities from Linearised Power-flow Model

Small changes $[\Delta y]$ in the independent parameters cause small changes $[\Delta x]$ in the dependent variables. The functional relationship can be obtained by using a Taylor series expansion around the power-flow solution point, with second-order and higher-order terms neglected:

$$\left[\frac{\partial g}{\partial x}\right] \cdot [\Delta x] = -\left[\frac{\partial g}{\partial u}\right][\Delta y]. \tag{10.5.1}$$

Equation (10.5.1) is in fact a linearisation of the power-flow equations around the solution point. Since Newton's method of power-flow solution produces $[\partial g/\partial x]$ in triangularised form as a by-product, it takes only one repeat solution to find the sensitivity $[\Delta x]/\Delta y_i$ with respect to one particular component Δy_i:

$$\left[\frac{\partial g}{\partial x}\right] \cdot \left[\frac{\Delta x}{\Delta y}\right] = -[r_i] \tag{10.5.2}$$

where $[r_i] = i$th column of $[\partial g/\partial y]$. As an example, the influence of a change ΔQ_{Ni} at a particular node i on all voltage magnitudes could be expressed as a sensitivity

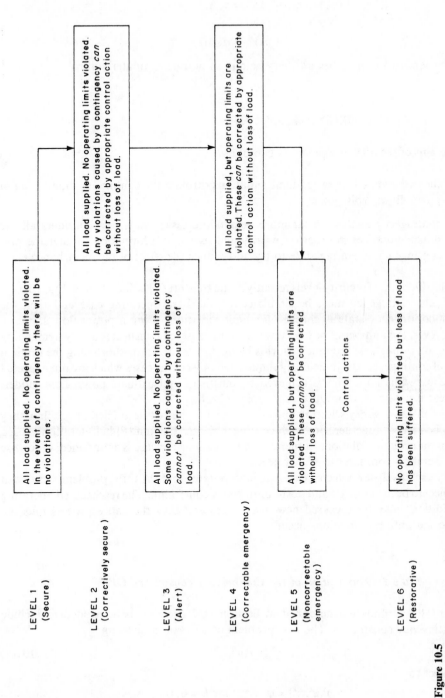

Figure 10.5
Power system static security levels [1]

vector $[Q]$:

$$\Delta V_1 = a_1 \Delta Q_i$$
$$\Delta V_2 = a_2 \Delta Q_i$$
$$\vdots \qquad \vdots$$
$$\Delta V_N = a_N \Delta Q_i.$$

Such sensitivities may become very helpful in on-line computer control.

10.6 SECURITY ASSESSMENT

The aim of security assessment is twofold:

(i) the detection of operating limit violations through the continuous monitoring of power flows, voltages, etc.

(ii) contingency analysis, a far more demanding task, which first considers all the possible outages in order of severity and uses that information to alter the pre-contingency operating state to try and reduce the effect of the disturbance.

A classification of security levels recently made by Stott [1] is illustrated in Fig. 10.5.

Each contingency must be simulated on the base operating case and then the post-contingency state is checked for limit violations using a power-flow solution. However, the number of power-flow solutions required constitutes a very demanding computational task and much effort is being devoted towards shortening the original list of contingencies by judiciously eliminating most of the cases which are not expected to cause violations. This is achieved by means of approximate power-flow models (linear if possible) to produce very rapid solutions.

When contingency selection and evaluation use the fast-decoupled power flow they can be merged together. For each contingency case the first (P) half-iteration is used to monitor limit violations. If there are no violations the case is abandoned; otherwise the iteration continues to higher accuracy.

The relatively few selected cases are incorporated into the OPF problem and solved subject to both base-case and post-contingency constraints. The rescheduled operating conditions may have caused new insecurities and thus the entire process must be repeated until no violations occur.

10.6.1 Formulation of the Contingency-constrained OPF

The OPF formulation described in Section 10.2 needs to be expanded to include contingency constraints. The new problem consists of minimising

$$f(u^0, x^0) \tag{10.6.1}$$

subject to

$$g^k(u^k, x^k) = 0 \qquad \text{for } k = 0, 1, \ldots, N_c$$

and

$$h^k(u^k, x^k) \geqslant 0 \qquad \text{for } k = 0, 1, \ldots, N_c \qquad (10.6.3)$$

where superscript '0' represents the pre-contingency (base-case) state being optimized, and superscript 'k' ($k > 0$) represents the post-contingency states for the N_c contingency cases selected for incorporation into the OPF analysis.

As a result of the outages the equality constraints g^0 change into g^k. Also the inequalities h^k will generally be different from h^0 as these may result from different monitored quantities or different limit conditions.

Regarding control variables the change from u^0 to u^k will depend on the security level. As explained in Fig. 10.5, security level 1 describes the conservative approach which prevents any post-contingency control action. In this case the control variable change at each state is

$$u^k = u^0 + \Delta u^k \qquad (10.6.4)$$

where Δu^k represents the automatic response of the system, e.g. generator inertia, AGC contribution, etc. For other controls, such as generator terminal voltage, u^k is generally equal to u^0.

On the other hand, security level 2 relies on post-contingency corrective rescheduling (δu^k) to remove any contingency limit violations and thus results in lower operating costs. Thus the control variable change at each state becomes

$$u^k = u^0 + \Delta u^k + \delta u^k. \qquad (10.6.5)$$

10.7   CHALLENGING PROBLEMS

The formulation of the optimal power flow is often regarded as a simple extension of a conventional power flow. However the application of general optimisation rules to the OPF solution is not yet well formalised. Some cases involve lack of uniqueness due to shortage of information about the desired operation of the power system. Such cases can lead to singularity or ill-conditioning.

Regarding the on-line implementation, one of the major problems is the interaction with the operator; the use of 'artificial intelligence' in the future will help to reduce this problem.

To take the system out of a bad operating condition it is critical to select the right sequence of control changes. This is a difficult topic in need of further investigation.

One of the most important questions on the implementation of on-line OPF is its interfacing with other system functions such as state estimation, contingency analysis, economic dispatch and automatic generation control which are not executed as often. Thus the OPF will normally receive outdated information from these other functions. Another formidable challenge is the communication and co-ordination of optimal secure solutions between geographically separated control centres.

Considering the practical nature of the problem the use of rigorous optimisation techniques is unwarranted. The decoupled characteristic of active and reactive power flows can be used to advantage in OPF, reducing the number of full (combined active and reactive optimisation) OPF's to the minimum needed to establish scheduling trajectories for economic, secure operation.

304

These and many other challenging problems and prospective ways of solving them are discussed in greater detail in reference [1].

10.8 REFERENCES

[1] B. Stott, O. Alsac and A. J. Monticelli, 1987. Security analysis and optimization, *Proc. IEEE* **75** 1623–1644.
[2] J. Carpentier, 1962. Contribution to the economic dispatch problem, (in French), *Bull. Soc. Franc. Elec.* **8** 431–447.
[3] H. V. Dommel and W. F. Tinney, 1968. Optimal power flow solutions, *IEEE Trans.* **PAS-87** 1866–1876.
[4] O. I. Elgerd, 1971. *Electric Energy Systems Theory: An Introduction.* McGraw-Hill, New York.
[5] H. W. Kuhn and A. W. Tucker, 1951. Non-linear programming, *Proc. Second Berkeley Symp. Math. Stat. Prob.* University of California Press.

11. A GRAPHICAL POWER SYSTEM ANALYSIS PACKAGE

11.1 INTRODUCTION

A large number of very versatile power system analysis programs have been developed during the last three decades. Most of those programs were originally written to run in batch mode on mainframe or minicomputers. With the introduction of multitasking/multiuser interactive computing environments, many of these programs have been upgraded to give them interactive user-friendly features. However, most of these programs are capable of analysing only one aspect of power systems operation such as load flow, faults, etc. Very often these programs require data input in different formats. They also need the help of separate presentation programs such as graph plotting, printing, etc., for result analysis and comprehension.

Most of these tasks can be dedicated to the computer, thus removing the tedious exercise of elaborate data preparation and processing. Tasks such as creating the design, analysis and result presentation can be integrated into one package so that less time is spent on switching between these important tasks. When several system analysis tasks are interdependent (performance of one analysis depends on the results of another), all of them can also be integrated to the same package [1].

This approach is very common in the CAD systems used for the design of electronics components and networks such as printed circuit boards and VLSI [2]. By using a similar approach in power system analysis, design turnover time can be considerably reduced [3].

This chapter describes a package named *Display Power*, developed at the University of Canterbury, New Zealand, mainly for educational purposes. The package integrates power system analysis programs under a single database, with the capability of switching between system editing, simulation and result analysis without leaving the environment.

In *Display Power*, power systems can be graphically constructed, modified, stored and any of the above simulations run at any time using CAD drawing and data editing facilities that are easy to use. A symbol editing program is used to permit custom design of the graphics displays to suit individual needs.

In the preceding chapters the algorithms have all been developed by the authors and their research group but the descriptions are quite general. When describing a complex CAD package like *Display Power* it is neither possible to be so general nor is it possible to be detailed. The information presented is specific to *Display Power*, but there are many ways of producing a similar package and it is the purpose of this chapter to give some assistance to programmers who wish to develop something similar, and to users who need to get the most out of a package.

11.2 PROGRAMMING CONCEPTS

The present power system software engineer has the problem of maintaining unstructured and nonmodular programs initially conceived many years ago. These programs are well tested and proven, but with the dramatic evolution in computer hardware and software, upgrading their capabilities is very difficult.

The development of versatile graphic windowing systems and networking facilities has created a new dimension to operator–computer interactions. Hardware has evolved to the extent that each user can have a personal computer or workstation at a relatively low cost. The computational capabilities of these workstations are equal to or higher than that of the minicomputers developed several years ago. They also provide faster response for the user interactions, so that interaction devices such as a mouse, graphic tablet, etc., can be used with them effectively.

All the simulations are from the existing stand alone FORTRAN programs described in earlier chapters. The philosophy adopted for *Display Power* is to change the well established FORTRAN algorithms as little as possible to avoid introducing errors. At the same time, the support structure for the simulations are all written using a language capable of exploiting the modern computing environment.

The selection of the language was based on the following requirements.

- It should support modern programming techniques such as modular and structured code, abstract data types, dynamic memory usage, etc.
- It should be sufficiently flexible to integrate with the existing FORTRAN programs.
- Language compilers should be readily available for running in mini- and micro-computer environments.
- It should be able to be used with the existing programming tools and operating systems.
- It should be easily adapted to the future developments of the computer software and hardware.

Fortran lacks the language support for easily creating modular and structured programs and has very limited data types and type checking capabilities. On the other hand, newer languages like C and Pascal provide necessary features to create more modular, structured and manageable programs. Ada and Modula-2, developed from Pascal, provide additional features and their popularity is still increasing.

Display Power has been written using Modula-2 for the graphics and data base. It is inherently a structured and modular programming language and supports abstract data structures, multitasking and certain aspects of object-oriented programming. The retention of FORTRAN for the algorithms, although possibly seen as an expediency, can be justified because of its efficiency in performing complex numerical tasks.

It is important to make the program flexible so that other algorithms and new graphic symbols together with associated data can be added in the future with minimum effort and disruption to the existing program.

11.3 PROGRAM OVERVIEW

11.3.1 Organisation

In *Display Power* a concept called the 'work sheet' is defined as the area where the one-line diagram of the system is drawn. A work sheet, which is the realisation of the data into useful form, is the environment which is entered on first starting the program. It is the environment in which a system can be built, modified and analysed. The work sheet is described in more detail later.

There are two modes of operation for the program.

- edit mode, in which all the drawing and data entry takes place
- simulation mode, in which the analysis takes place and the results are observed.

Pop-up menus are used throughout instead of permanently displayed menus. The advantage of this is that the graphics display never becomes cluttered.

All operations wherever possible, are carried out using a mouse and mouse-driven menus. The mouse button philosophy, which is consistent throughout the program, uses the left button for operations associated with the work sheet as a whole, the right button for component or simulation selections and the centre button for utility operations. Where there is no conflict any button can be used.

A schematic representation of all the major tasks associated with *Display Power* is shown in Fig. 11.1. The work sheet handler reads the network data from a pre-stored file (unless a new system is to be drawn) and restores the network.

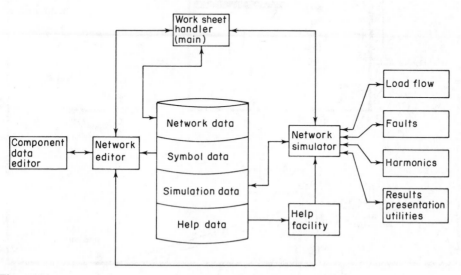

Figure 11.1
A schematic representation of the program

11.3.2 Network Display and Data Editing

At start up, the user may either retrieve an existing pre-saved system or start a new system. *Display Power* always commences by drawing the one-line diagram then entering the edit mode to allow any necessary changes to be made. The components are selected from a menu and placed anywhere on the screen. Once in place, the components can be easily manipulated. Rubber banding, which is the stretching or contracting of line elements, allows components to be moved and the effect of the move seen before its acceptance. Components can also be rotated, translated and deleted provided no conflict occurs with other components. To allow for large systems, zooming and panning features are provided.

The user is free to either include data concerning the components while drawing or at some later time. The busbar name is the only piece of data which is displayed along with the component. The name is treated like a component and can be moved, modified or deleted as necessary. The user can deliberately choose not to name busbars if necessary.

The window displaying a component's parameters is popped up by clicking a mouse button when pointing to the component. Fig. 11.2 shows an example of a data

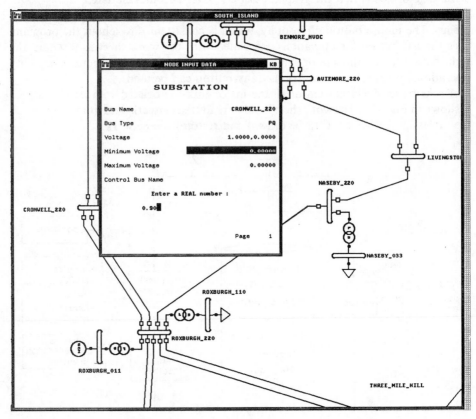

Figure 11.2
Example of the screen display while in edit mode, showing data editing window

editing window. Each type of component has its own data structure and this is reflected in the window layout. Despite being in the edit mode, the type of simulation to be performed can be specified (the default is load flow or the previous simulation). This allows the window layout to show only the relevant data. Windows can be paged if the amount of data necessary to fully specify a component is large. The data are readily modified using the keyboard. The load-flow slack busbar, the fault position or the harmonic injection point can also be specified in the edit mode.

11.3.3 Simulation

Once sufficient data have been entered, the program can be toggled to the simulation mode from which the load flow, faults and harmonics programs can be run. A data input file is created from the work sheet in a suitable form for the analysis program, thus ensuring the minimum changes to the FORTRAN program. On completion of a run the analysis program generates an output listing file and automatically updates the worksheet to reflect results where necessary.

The system can be stored at any time and another system retrieved from the data base. In fact, several systems can be retrieved into work sheets and operated on although only one is visible at any time. Toggling between work sheets is rapid. This allows data developed in one system to be transferred easily to another system.

11.3.4 Output

The output stage of this package is very important because it is required to perform a large number of tasks in a user-friendly manner so that the results can be comprehended quickly. The results can be viewed in several different ways.

- A window can be opened to view the output listing directly on the screen.
- The output listing can be saved to be printed out later.
- Where relevant, a window can be opened and the results graphed. At present this is limited to harmonic locus diagrams, but it can be extended whenever necessary.
- On returning to the edit mode, relevant data will be seen to have been updated. These can be saved with the system data if necessary, for future use.
- Results such as overloaded circuits or voltage profiles may be displayed by drawing the components in different colours. For example, the voltage profile of the network can be indicated by giving the busbars different colours to show different per unit voltage levels.
- The user may choose to permanently display a quantity of user-defined variables on the screen.

Harmonic studies demand more versatile result comprehension methods than load flow or faults because of the large amount of harmonic data associated with each component. Due to the large quantity of results produced by the harmonic penetration program, the results are stored in a file rather than in the database. Any information relevant to a specified component can be extracted from the file and displayed whenever necessary.

The total harmonic voltage distortion and the equivalent disturbing voltage of the busbars and the respective current quantities for branch components are calculated and can be viewed. Also the spectrum of harmonic voltages at various points of the system and harmonic currents in branch components can be viewed in list form and graphical form. Graphs are plotted as continuous curves, since features like the rate of change and trend of change are more easily understood than from a discrete graph (such as a bar chart). The intermediate points are obtained by interpolating the harmonic results and approximate the value of the variable for noninteger harmonics.

Polar plots are more useful than cartesian co-ordinate plots for the interpretation of equivalent system harmonic impedance. Therefore an option is provided to view the impedances as a harmonic loci diagram, an example of which is given in Fig. 11.3.

Comparison of two or more harmonic graphs associated with a component or several components is possible by opening several windows simultaneously. All the harmonic data associated with components can be stored in different files for later use. With this option, up to four graphs can be viewed simultaneously, by specifying a pre-stored data file for each channel.

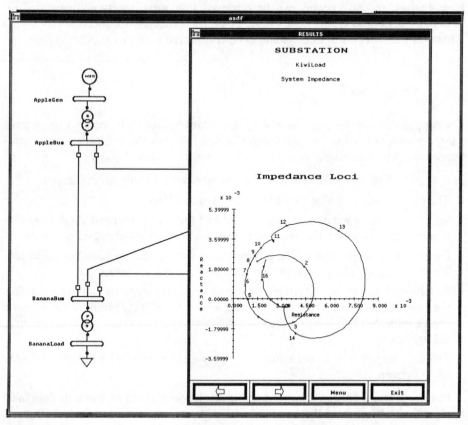

Figure 11.3
Example of the screen display while in simulation mode, showing harmonic locus diagram

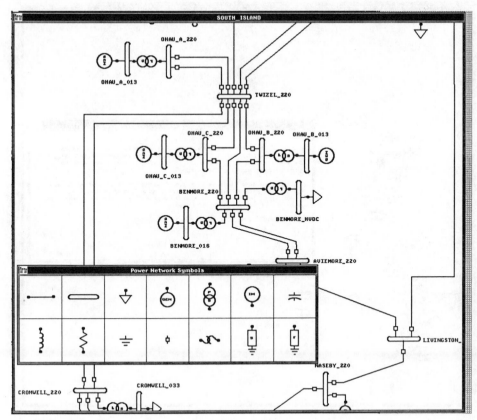

Figure 11.4
Example of the screen display while in edit mode, showing component menu window

Two examples of the display showing the power system component selection window and a help window are given in Figs. 11.4 and 11.5.

11.4 DATA STRUCTURE

Display Power is built in several program layers so that the lower layers are associated with primitive object definitions and manipulations and the upper layers are associated with the definition and manipulation of more complex objects which are in turn made out of the primitive objects. In Fig. 11.6, several levels of objects are defined to demonstrate the approach as applied to *Display Power*. These are explained below.

(i) The *graphic primitive* is a basic element such as a line, circle or text and has co-ordinates which define its size. The origin for the co-ordinates is the centre of a bounded box of the symbol.

(ii) A *symbol* consists of one or more graphic primitives forming a power system symbol such as a transformer or busbar. These objects are stored in the library.

312

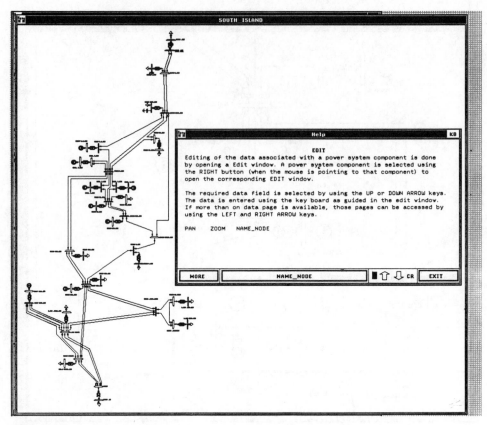

Figure 11.5
Example of the screen display showing the on-line help facility

(iii) A *power symbol* is an object which can be drawn on the screen and consists of a symbol, obtained from the library, plus its location, rotation (orientation) and image (left or right handed).

(iv) Each *power system component* consists of a power symbol, data and, if necessary, results associated with the component. Its association with other power system components is also recorded.

(v) A *network* is the total set of power system components necessary to suitably describe a power system.

(vi) The *work sheet* consists of the network plus other global data.

(vii) *Display Power* itself may be considered as the overall object containing the work sheets and their tasks.

Fig. 11.7 shows some source code for data abstraction types for objects in the symbol local co-ordinate frame. The listing starts with some basic data structures that are used to define these objects. The listing is not complete but intended to

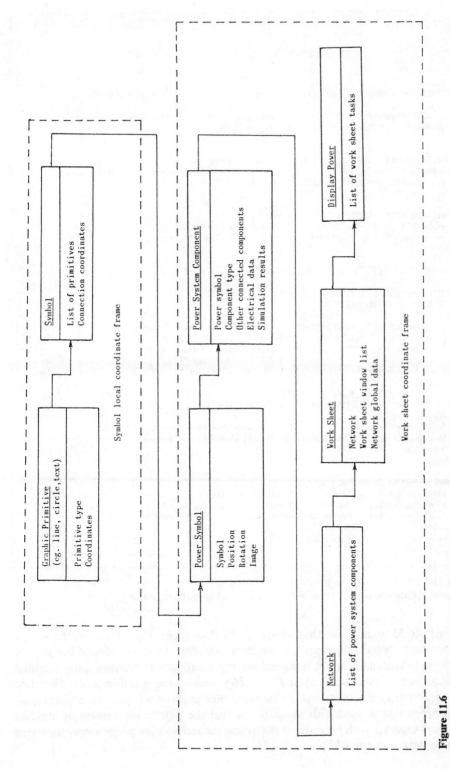

Figure 11.6
Hierarchical implementation of objects

```
TYPE

    POINT = RECORD      (* coordinates of a point *)
      X,Y : REAL;
    END; (* POINT *)

    CIRCUITRECORD = POINTER TO RECORD    (* information about one circuit of a multiple circuit
                                            transmission line *)
      numOfLineSegments : CARDINAL;    (* number of line segments required to compose one circuit *)
      pointArray : POINTER TO ARRAY [1..MaxSeg] OF POINT;
    END;

    CONNECTIONRECORD = RECORD      (* connection information *)
      NumOfConnections : CARDINAL;      (* number of connection points in one polarity group *)
      ConnectionPts : POINTER TO ARRAY [1..MaxCon] OF POINT;
    END;

    CONELEMENTRECORD = RECORD      (* connection element *)
      numberOfConnections  : CARDINAL;    (* number of connections in a polarity group *)
      connections          : POINTER TO ARRAY [1..MaxCon] OF PSELEMENT;
    END;

    PRIMITIVETYPE = (Circle, Arc, Line, Rectangle, Polygon, Text);

    PRIMITIVE = POINTER TO RECORD      (* Graphic Primitive *)
      CASE type : PRIMITIVETYPE OF
        Text:
            textarray   : POINTER TO ARRAY [0..HighChar] OF CHAR; (* text string*)
            textpath    : TEXTPATH;    (* horizontal or vertical text *)
            textfont    : FONT;        (* font to be used to draw text *)
            textfontsizeX,             (* width of a character *)
            textfontsizeY : REAL;      (* height of a character *)
        ELSE
            linewidth   : REAL;        (* line width *)
            objectfill  : FILLSTATUS;  (* filled polygons or not *)
      END; (* case *)
      NumOfPoints   : CARDINAL;        (* number of points to define entity *)
      CoordPointer  : POINTER TO ARRAY [1..MaxPts] OF POINT;
      ObjectList    : OBJECT;
    END;

    SYMBOL = POINTER TO RECORD    (* Symbol *)
      width, height     : REAL;        (* bounded box size *)
      polarityDimension : CARDINAL;    (* number of polarity groups *)
      connectionPointer : POINTER TO ARRAY [1..MaxDem] OF CONNECTIONRECORD;
      colour            : COLOUR;
      object            : PRIMITIVE;
    END;
```

Figure 11.7
Examples of data abstraction types in the symbol local co-ordinate frame

demonstrate Modula-2's method of specifying data types. Fig. 11.8 gives the source code for data abstraction types for objects in the work sheet coordinate frame.

Although Modula-2 is not designed strictly as an object oriented programming language such as Smalltak [5] or C++ [6], which have a self-imposed discipline on programming, certain aspects of object-oriented programming can be implemented. The concept that is used with Modula-2 is that the *objects* are defined as abstract data types together with procedures describing the *methods* for performing operations on these data structures [7].

```
POWERSYMBOL = POINTER TO RECORD        (* Power Symbol *)
    symbol       : SYMBOL;
    CASE type    : SYMBOLTYPE OF
        PictureSymbol :
            posX, posY  : REAL;              (* position of symbol *)
            rotation    : CARDINAL;
            image       : IMAGEFLAG;|
        LineSymbol:
            NumOfCircuits  : CARDINAL;     (* number of circuits in transmission line *)
            CircuitPointer : POINTER TO ARRAY [1..MaxCct] OF CIRCUITRECORD;
    END; (* case *)
END;


PSELEMENT = POINTER TO RECORD      (* Power Symbol Component *)
    Type               : ELEMENTTYPE;                    (* power system element type identity *)
    Name               : POINTER TO ARRAY [0..HighChar] OF CHAR;
    NamePosition       : POINT;
    NodeNumber         : CARDINAL;
    PolarityDimension  : CARDINAL;            (* number of polarity groups *)
    ElementPointer     : POINTER TO ARRAY [0..MaxCon] OF ConElementRecord;
    Data               : POINTER TO DATARECORD;     (* power system data of the component *)
    Results            : POINTER TO RESULTSRECORD;  (* results of last analysis *)
    Symbol             : POWERSYMBOL;
    List               : PSElement;
END;
```

Figure 11.8
Examples of data abstraction types in the work sheet co-ordinate frame

The data elements of the objects may not be operated on directly but only by using provided procedures or by asking the object to perform operations on itself. This concept is supported by the language by providing a facility to define opaque or hidden data types in modules. In *Display Power* hidden data types are used as much as possible to define the objects. Sometimes the objects are declared as visible data structures, allowing a set of modules to operate on data elements. This is done for the reasons of reducing the modules to easily manageable sizes, and reducing the program effort. Any module not included in this set treats the object as a hidden data type.

11.5   PROGRAM STRUCTURE

The programs have been written to operate with objects as whole entities or to pass messages to them asking them to operate on themselves. Figure 11.9 shows the hierarchical structure of the implementation.

The system-dependent features are pushed to the lowest layers making the upper layer functions computer independent. To transfer to another operating system, only a few low-level routines needed to be altered.

The language-supported multitasking feature makes the programming relatively easy and the code more elegant. To obtain the maximum benefit from this feature, *Display Power* is designed with its own non pre-emptive scheduler. The concept of

```
MODULE NetworkEditor
       .
       .
    RotateSymbol (PowerSymbol, Angle);
       .
       .
```

```
MODULE PowerSymbols
    PROCEDURE RotateSymbol (pSymb: POWERSYMBOL; ang: INTEGER);
       .
       .
       GetSymbolPosition (pSymb, posX, posY);
       CreateRotation Matrix (posX, posY, ang, matrix);
       DrawSymbol (pSymb.symb, matrix);
       .
       .
```

```
MODULE Symbols
    PROCEDURE DrawSymbol (sym: SYMBOL; m:MATRIX);
       FOR allComponentsInSymbol DO
          DrawComponent (comp, M);
       END;
    END DrawSymbol;
```

Figure 11.9
Hierarchical implementation of objects and methods

creating several concurrent processes and the ability to pass messages between them
has been extensively used in the design of *Display Power*.

Multitasking is used to perform tasks such as simultaneous monitoring of several
input devices (or several windows) for user inputs, rubber banding, dragging etc.
Dragging is moving a screen object (a power symbol in this case) on the screen. In
some circumstances, dragging can be used to change the shape or size of a screen
object by moving one side or corner of the object, and in this case the other sides
will be rubber banded. An example is shown in Fig. 11.10 where two processes—'Get
command from keyboard' and 'Get command from mouse'—work concurrently with
other processes. In both these processes, the major period of running time is spent
waiting for an event and if the event occurs the global variable 'com' is set to indicate

```
VAR com : COMMAND;

PROCEDURE GetCommandFromKeyboard ( VAR com : COMMAND );
                                         (* PROCESS *)

   VAR
      key : CHAR;
   BEGIN
      LOOP
         WaitForKeyInput( key );
         com := DecodeKey ( key );
         Yield;
      END;
END GetCommandFromKeyboard;

PROCEDURE GetCommandFromMouse ( VAR com : COMMAND );
                                         (* PROCESS *)

   VAR
      region : REGION;
   BEGIN
      LOOP
         WaitForRegionSelect ( region );
         com := DecodeRegion ( region );
         Yield;
      END;
END GetCommandFromMouse;

PROCEDURE ExecuteCommand ( VAR com : COMMAND );
   VAR
      procId1, procId2 : PROCESSID;
   BEGIN
      com := None;
      procId1 := StartProcess ( GetCommandFromKeyboard );
      procId2 := StartProcess ( GetCommandFromMouse );
      LOOP
         CASE com OF
            LEFTARROW :    (* select previous page     *)
                .          (* as in Help command line  *)
                ... |      (* shown in Fig. 2          *)
                .

                .
            NONE : |       (* do nothing *)
                .

                .
            Exit : EXIT;   (* exit loop *)
         END;
         com := None;
         Yield;
      END; (* loop *)
      DeleteProcess ( procId1 );
      DeleteProcess ( procId2 );
END ExecuteCommand;
```

Figure 11.10
An example of multitasking

```
PROCEDURE ErrorHandler ( msg : ARRAY OF CHAR );
   BEGIN
      InformError ( msg );
      GetAcknowledgement;
      DeleteProcess ( GetMyProcId() );
END ErrorHandler;


PROCEDURE  LoadFlow ( VAR w : WORKSHEETPTR ); (* PROCESS *)
   BEGIN
      .
      .
      GetDataFromDataBase;
      .
      .
      IF (error) THEN
         ErrorHandler ( 'Error....Encountered' );
      END; (* if *)
      .
      .
      CalcLoadFlow;
      .
      .
END LoadFlow;

PROCEDURE Simulation ( VAR w : WORKSHEETPTR );
   VAR
      procId : PROCESSID;
   BEGIN
      CASE selectedSimulation OF
         loadflow :
            procId := StartProcess ( LoadFlow );
         faults :
            .
            .
      END; (* case *)
END Simulation;
```

Figure 11.11
An example of exception handling

the event. A procedure 'Execute command', issues the command to start the other two processes and then continuously loops until a concurrent process reports an event via 'com'. The procedure 'Yield' transfers control to the process scheduler which determines which process to activate.

Exception handling is also made relatively easy by creating processes. When an exception occurs in a created process, it can be detected and the process stopped. This allows the user to return to the main process and rectify the error, rather than abort the whole session. This is a very useful way to exit from a program after errors have occurred in deeply nested procedures. Instead of passing an error flag across all the procedure calls, the error can be identified 'in situ' and the process can be terminated. In the example of exception handling, shown in Fig. 11.11, the load-flow algorithm is a process which is started by the simulation procedure.

11.6   CONCLUSIONS AND FUTURE DEVELOPMENTS

The *Display Power* package containing load flow, faults and single-phase harmonic analysis, is already used in undergraduate laboratory exercises. Depending on the particular interests of the researchers and also the time available further programs can be integrated, such as transient stability, three-phase harmonic penetration [8], iterative harmonic analysis, etc. It is unlikely to be ever considered in a final form but the existing simple version may well remain in use for quite some time.

Modula-2 was chosen as the language to develop the data base and graphic handling part of the package. This was not the only possibility and in the future the development of more powerful languages will make a decision more difficult unless a single universal language can be accepted. The data structure has been carefully designed to be object oriented. This allows the future addition of new power system components, different simulation algorithms or even new graphical symbols to be made without disruption to the existing package. The decision to keep FORTRAN as the language for the simulation algorithms made the production of the program quicker. In time these algorithms may be converted to Modula-2 or some other modern language when it can be demonstrated to be at least as efficient as Fortran.

Display Power has been developed in a VAX workstation environment with the intention of producing a PC-AT version when the successor to DOS has been finally decided. It has already been downloaded, compiled and run in parts on a PC-AT but the 640 K byte limit of DOS prevents a satisfactory overall program to be constructed. The program structure allows for multitasking, which although not yet supported on many workstations, will give the package a natural advantage when multiprocessor computers become more readily available.

11.7   REFERENCES

[1] R. C. Dugan, J. A. King and M. A. Barel, 1983. The personal scientific computing environment: A new approach to power industry computer applications, *IEEE Trans.* **PAS-102** 3522–3528.

[2] *Daisy Systems Corp. Personal Logican—Systems Manual* 1986 Daisy Systems Corp., USA.

[3] R. E. Hursch, 1982. Starting and maintaining a computer aided design system, *IEEE Trans.* **CGA-2** 15–22.

[4] N. C. Pahalawaththa, C. P. Arnold, M. Shurety and N. R. Watson, 1990. Display power—An advanced graphic interface for power system analysis IPENZ Conf., Wellington, New Zealand, February 1990.

[5] A. Goldberg and D. Robson, 1984. *Smalltalk-80: The Language and its Implementation* Addison-Wesley, USA.

[6] B. Stroustrup, 1986. *The C++ Reference Manual* Addison-Wesley, USA.

[7] D. F. Stubbs, 1985. *Data Structures with Abstract Data Types and Pascal* Brooks Cole Publ. Co., USA.

[8] J. Arrillaga, D. Bradley and P. S. Bodger, 1985. *Power System Harmonics* Wiley and Sons, London.

APPENDIX I.
LINEAR TRANSFORMATION
TECHNIQUES

I.1 INTRODUCTION

Linear transformation techniques are used to enable the admittance matrix of any network to be found in a systematic manner. Consider, for the purpose of illustration, the network drawn in Fig. I.1.

Five steps are necessary to form the network admittance matrix by linear transformations.

(i) Label the nodes in the original network.

(ii) Number, in any order, the branches and branch admittances.

(iii) Form the primitive network admittance matrix by inspection. This matrix relates the nodal injected currents to the node voltages of the primitive network. The primitive network is also drawn by inspection of the actual network. It consists of the unconnected branches of the original network with a current equal to the original branch current injected into the corresponding node of the primitive network. The voltages across the primitive network branches then equal those across the same branch in the actual network.

The primitive network for Fig. I.1. is shown in Fig. I.2.

The primitive admittance matrix relationship is

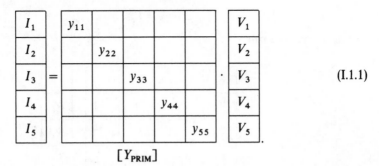

$$[Y_{\text{PRIM}}]$$

$$(\text{I.1.1})$$

Off-diagonal terms are present where mutual coupling between branches is present.

(iv) Form the connection matrix $[C]$. This relates the nodal voltages of the actual

321

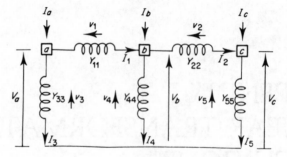

Figure I.1
Actual connected network

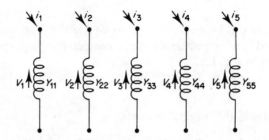

Figure I.2
Primitive or unconnected network

network to the nodal voltages of the primitive network. By inspection of Fig. I.1,

$$V_1 = V_a - V_b$$
$$V_2 = V_b - V_c$$
$$V_3 = V_a \qquad \qquad \text{(I.1.2)}$$
$$V_4 = V_b$$
$$V_5 = V_c$$

or in matrix form

$$
\begin{array}{|c|}
\hline V_1 \\ \hline V_2 \\ \hline V_3 \\ \hline V_4 \\ \hline V_5 \\ \hline
\end{array}
=
\begin{array}{|c c c|}
\hline 1 & -1 & \\ \hline & 1 & -1 \\ \hline 1 & & \\ \hline & 1 & \\ \hline & & 1 \\ \hline
\end{array}
\begin{array}{|c|}
\hline V_a \\ \hline V_b \\ \hline V_c \\ \hline
\end{array}
\qquad \text{(I.1.3)}
$$

(v) The actual network admittance matrix which relates the nodal currents to the voltages by

$$
\begin{array}{|c|}
\hline I_a \\
\hline I_b \\
\hline I_c \\
\hline
\end{array}
=
\begin{array}{|c|}
\hline \\
\left[Y_{abc} \right] \\
\\
\hline
\end{array}
\cdot
\begin{array}{|c|}
\hline V_a \\
\hline V_b \\
\hline V_c \\
\hline
\end{array}
\tag{I.1.4}
$$

can now be derived from

$$
\underset{3\times3}{[Y_{abc}]} = \underset{3\times5}{[C]^{\mathrm{T}}} \cdot \underset{5\times5}{[Y_{\mathrm{PRIM}}]} \cdot \underset{5\times3}{[C]}
\tag{I.1.5}
$$

which is a straightforward matrix multiplication.

I.2   THREE-PHASE SYSTEM ANALYSIS

I.2.1   Discussion of the Frame of Reference

Sequence components have long been used to enable convenient examination of the balanced power system under both balanced and unbalanced loading conditions.

The symmetrical component transformation is a general mathematical technique developed by Fortescue whereby any 'system of n vectors or quantities may be resolved, when n is prime, into n different symmetrical n phase systems'. Any set of three-phase voltages or currents may therefore be transformed into three symmetrical systems of three vectors each. This, in itself, would not commend the method and the assumptions, which lead to the simplifying nature of symmetrical components, must be examined carefully.

Consider, as an example, the series admittance of a three-phase transmission line, shown in Fig. I.3, i.e. three mutually coupled coils. The admittance matrix relates the illustrated currents and voltages by

$$
[I_{abc}] = [Y_{abc}][V_{abc}]
\tag{I.2.1}
$$

where

$$
[I_{abc}] = [I_a I_b I_c]^{\mathrm{T}}
$$
$$
[V_{abc}] = [V_a V_b V_c]^{\mathrm{T}}
$$

and

$$
[Y_{abc}] =
\begin{array}{|c|c|c|}
\hline
y_{aa} & y_{ab} & y_{ac} \\
\hline
y_{ba} & y_{bb} & y_{bc} \\
\hline
y_{ca} & y_{cb} & y_{cc} \\
\hline
\end{array}
\tag{I.2.2}
$$

By the use of symmetrical component transformation the three coils of Fig. I.3

324

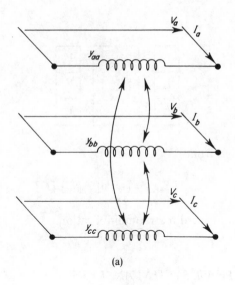

(a)

$$\begin{array}{|c|} I_a \\ I_b \\ I_c \end{array} = \begin{array}{|ccc|} y_{aa} & y_{ab} & y_{ac} \\ y_{ba} & y_{bb} & y_{bc} \\ y_{ca} & y_{cb} & y_{cc} \end{array} \begin{array}{|c|} V_a \\ V_b \\ V_c \end{array}$$

(b)

Figure I.3
Admittance representation of a three-phase series element: (a) series admittance element; (b) admittance matrix representation

can be replaced by three uncoupled coils. This enables each coil to be treated separately with a great simplification of the mathematics involved in the analysis.

The transformed quantities (indicated by subscripts 012 for the zero, positive and negative sequences respectively) are related to the phase quantities by

$$[V_{012}] = [T_s]^{-1} \cdot [V_{abc}] \tag{I.2.3}$$

$$[I_{012}] = [T_s]^{-1} \cdot [I_{abc}] \tag{I.2.4}$$

$$= [T_s]^{-1} \cdot [Y_{abc}] \cdot [T_s] \cdot [V_{012}] \tag{I.2.5}$$

where $[T_s]$ is the transformation matrix.

The transformed voltages and currents are thus related by the transformed admittance matrix

$$[Y_{012}] = [T_s]^{-1} \cdot [Y_{abc}] \cdot [T_s]. \tag{I.2.6}$$

Assuming that the element is balanced, we have

$$y_{aa} = y_{bb} = y_{cc}$$
$$y_{ab} = y_{bc} = y_{ca} \tag{I.2.7}$$
$$y_{ba} = y_{cb} = y_{ac}$$

and a set of invariant matices $[T]$ exist. Tranformation (I.2.6) will then yield a diagonal matrix $[y]_{012}$.

In this case the mutually coupled three-phase system has been replaced by three uncoupled symmetrical systems. In addition, if the generation and loading are balanced, or may be assumed balanced, then only one system, the positive-sequence system, has any current flow and the other two sequences may be ignored. This is essentially the situation with the single-phase load flow.

If the original phase admittance matrix $[y_{abc}]$ is in its natural unbalanced state then the transformed admittance matrix $[y_{012}]$ is full. Therefore, current flow of one sequence will give rise to voltages of all sequences, i.e. the equivalent circuits for the sequence networks are mutually coupled. In this case the problem of analysis is no simpler in sequence components than in the original phase components and symmetrical components should not be used.

From the above considerations it is clear that the asymmetry inherent in all power systems cannot be studied with any simplification by using the symmetrical component frame of reference. Data in the symmetrical component frame should only be used when the network element is balanced, for example, synchronous generators.

In general, however, such an assumption is not valid. Unsymmetrical interphase coupling exists in transmission lines and to a lesser extent in transformers and this results in coupling between the sequence networks. Furthermore, the phase shift introduced by transformer connections is difficult to represent in sequence component models.

With the use of phase co-ordinates the following advantages become apparent.

- Any system element maintains its identity.

- Features such as asymmetric impedances, mutual couplings between phases and between different system elements, and line transpositions are all readily considered.

- Transformer phase shifts present no problem.

I.2.2 The Use of Compound Admittances

When analysing three-phase networks, where the three nodes at a busbar are always associated together in their interconnections, the graphical representation of the network is greatly simplified by means of 'compound admittances', a concept which is based on the use of matrix quantities to represent the admittances of the network.

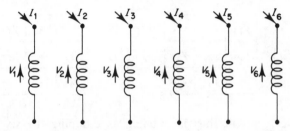

Figure I.4
Primitive network of six coupled admittances

The laws and equations of ordinary networks are all valid for compound networks by simply replacing single quantities by appropriate matrices.

Consider six mutually coupled single admittances, the primitive network of which is illustrated in Fig. I.4.

The primitive admittance matrix relates the nodal injected currents to the branch voltages as follows:

$$
\begin{bmatrix} I_1 \\ I_2 \\ I_3 \\ I_4 \\ I_5 \\ I_6 \end{bmatrix}
=
\begin{bmatrix}
y_{11} & y_{12} & y_{13} & y_{14} & y_{15} & y_{16} \\
y_{21} & y_{22} & y_{23} & y_{24} & y_{25} & y_{26} \\
y_{31} & y_{32} & y_{33} & y_{34} & y_{35} & y_{36} \\
y_{41} & y_{42} & y_{43} & y_{44} & y_{45} & y_{46} \\
y_{51} & y_{52} & y_{53} & y_{54} & y_{55} & y_{56} \\
y_{61} & y_{62} & y_{63} & y_{64} & y_{65} & y_{66}
\end{bmatrix}
\cdot
\begin{bmatrix} V_1 \\ V_2 \\ V_3 \\ V_4 \\ V_5 \\ V_6 \end{bmatrix}
\qquad (I.2.8)
$$

$6 \times 1 \qquad\qquad 6 \times 6$

Partitioning equation (I.2.8) into 3×3 matrices and 3×1 vectors, the equation becomes

$$
\begin{bmatrix} [I_a] \\ [I_b] \end{bmatrix}
=
\begin{bmatrix} [Y_{aa}] & [Y_{ab}] \\ [Y_{ba}] & [Y_{bb}] \end{bmatrix}
\cdot
\begin{bmatrix} [V_a] \\ [V_b] \end{bmatrix}
\qquad (I.2.9)
$$

where

$$[I_a] = [I_1\, I_2\, I_3]^T$$
$$[I_b] = [I_4\, I_5\, I_6]^T$$

$$
[Y_{aa}] =
\begin{bmatrix}
y_{11} & y_{12} & y_{13} \\
y_{21} & y_{22} & y_{23} \\
y_{31} & y_{32} & y_{33}
\end{bmatrix}
\qquad
[Y_{bb}] =
\begin{bmatrix}
y_{44} & y_{45} & y_{46} \\
y_{54} & y_{55} & y_{56} \\
y_{64} & y_{65} & y_{66}
\end{bmatrix}
$$

$$
\qquad (I.2.10)
$$

$$
[Y_{ab}] =
\begin{bmatrix}
y_{14} & y_{15} & y_{16} \\
y_{24} & y_{25} & y_{26} \\
y_{34} & y_{35} & y_{36}
\end{bmatrix}
\qquad
[Y_{ba}] =
\begin{bmatrix}
y_{41} & y_{42} & y_{43} \\
y_{51} & y_{52} & y_{53} \\
y_{61} & y_{62} & y_{63}
\end{bmatrix}
$$

Graphically we represent this partitioning as grouping the six coils into two compound coils (a) and (b), each composed of three individual admittances. This is illustrated in Fig. I.5.

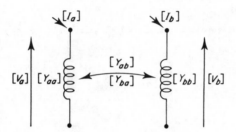

Figure I.5
Two coupled compound admittances

On examination of $[Y_{ab}]$ and $[Y_{ba}]$ it can be seen that $[Y_{ba}] = [Y_{ab}]^T$ if, and only if $y_{ik} = y_{ki}$ for $i = 1$ to 3 and $k = 4$ to 6; that is, if and only if the couplings between the two groups of admittances are bilateral. In this case equation (I.2.9) may be written

$$
\begin{array}{|c|c|c|c|}
\hline
[I_a] & [Y_{aa}] & [Y_{ab}] & [V_a] \\
\hline
[I_b] & [Y_{ab}]^T & [Y_{bb}] & [V_b] \\
\hline
\end{array}
\qquad (I.2.11)
$$

The primitive network for any number of compound admittances is formed in exactly the same manner as for single admittances, except in that all quantities are matrices of the same order as the compound admittances.

The actual admittance matrix of any network composed of the compound admittances can be formed by the usual method of linear transformation; the elements of the connection matrix are now $n \times n$ identity matrices where n is the dimension of the compound admittances.

If the connection matrix of any network can be partitioned into identity elements of equal dimensions greater than one, the use of compound admittances is advantageous.

As an example, consider the network shown in Figs I.6 and I.7, which represent a simple line section. The admittance matrix will be derived using single and compound admittances to show the simple correspondence. The primitive networks and

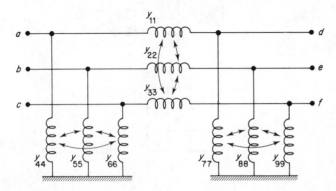

Figure I.6
Sample network represented by single admittances

328

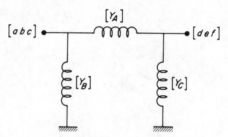

Figure I.7
Sample network represented by compound admittances

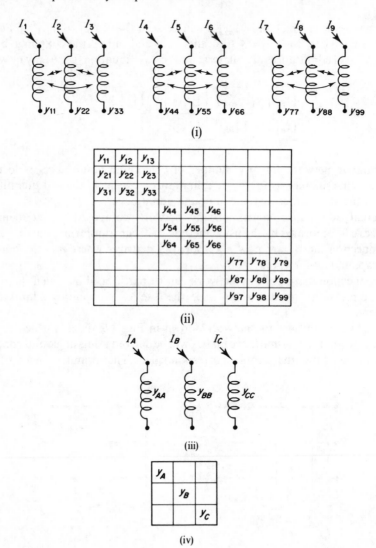

(i)

(ii)

(iii)

(iv)

Figure I.8
Primitive networks and corresponding admittance matrices: (i) primitive network using single admittances; (ii) primitive admittance matrix; (iii) primitive network using compound admittances; (iv) primitive admittance matrix

associated admittance matrices are drawn in Fig. I.8. The connection matrices for the single and compound networks are illustrated by equations (I.2.12) and (I.2.13) respectively:

$$
\begin{bmatrix} V_1 \\ V_2 \\ V_3 \\ V_4 \\ V_5 \\ V_6 \\ V_7 \\ V_8 \\ V_9 \end{bmatrix} =
\begin{bmatrix}
-1 & & & 1 & & \\
& -1 & & & 1 & \\
& & -1 & & & 1 \\
1 & & & & & \\
& 1 & & & & \\
& & 1 & & & \\
& & & 1 & & \\
& & & & 1 & \\
& & & & & 1
\end{bmatrix}
\begin{bmatrix} V_a \\ V_b \\ V_c \\ V_d \\ V_e \\ V_f \end{bmatrix}
\qquad (I.2.12)
$$

$$
\begin{bmatrix} [V_A] \\ [V_B] \\ [V_c] \end{bmatrix} =
\begin{bmatrix}
-I & I \\
I & \\
& I
\end{bmatrix} \cdot
\begin{bmatrix} [V_{abc}] \\ [V_{def}] \end{bmatrix} .
\qquad (I.2.13)
$$

The exact equivalence, with appropriate matrix partitioning, is clear.

The network admittance matrix is given by the linear transformation equation

$$[Y_{\text{NODE}}] = [C]^{\text{T}} \cdot [Y_{\text{PRIM}}] \cdot [C].$$

This matrix multiplication can be executed using the full matrices or in partitioned form. The result in partitioned form is

$$
[Y_{\text{NODE}}] =
\begin{bmatrix}
[Y_A] + [Y_B] & -[Y_A] \\
-[Y_A] & [Y_A] + [Y_C]
\end{bmatrix} .
$$

I.2.3   Rules for Forming the Admittance Matrix of Simple Networks

The method of linear transformation may be used to obtain the admittance matrix of any network. For the special case of networks where there is no mutual coupling, simple rules may be used to form the admittance matrix by inspection. These rules, which apply to compound networks with no mutual coupling between the compound admittances, may be stated as follows.

(i) Any diagonal term is the sum of the individual branch admittances connected to the node corresponding to that term.

(ii) Any off-diagonal term is the negated sum of the branch admittances which are connected between the two corresponding nodes.

I.2.4   Network Subdivision

To enable the transmission system to be modelled in a systematic, logical and convenient manner the system must be subdivided into more manageable units. These units, called subsystems, are defined as follows.

A SUBSYSTEM is the unit into which any part of the system may be divided such that no subsystem has any mutual couplings between its constituent branches and those of the rest of the system.

This definition ensures that the subsystems may be combined in an extremely straightforward manner.

The system is first subdivided into the most convenient subsystems consistent with the definition above. The most convenient unit for a subsystem is a single network element. In previous sections the nodal admittance matrix representation of all common elements has been derived.

The subsystem unit is retained for input data organisation. The data for any subsystem is input as a complete unit, the subsystem admittance matrix is formulated and stored and then all subsystems are combined to form the total system admittance matrix.

I.3   LINE SECTIONALISATION

A line may be divided into sections to account for features such as the following:

- transposition of line conductors
- change of type of supporting towers
- variation of soil permitivity
- improvement of line representation (series of two or more equivalent-π networks)
- series capacitors for line compensation
- Lumping of series elements not central to a particular study.

An example of a line divided into a number of sections is shown in Fig. I.9. The network shown is considered to form a single subsystem. The resultant admittance matrix between bus A and bus B may be derived by finding, for each section, the $ABCD$ or transmission parameters, then combining these by matrix multiplications to give the resultant transmission parameters. These are then converted to the required admittance parameters.

This procedure involves an extension of the usual two-port network theory to multi-two-port networks. Currents and voltages are new matrix quantities and are defined in Fig. I.10. The $ABCD$ matrix parameters are also shown.

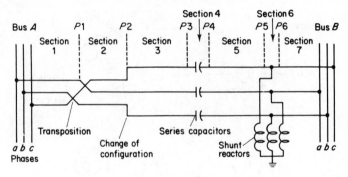

Figure I.9
Example of a transmission line divided into sections

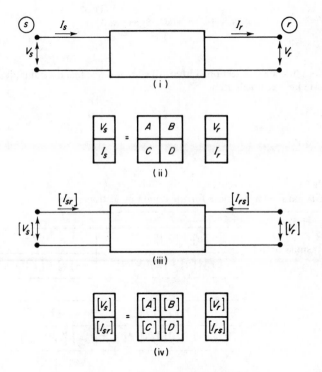

Figure I.10
Two-part network transmission parameters: (i) normal two-port network; (ii) transmission parameters; (iii) multi-two-port network; (iv) matrix transmission parameters

The dimensions of the parameters matrices correspond to those of the section being considered, i.e. 3, 6, 9, or 12 for 1, 2, 3 or 4 mutually coupled three-phase elements respectively. All sections must contain the same number of mutually coupled three-phase elements, ensuring that all the parameter matrices are of the same order and that the matrix multiplications are executable. To illustrate this feature, consider the example of Fig. I.11. Features of interest are as follows.

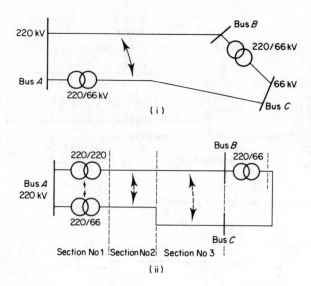

Figure I.11
Sample system to illustrate line sectionalisation: (i) system single-line diagram; (ii) system redrawn
to illustrate line sectionalisation

Table I.1
ABCD parameter matrices for the common section types

Transmission line	$[u] + [Z][Y]/2$	$-[Z]$
	$[Y]\{[u] + [Z)[Y]/4\}$	$-\{[u] + [Y][Z]/2\}$

Transfomer	$-[Y_{SP}]^{-1}[Y_{SS}]$	$[Y_{SP}]^{-1}$
	$[Y_{PS}] - [Y_{PP}][Y_{SP}]^{-1}[Y_{SS})$	$[Y_{PP}][Y_{SP}]^{-1}$

Shunt element	$[u]$	$[0]$
	$[Y_{SH}]$	$-[\partial]$

Series element	$[u]$	$-[Y_{SE}]^{-1}$
	$[0]$	$-[u]$

(i) As a matter of programming convenience an ideal transformer is created and included in section 1.

(ii) The dotted coupling represents coupling which is zero. It is included to ensure correct dimensionality of all matrices.

(iii) In the p.u. system the mutual coupling between the 220 kV and 66 kV lines is expressed to a voltage base given by the geometric mean of the base line-neutral voltages of the two parallel circuits.

In Table I.1 $[u]$ is the unit matrix, $[0]$ is a matrix of zeros, and all other matrices have been defined in their respective sections. Note that all these matrices have dimensions corresponding to the number of coupled three-phase elements in the section.

Once the resultant $ABCD$ parameters have been found the equivalent nodal admittance matrix for the subsystem can be calculated from the equation

$$[Y] = \begin{array}{|c|c|} \hline [D][B]^{-1} & [C] - [D][B]^{-1}[A] \\ \hline [B]^{-1} & -[B]^{-1}[A] \\ \hline \end{array} \tag{I.3.1}$$

I.4 FORMATION OF THE SYSTEM ADMITTANCE MATRIX

It has been shown that the element (and subsystem) admittance matrices can be manipulated efficiently if the three nodes at the busbar are associated together. This association proves equally helpful when forming the admittance matrix for the total system.

The subsystem, as defined in Section I.2, may have common busbars with other subsystems, but may not have mutual coupling terms to the branches of other subsystems. Therefore the subsystem admittance matrices can be combined to form the overall system admittance matrix as follows.

• The self-admittance of any busbar is the sum of all the individual self-admittance matrices at that busbar.

• The mutual admittance between any two busbars is the sum of the individual mutual-admittance matrices from all the subsytems containing those two nodes.

APPENDIX II.
MODELLING OF STATIC
A.C.–D.C. CONVERSION PLANT

II.1 INTRODUCTION

Although the power electronic device is basically a switch, it is only explicitly represented as such in dynamic studies. The periodicity of switching sequences can be used in steady-state studies to model the active and reactive power loading conditions of a.c.–d.c. converters at the relevant busbars. Such modelling is discussed here with reference to the most common configuration used in power systems, i.e. the three-phase bridge rectifier shown in Fig. II.1.

For large power ratings static converter units generally consist of a number of series and/or parallel connected bridges, some or all bridges being phase-shifted relative to the others. With these configurations twelve-pulse and higher pulse numbers can be achieved to reduce the distortion of the supply current with limited or no filtering. A multiple bridge rectifier can therefore be modelled as a single equivalent bridge with a sinusoidal supply voltage at the terminals.

The following basic assumptions are normally made in the development of the steady-state model.

(i) The forward voltage drop in a conducting valve is neglected so that the valve may be considered as a switch. This is justified by the fact that the voltage drop is very small in comparison with the normal operating voltage. It is, further, quite

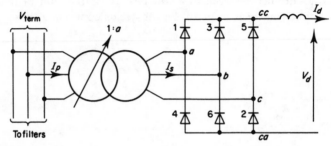

Figure II.1
Basic three-phase rectifier bridge

independent of the current and should therefore play an insignificant part in the commutation process since all valves commutating on the same side of the bridge suffer similar drops. Such a voltage drop is taken into account by adding it to the d.c. line resistance. The transformer windings resistance is also ignored in the development of the equations, though it should also be included to calculate the power loss.

(ii) The converter transformer leakage reactances as viewed from the secondary terminals are identical for the three phases, and variations of leakage reactance caused by on-load tap-changing are ignored.

(iii) The direct current ripple is ignored, i.e. sufficient smoothing inductance is assumed on the d.c. side.

II.2   RECTIFICATION

Rectifier loads can use diode and thyristor elements in full or half-bridge configurations. In some cases the diode bridges are complemented by transformer on-load tap-changer and saturable reactor control. Saturable reactors produce the same effect as thyristor control over a limited range of delay angles.

Referring to the voltage waveforms in Fig. II.2. and using as time reference the

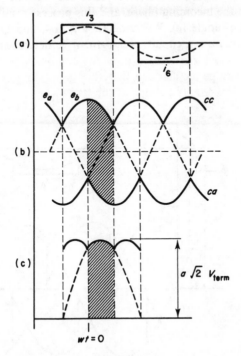

Figure II.2
Diode rectifier waveforms: (a) alternating current in phase '*b*'; (b) common anode (*ca*) and cathode (*cc*) voltage waveform; (c) rectified voltage

instant when the phase to neutral voltage in phase 'b' is a maximum, the commutating voltage of valve 3 can be expressed as:

$$e_b - e_a = \sqrt{2}aV_{\text{term}}\sin\left(\omega t + \frac{\pi}{3}\right)$$

where 'a' is the off-nominal tap-change position of the converter transformer. The shaded area in Fig. II.2(b) indicates the potential difference between the common cathode (cc) and common anode (ca) bridge poles for the case of uncontrolled rectification. The maximum average rectified voltage is therefore

$$V_0 = \frac{1}{\pi/3}\int_0^{\pi/3}\sqrt{2}aV_{\text{term}}\sin\left(\omega t + \frac{\pi}{3}\right)d(\omega t) = \frac{3\sqrt{2}}{\pi}aV_{\text{term}}. \tag{II.2.1}$$

However, uncontrolled rectification is rarely used in large power conversion. Controlled rectification is achieved by phase-shifting the valve conducting periods with respect to their corresponding phase voltage waveforms.

With delay angle control the average rectified voltage (shown in Fig. II.3) is thus

$$V_d = \frac{1}{\pi/3}\int_\alpha^{\pi/3+\alpha}\sqrt{2}aV_{\text{term}}\sin\left(\omega t + \frac{\pi}{3}\right)d(\omega t) = V_0\cos\alpha. \tag{II.2.2}$$

In practice the voltage waveform is that of Fig. II.4, where a voltage area (δA) is lost due to the reactance (X_c) of the a.c. system (as seen from the converter), referred to as commutation reactance. The energy stored in this reactance has to be transferred from the outgoing to the incoming phase, and this process results in a commutation or conduction overlap angle (u). Referring to Fig. II.4, and ignoring the effect of resistance in the commutation circuit, area δA can be determined as follows:

$$e_b - e_a = 2\frac{X_c}{\omega}\frac{di_c}{dt} \tag{II.2.3}$$

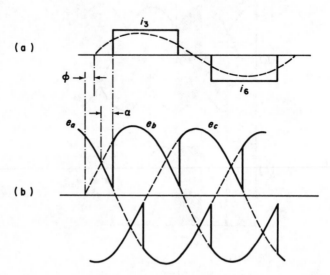

Figure II.3
Thyristor-controlled waveform: (a) alternating current in phase 'b'; (b) rectified d.c. voltage waveforms

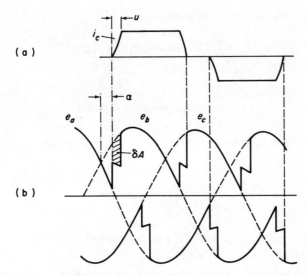

Figure II.4
Effect of commutation reactance: (a) alternating current; (b) d.c. voltage waveforms

where e_a, e_b are the instantaneous voltages of phases a and b respectively, and i_c is the incoming valve (commutating) current. Hence

$$\delta A = \int_{\alpha}^{\alpha+\gamma} \frac{e_b - e_a}{2} d(\omega t) = X_c \int_0^{I_d} di_c = X_c I_d. \tag{II.2.4}$$

Finally, by combining equations (II.2.1), (II.2.2) and (II.2.4) the following a.c.–d.c. voltage relationship is obtained:

$$V_d = V_0 \cos \alpha - \frac{\delta A}{\pi/3} = \frac{3\sqrt{2}}{\pi} a V_{\text{term}} \cos \alpha - \frac{3}{\pi} X_c I_d. \tag{II.2.5}$$

It must be emphasised that the commutating voltage (V_{term}) is the a.c. voltage at the closest point to the converter bridge where sinusoidal waveforms can be assumed. The commutation reactance (X_c) is the reactance between the point at which V_{term} exists and the bridge. Where filters are installed the filter busbar voltage can be used as V_{term}. In the absence of filters, V_{term} must be established at some remote point and X_c must be modified to include the system impedance from the remote point to the converter.

With perfect filtering, only the fundamental component of the current waveform will appear in the a.c. system. This component is obtained from the Fourier analysis of the current waveform in Fig. II.4, and requires information of i_c and u.

Taking as a reference the instant when the line voltage ($e_b - e_a$) is zero, equation (II.2.3) can be written as

$$\sqrt{2} a V_{\text{term}} \sin \omega t = 2 \frac{X_c}{\omega} \frac{di_c}{dt}$$

and integrating with respect to ωt gives

$$\int \frac{aV_{\text{term}}}{\sqrt{2}} \sin \omega t \, d(\omega t) = X_c \int di_c$$

or

$$-\frac{1}{\sqrt{2}} aV_{\text{term}} \cos \omega t + K = X_c i_c.$$

From the initial condition, that $i_c = 0$ at $\omega t = \alpha$, the following expressions for K and i_c are obtained:

$$K = \frac{1}{\sqrt{2}} aV_{\text{term}} \cos \alpha$$

$$i_c = \frac{aV_{\text{term}}}{\sqrt{2} X_c} [\cos \alpha - \cos \omega t]. \tag{II.2.6}$$

From the final condition $i_c = I_d$ at $\omega t = \alpha + u$, the following expressions for I_d and u are obtained:

$$I_d = \frac{aV_{\text{term}}}{\sqrt{2} X_c} [\cos \alpha - \cos(\alpha + u)] \tag{II.2.7}$$

$$u = \cos^{-1} \left[\cos \alpha - \frac{\sqrt{2} X_c I_d}{aV_{\text{term}}} \right] - \alpha. \tag{II.2.8}$$

Equation (II.2.6) provides the time-varying commutating current and equation (II.2.8) the limits for the Fourier analysis.

Fourier analysis of the a.c. current waveform, including the effect of commutation (Fig. II.4), leads to the following relationship between the r.m.s. of the fundamental component and the direct current:

$$I_s = k \frac{\sqrt{6}}{\pi} I_d \tag{II.2.9}$$

where

$$k = \frac{\sqrt{[\cos 2\alpha - \cos 2(\alpha + u)]^2 + [2u + \sin 2\alpha - \sin 2(\alpha + u)]^2}}{4[\cos \alpha - \cos(\alpha + u)]}$$

for values of u not exceeding $60°$.

The values of k are very close to unity under normal operating conditions, i.e., when the voltage and currents are close to their nominal values and the a.c. voltage waveforms are symmetrical and undistorted. Alternative steady-state models for operating conditions deviating from the above are described in Chapters 4 and 7.

Taking into account the transformer tap position the current on the primary side becomes

$$I_p = k \frac{\sqrt{6}}{\pi} aI_d. \tag{II.2.10}$$

When using per unit values based on a common power and voltage base on both sides of the converter, the direct current base has to be $\sqrt{3}$ times larger than the a.c. current base (as explained in Section 4.3) and equation (II.2.10) becomes

$$I_p = k\frac{3\sqrt{2}}{\pi}aI_d. \tag{II.2.11}$$

Using the fundamental components of voltage and current and assuming perfect filtering at the converter terminals the power factor angle at the converter terminals is ϕ (the displacement between fundamental voltage and current waveforms) and we may write

$$P = \sqrt{3}V_{\text{term}}I_p\cos\phi = V_dI_d \tag{II.2.12}$$

or

$$\cos\phi = \frac{1}{2k}(\cos\alpha + \cos(\alpha + u)) \tag{II.2.13}$$

and

$$Q = \sqrt{3}V_{\text{term}}I_p\sin\phi. \tag{II.2.14}$$

II.3   INVERSION

Owing to the unidirectional nature of current flow through the converter valves, power reversal (i.e. power flow from the d.c. to the a.c. side) requires direct voltage polarity reversal. This is achieved by delay angle control, which, in the absence of commutation overlap produces rectification between $0° < \alpha < 90°$ and inversion between $90° < \alpha < 180°$. In the presence of overlap, the value of 'α' at which inversion begins is always less than $90°$. Moreover, unlike with rectification, full inversion (i.e. $\alpha = 180°$) can not be achieved in practice. This is due to the existence of a certain deionisation angle γ at the end of the conducting period, before the voltage across the commutating valve reverses, i.e.

$$\alpha + u \leqslant 180 - \gamma_0.$$

If the above condition is not met (γ_0 being the minimum required extinction angle) a commutation failure occurs; this event would upset the normal conducting sequence and preclude the use of the steady-state model derived in this appendix.

The inverter voltage, although of opposite polarity with respect to the rectifier, is usually expressed as positive when considered in isolation.

Typical inverter voltage and current waveforms are illustrated in Fig. II.5. By similarity with the waveforms of Fig. II.4, the following expression can be written for the inverter voltage in terms of the extinction angle:

$$V_d = \frac{3\sqrt{2}}{\pi}aV_{\text{term}}\cos\gamma - \frac{3X_c}{\pi}I_d \tag{II.3.1}$$

which is the same as equation (II.2.5) substituting γ for α.

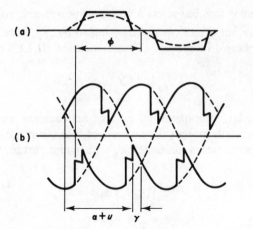

Figure II.5
Inverter waveforms: (a) alternating current; (b) d.c. voltage waveforms

It should by now be obvious that inverter operation requires the existence of three conditions.

(i) An active a.c. system which provides the commutating voltages.

(ii) A d.c. power supply of opposite polarity to provide continuity for the unidirectional current flow (i.e. from anode to cathode through the switching devices).

(iii) Fully controlled rectification to provide firing delays beyond 90°.

When these three conditions are met, a negative voltage of a magnitude given by equation (II.3.1), is impressed across the converter bridge and power $(-V_d I_d)$ is inverted. Note that the power factor angle (ϕ) is now larger than 90°, i.e.

$$P = \sqrt{3} V_{\text{term}} I_p \cos\phi = -\sqrt{3} V_{\text{term}} I_p \cos(\pi - \phi) \qquad \text{(II.3.2)}$$

$$Q = \sqrt{3} V_{\text{term}} I_p \sin\phi = \sqrt{3} V_{\text{term}} I_p \sin(\pi - \phi). \qquad \text{(II.3.3)}$$

Equations (II.3.2) and (II.3.3) indicate that the inversion process still requires reactive power supply from the a.c. side. The vector diagram of Fig. II.6 illustrates the sign of P and Q for rectification and inversion.

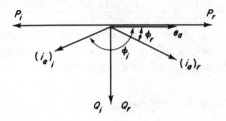

Figure II.6
P and Q vector diagram

II.4 COMMUTATION REACTANCE

Fig. II.7 shows the general case of n bridges connected in parallel on the a.c. side. In the absence of filters the pure sinusoidal voltages exist only behind the system source impedance (X_{ss}) and the commutation reactance (X_{c_j}) for the jth bridge is thus

$$X_{cj} = X_{ss} + X_{tj}. \tag{II.4.1}$$

However, if the bridges are under the same controller or under identical controllers then it is preferable to create a single equivalent bridge. The commutation reactance of such an equivalent bridge depends upon the d.c. connections and also the phase shifting between bridges.

If there are k bridges with the same phase shift then they will commutate at the same time and the equivalent reactance must reflect this. For a series connection of bridges the commutation reactance of the equivalent bridge is

$$X_{c_{\text{series}}} = kX_{ss} + X_{tj} \tag{II.4.2}$$

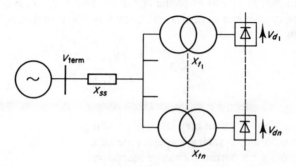

Figure II.7
'n' bridges connected in series on the d.c. side and in parallel on the a.c. side

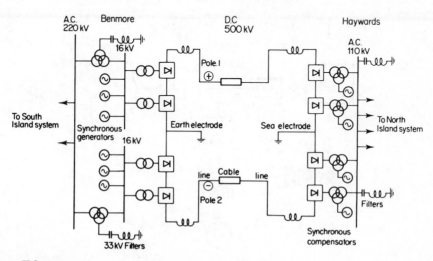

Figure II.8
Simplified diagram of the New Zealand h.v.d.c. interconnection

where j represents any of the n bridges. For bridges connected in parallel on the d.c. side the equivalent bridge commutation reactance is

$$X_{c_{parallel}} = X_{ss} + \frac{1}{k} X_{tj}. \qquad (II.4.3)$$

It should be noted that with perfect filtering or when many bridges are used with different transformer phase shifts the voltage on the a.c. side of the converter transformers may be assumed to be sinusoidal and hence X_{ss} has no influence on the commutation.

Moreover, the presence of local plant components at the converter terminals may affect the commutation reactance. By way of example, let us consider the two ends of the New Zealand h.v.d.c. link (with reference to Fig. II.8). It must be noted that h.v.d.c. schemes are normally designed for twelve-pulse operation and that filters are always provided (i.e. the system impedance can be ignored).

(i) At Haywards the effect of the subtransient reactance of the synchronous compensators on the tertiaries of the converter transformers must be taken into account. The approximate equivalent circuit is illustrated in Fig. II.9 and the commutation reactance is

$$X_c = X_s + \frac{X_p \cdot (X_t + X_d'')}{X_p + X_t + X_d''} \qquad (II.4.4)$$

where

X_s is the transformer secondary leakage reactange
X_p is the transformer primary leakage reactance
X_t is the transformer tertiary leakage reactance
X_d'' is the subtransient reactance of the synchronous condenser unit.

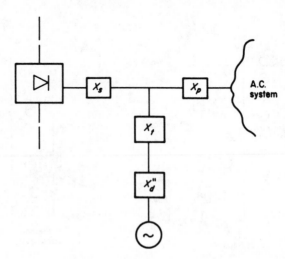

Figure II.9
Equivalent circuit for the calculation of the commutation reactance at the Haywards end

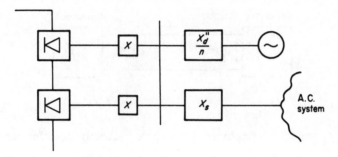

Figure II.10
Equivalent circuit for the calculation of the commutation reactance at the Benmore end

(ii) At the Benmore end the subtransient reactance of the generators is combined in parallel with the secondary reactance of the interconnecting transformer. (The primary reactance is beyond the filters and can thus be neglected.) The approximate equivalent circuit is illustrated in Fig. II.10. Although there are two converter groups commutating on this reactance, the commutations are not simultaneous due to the 30° phase shift of their respective transformers. Thus the effective commutation reactance per group is

$$X_c = X + \frac{X_d'' \cdot X_s}{X_d'' + nX_s} \tag{II.4.5}$$

where

X is the two-winding transformer leakage reactance
X_s is the interconnecting transformer secondary leakage reactance (note filters connected to tertiary winding)
X_d'' is the generator subtransient reactance
n is the number of generators connected.

II.5   D.C. TRANSMISSION

The sending and receiving ends of a two-terminal d.c. transmission link such as that illustrated in Fig. II.8 can be modelled as single equivalent bridges with terminal voltages V_{d_r} and V_{d_i} respectively. The direct current is thus given by

$$I_d = I_{d_r} = I_{d_i} = \frac{V_{d_r} - V_{d_i}}{R_d} \tag{II.5.1}$$

where R_d is the resistance of the link and includes the loop transmission resistance (if any), the resistance of the smoothing reactors and the converter valves.

    The prime considerations in the operation of a d.c. transmission system are to minimise the need for reactive power at the terminals and reduce system losses. These objectives require maintaining the highest possible transmission voltage and this is achieved by minimising the inverter end extinction angle, i.e. operating the inverter on constant extinction angle (e.a.) control while controlling the direct current at the

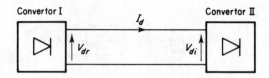

Figure II.11
Two-terminal d.c. link

rectifier end by means of temporary delay angle backed by transformer tap-change control.

When e.a. control is applied to the inverter it automatically varies the firing angle of advance to maintain the extinction angle γ at a constant value. Deionisation imposes a definite minimum limit on γ, and the e.a. control usually maintains it at this limit.

Constant current (c.c.) control applied to the rectifier regulates the firing angle α to maintain a pre-specified link current I_d^{sp}, within the range of α. If the value of α required to maintain I_d^{sp} falls below its minimum limit, current control is transferred to the inverter, i.e. α is fixed on its minimum limit, and the inverter firing angle is advanced to control the current.

The converter-transformer tap-change is a composite part of this control. The rectifier transformer attempts to maintain α within its permitted range. The inverter transformer attempts to regulate the d.c. voltage at some point along the line to a specified level. For minimum loss and minimum reactive-power absorption, this voltage is required to be as high as possible, and the firing angle of the rectifier should be as low as possible.

Fig. II.12 shows the d.c. voltage/current characteristics at the rectifier and inverter ends (the latter have been drawn with reverse polarity in order to illustrate the operating point). The current controller gains are very large and for all practical purposes the slopes of the constant current characteristics can be ignored. Consequently the operating current is equal to the relevant current setting, i.e. $I_{d_{sr}}$ and $I_{d_{si}}$ for rectifier and inverter constant current control respectively.

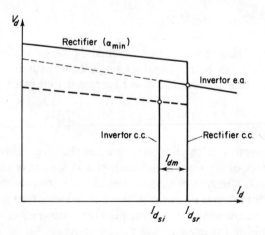

Figure II.12
Normal control characteristics

The direction of power flow is determined by the current settings, the rectifier end always having the larger setting. The difference between the settings is the current margin I_{d_m} and is given by

$$I_{d_m} = I_{d_{sr}} - I_{d_{si}} > 0. \tag{II.5.4}$$

Many d.c. transmission schemes are bidirectional, i.e. each converter operates sometimes as a rectifier and sometimes as an inverter. Moreover, during d.c. line faults, both converters are forced into the inverter mode in order to de-energise the line faster. In such cases each converter is provided with a combined characteristic as shown in Fig. II.13 which includes natural rectification, constant current control and constant extinction angle control.

With the characteristics shown by solid lines (i.e. operating at point A), power is transmitted from converter I to converter II. Both stations are given the same current command but the current margin setting is subtracted at the inverter end. When power reversal is to be implemented the current settings are reversed and the broken line characteristics apply. This results in operating point B, with direct voltage reversed and no change in direct current.

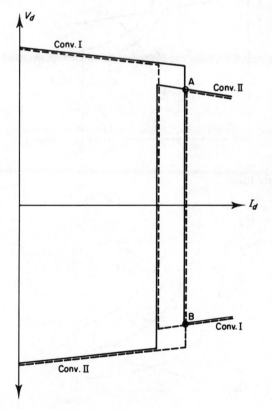

Figure II.13
Control characteristics and power-flow reversal

II.5.1 Alternative Forms of Control

A common used operating mode is constant power (c.p.) control. As with constant current control either converter can control power. The power setting at the rectifier terminal $P_{d_{sr}}$ must be larger than that at the invertor terminal $P_{d_{si}}$ by a suitable power margin P_{d_m}, that is

$$P_{d_m} = P_{d_{sr}} - P_{d_{si}} > 0. \tag{II.5.5}$$

The c.p. controller adjusts the c.c. control setting I_d^{sp} to maintain a specified power flow P_d^{sp} through the link, which is usually more practical than c.c. control from a system operation point of view. The voltage/current loci now become nonlinear, as shown in Fig. II.14.

Several limits are added to the c.p. characteristics as shown in Fig. II.15. These are:

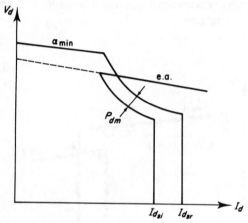

Figure II.14
Constant power characteristics

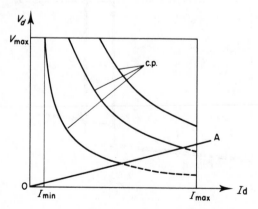

Figure II.15
Voltage and currents limits

- a maximum current limit with the purpose of preventing thermal damage to the converter valves; normally between 1 and 1.2 times the nominal current
- a minimum current limit (about 10 % of the nominal value) in order to avoid possible current discontinuities which can cause overvoltages
- voltage-dependent current limit (line OA in the figure) in order to reduce the power loss and reactive power demand.

In cases where the power rating of the d.c. link is comparable with the rating of either the sending or receiving a.c. system interconnected by the link, the frequency of the smaller a.c. system is often controlled to a large extent by the d.c. link. With power frequency (p.f.) control if the frequency goes out of pre-specified limits, the output power is made proportional to the deviation of frequency from its nominal value. Frequency control is analogous to the current control described earlier, i.e. the converter with lower voltage determines the direct voltage of the line and the one with higher voltage determines the frequency. Again, current limits have to be imposed, which override the frequency error signal.

The c.p./e.a. and c.c./e.a. controls were evolved principally for bulk point-to-point power transmission over long distances or submarine crossings and are still the main control modes in present use.

Multiterminal d.c. schemes are also being considered, based on the basic controls already described. Two alternatives are possible, i.e. constant voltage parallel and constant current series schemes.

APPENDIX III.
MODAL ANALYSIS
OF MULTICONDUCTOR LINES

The steady-state behaviour of a multiconductor line at a discrete frequency is described by the equations

$$-\left[\frac{dV}{dx}\right] = [Z'] \cdot [I] \tag{III.1}$$

$$-\left[\frac{dI}{dx}\right] = [Y'] \cdot [V] \tag{III.2}$$

where $[Z']$ and $[Y']$ are the series impedance and shunt admittance matrices per unit distance and $[V]$ and $[I]$ are the vectors of voltage and current phasors in the various conductors.

Differentiating equations (III.1) and (III.2) again with respect to x gives

$$\left[\frac{d^2V}{dx^2}\right] = [Z'] \cdot [Y'] \cdot [V] \tag{III.3}$$

$$\left[\frac{d^2I}{dx^2}\right] = [Y'] \cdot [Z'] \cdot [I]. \tag{III.4}$$

It should be noted that in this case the matrix products $[Z'] \cdot [Y']$ and $[Y'] \cdot [Z']$ are not equal, except in special cases.

These equations are still difficult to solve because all phases are coupled. However, just as three-phase equations with balanced matrices can be transformed into decoupled single-phase equations using symmetrical components, it is possible to transform equations (III.3) and (III.4) into decoupled equations as well. By transforming phase voltages to 'modal' voltages,

$$[V] = [T_v] \cdot [V_{\text{mode}}] \quad \text{and} \quad [V_{\text{mode}}] = [T_v]^{-1} \cdot [V] \tag{III.5}$$

and by choosing the proper transformation matrix $[T_v]$, equation (III.3) can be changed to

$$\left[\frac{d^2V_{\text{mode}}}{dx^2}\right] = [\Lambda][V_{\text{mode}}] \tag{III.6}$$

348

where $[\Lambda]$ is now a diagonal matrix. This diagonalisation is a well defined procedure in matrix algebra; the elements of $[\Lambda]$ are the eigenvalues of the matrix product $[Z']\cdot[Y']$, and the transformation matrix $[T_v]$ is the matrix of eigenvectors of that matrix product. Equation (III.4) can be diagonalised as well, with the same diagonal matrix $[\Lambda]$, i.e.

$$\left[\frac{d^2 I_{\text{mode}}}{dx^2}\right] = [\Lambda][I_{\text{mode}}] \tag{III.7}$$

but the transformation matrix for currents differs from that used for voltages (in contrast to symmetrical components):

$$[I] = [T_i]\cdot[I_{\text{mode}}] \quad \text{and} \quad [I_{\text{mode}}] = [T_i]^{-1}\cdot[I] \tag{III.8}$$

though both are related by

$$[T_i]^t = [T_v]^{-1} \tag{III.9}$$

where the subscript 't' indicates a transposed matrix.

With the diagonalised equations (III.6) and (III.7), an m-phase line can now be studied as if it consisted of m single-phase lines, similar to the symmetrical component approach, except that the zero-, positive- and negative-sequence networks now become the mode 1, mode 2 and mode 3 networks. The modal series impedance and shunt admittance are not directly available but must be computed from

$$[Z'_{\text{mode}}] = [T_v]^{-1}\cdot[Z']\cdot[T_i] \tag{III.10a}$$

and

$$[Y'_{\text{mode}}] = [T_i]^{-1}\cdot[Y']\cdot[T_v] \tag{III.10b}$$

with both modal matrices being diagonal. $[Y'_{\text{mode}}]$ may no longer be purely imaginary even though only shunt capacitance is modelled. This will depend on how the transformation matrices were normalised. For steady-state analysis at one particular frequency, this causes no problems. Once Z_{series} and Y_{shunt} have been calculated for each mode, the representation in phase quantities is easily obtained by transforming back, with

$$[Z_{\text{series}}] = [T_v]\cdot[Z_{\text{series-mode}}]\cdot[T_i]^{-1} \tag{III.11a}$$

and

$$[Y_{\text{shunt}}] = [T_i]\cdot[Y_{\text{shunt-mode}}]\cdot[T_v]^{-1} \tag{III.11b}$$

becoming the values of the equivalent-π model which will accurately represent the untransposed line.

In expanded form the following are expressions for the series impedance and shunt admittance of the equivalent-π model:

$$[Z]_{\text{EPM}} = l[Z']\cdot[M]\left[\frac{\sinh \gamma l}{\gamma l}\right][M]^{-1} \tag{III.12}$$

350

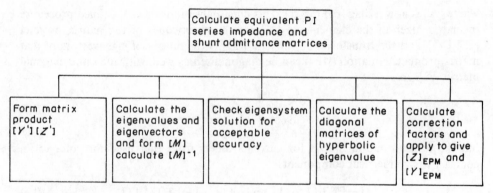

Figure III.1
Structure diagram for calculation of the equivalent-π model

where l is the transmission line length, $[Z]_{EPM}$ is the equivalent-π series impedance matrix, $[M]$ is the matrix of normalised eigenvectors, and

$$\left[\frac{\sinh \gamma l}{\gamma l}\right] = \begin{bmatrix} \dfrac{\sinh_{\gamma_1} l}{\gamma_1 l} & 0 & \cdots & 0 \\ 0 & \dfrac{\sinh_{\gamma_2} l}{\gamma_2 l} & \cdots & 0 \\ \vdots & \vdots & & \vdots \\ 0 & 0 & \cdots & \dfrac{\sinh \gamma_j l}{\gamma_j l} \end{bmatrix}$$
(III.13)

and γ_j is the jth eigenvalue for $j/3$ mutually coupled circuits. Similarly

$$[Y]_{EPM} = l[M]\left[\frac{\tanh(\gamma l/2)}{\gamma l/2}\right][M]^{-1} \cdot [Y']$$
(III.14)

where $[Y]_{EPM}$ is the equivalent-π shunt admittance matrix.

Computer derivation of the correction factors for conversion from the nominal-π to the equivalent-π model, and their incorporation into the series impedance and shunt admittance matrices, is carried out as indicated in the structure diagram of Fig. III.1. The LR2 algorithm of Wilkinson and Reinsch[†] is used with due regard for accurate calculations in the derivation of the eigenvalues and eigenvectors.

[†]J. H. Wilkinson and C. Reinsch, (1971). '*Handbook for Automatic Computations*' Vol. II (*Linear Algebra*) Springer-Verlag, Berlin.

APPENDIX IV.
NUMERICAL INTERGRATION METHODS

IV.1 INTRODUCTION

Basic to the computer modelling of power system transients is the numerical integration of the set of differential equations involved. Many books have been written on the numerical solution of ordinary differential equations, but this appendix is restricted to the techniques in common use for the dynamic simulation of power system behaviour.

It is therefore appropriate to start by identifying and defining the properties required from the numerical integration method in the context of power system analysis.

IV.2 PROPERTIES OF THE INTEGRATIONS METHODS

IV.2.1 Accuracy

This property is limited by two main causes, i.e. round-off and truncation errors. Round-off error occurs while performing arithmetic operations and is due to the inability of the computer to represent numbers exactly. A word length of 48 bits is normally sufficient for scientific work and is certainly acceptable for transient stability analysis. When the stability studies are carried out on computers with a 32-bit word length, it is necessary to use double precision on certain areas of the storage to maintain adequate accuracy.

The difference between the true and calculated results is mainly determined by the truncation error, which is due to the numerical method chosen. The true solution at any one point can be expressed as a Taylor series based on some initial point and by substituting these into the formulae, the order of accuracy can be found from the lowest power of step length (h) which has a nonzero coefficient. In general terms, the truncation error $T(h)$ of a method using a step length h is given by

$$T(h) = O(h^{p+1}) \tag{IV.2.1}$$

where superscript p represents the order of accuracy of the method.

The true solution $y(t_n)$ at t_n is thus

$$y(t_n) = y_n + O(h^{p+1}) + \varepsilon_n \tag{IV.2.2}$$

where y_n is the value of y calculated by the method after n steps, and ε_n represents other possible errors.

IV.2.2 Stability

Two types of instability occur in the solution of ordinary differential equations, i.e. inherent and induced instability.

Inherent instability occurs when, during a numerical step-by-step solution, errors generated by any means (truncation or round-off) are magnified until the true solution is swamped. Fortunately transient stability studies are formulated in such a manner that inherent instability is not a problem.

Induced instability is related to the method used in the numerical solution of the ordinary differential equation. The numerical method gives a sequence of approximations to the true solution and the stability of the method is basically a measure of the difference between the approximate and true solutions as the number of steps becomes large.

Consider the ordinary differential equation

$$py = \lambda y \qquad (IV.2.3)$$

with the initial conditions $y(0) = y_0$ which has the solution

$$y(t) = y_0 e^{\lambda t}. \qquad (IV.2.4)$$

Note that λ is the eigenvalue [1] of the single-variable system given by the ordinary differential equation (IV.2.3). This may be solved by a finite difference equation of the general multistep form:

$$\sum_{i=0}^{k} \alpha_i y_{n-i+1} - h \sum_{i=0}^{k} \beta_i p y_{n-i+1} = 0 \qquad (IV.2.5)$$

where α_i and β_i are constants.

Letting

$$m(z) = \sum_{i=0}^{k} \alpha_i (z)^i \qquad (IV.2.6)$$

and

$$\sigma(z) = \sum_{i=0}^{k} \beta_i (z)^i$$

and constraining the difference scheme to be stable when $\lambda = 0$, then the remaining part of (IV.2.5) is linear and the solutions are given by the roots z_i (for $i = 1, 2, \ldots, k$) of $m(z) = 0$. If the roots are all different, then

$$y_n = A_1 (z_1)^n + A_2 (z_2)^n + \cdots A_k (z_k)^n \qquad (IV.2.7)$$

and the true solution in this case ($\lambda = 0$) is given by

$$y(t_n) = A_1 (z_1)^n + O(h^{p+1}) = y_0 \qquad (IV.2.8)$$

where superscript p is the order of accuracy.

The principal root z_1, in this case, is unity and instability occurs when $|z_i| \geqslant 1$ (for $i = 2, 3, \ldots, k, i \neq 1$) and the true solution will eventually be swamped by this root as n increases.

If a method satisfies the above criteria, then it is said to be stable but the degree of stability requires further consideration.

Weak stability occurs where a method can be defined by the above as being stable, but because of the nature of the differential equation, the derivative part of (IV.2.5) gives one or more roots which are greater than or equal to unity. It has been shown by Dalquist [2] that a stable method which has the maximum order of accuracy is always weakly stable. The maximum order or accuracy of a method is either $k + 1$ or $k + 2$ depending on whether k is odd or even, respectively.

Partial stability occurs when the step length (h) is critical to the solution and is particularly relevant when considering Runge–Kutta methods. In general, the roots z_i of (IV.2.7) are dependent on the product $h\lambda$ and also on equations (IV.2.6). The stability boundary is the value of $h\lambda$ for which $|z_i| = 1$, and any method which has this boundary is termed conditionally stable.

A method with an infinite stability boundary is known as A-stable (unconditionally stable). A linear multistep method is A-stable if all solutions of (IV.2.5) tend to zero as $n \to \infty$ when the method is applied with fixed $h > 0$ to (IV.2.3) where λ is a complex constant with $\mathrm{Re}(\lambda) < 0$.

Dalquist has demonstrated that for a multistep method to be A-stable the order of accuracy cannot exceed $p = 2$, and hence the maximum k is unity, that is, a single-step method. Backward Euler and the trapezoidal method are A-stable, single-step methods. Other methods not based upon the multistep principle may be A-stable and also have high orders of accuracy. In this category are implicit Runge–Kutta methods in which $p < 2r$, where r is the number of stages in the method.

A further definition of stability has been introduced recently [3], i.e. Σ-stability which is the multivariable version of A-stability. The two are equivalent when the method is linear but may not be equivalent otherwise. Backward Euler and the trapezoidal method are Σ-stable single-step methods.

The study of scalar ordinary differential equations of the form (IV.2.3) is sufficient for the assessment of stability in coupled equations, provided that λ are the eigenvalues of the ordinary differential equations. Unfortunately, not all the equations used in transient stability analysis are of this type.

IV.2.3 Stiffness

A system or ordinary differential equations in which the ratio of the largest to the smallest eigenvalue is very much greater than one is usually referred to as being stiff. Only during the initial period of the solution of the problem are the largest negative eigenvalues significant, yet they must be accounted for during the whole solution.

For methods which are conditionally stable, a very small step length must be chosen to maintain stability. This makes the method very expensive in computing time.

The advantages of Σ-stability thus become apparent for this type of problem as the step length need not be adjusted to accommodate the smallest eigenvalues.

In an electrical power system the differential equations which describe its behaviour in a transient state have greatly varying eigenvalues. The largest negative eigenvalues are related to the network and the machine stators but these are ignored by establishing algebraic equations to replace the differential equations. The associated variables are then permitted to vary instantaneously.

However, the time constants of the remaining ordinary differential equations are still sufficiently varied to give a large range of eigenvalues. It is therefore important that if the fastest remaining transient are to be considered and not ignored, as so often done in the past, a method must be adopted which keeps the computation to a minimum.

IV.3 PREDICTOR–CORRECTER METHODS

These methods for the solution of the differential equation

$$pY = F(Y, X) \tag{IV.3.1}$$

with $Y(0) = Y_0$ and $X(0) = X_0$ have all been developed from the general k-step finite difference equation

$$\sum_{i=0}^{k} \alpha_i Y_{n-i+1} - h \sum_{i=0}^{k} \beta_i F_{n-i+1} = 0. \tag{IV.3.2}$$

Basically the methods consist of a pair of equations, one being explicit ($\beta_0 = 0$) to give a prediction of the solution at t_{n+1} and the other being implicit ($\beta_0 \neq 0$) which corrects the predicted value. There are a great variety of methods available, the choice being made by the requirements of the solution. It is usual for simplicity to maintain a constant step length with these methods if $k > 2$.

Each application of a correcter method improves the accuracy of the method by one order, up to a maximum given by the order of accuracy of the correcter. Therefore, if the correcter is not to be iterated, it is usual to use a predictor with an order of accuracy one less than that of the correcter. The predictor is thus not essential, as the value at the previous step may be used as a first crude estimate, but the number of iterations of the correcter may be large.

While, for accuracy, there is a fixed number of relevant iterations, it is desirable for stability purposes to iterate to some predetermined level of convergence. The characteristic root (z_1) of a predictor or corrector when applied to the single-variable problem

$$py = \lambda y \tag{IV.3.3}$$

with $y(0) = y_0$ may be found from

$$\sum_{i=0}^{k} (\alpha_i - h\lambda\beta_i) z^{(k-i)} = 0. \tag{IV.3.4}$$

When applying a correcter to the problem defined by equation (IV.3.3) and rearranging equation (IV.3.2) to give

$$y_{n+1} = \frac{-\sum_{i=1}^{k} (\alpha_i - h\lambda\beta_i) y_{n-i+1}}{(\alpha_0 - h\lambda\beta_0)} \tag{IV.3.5}$$

the solution to the problem becomes direct. The predictor is now not necessary as the solution only requires information of y at the previous steps, i.e. at $y_{n=i+1}$ for $i = 1, 2, \ldots, k$.

Where the problem contains two variables, one nonintegrable, such that

$$py = \lambda y + \mu x \qquad \text{(IV.3.6)}$$

with $y(0) = y_0$, $x(0) = x_0$, and

$$0 = g(y, x) \qquad \text{(IV.3.7)}$$

then

$$y_{n+1} = c_{n+1} + m_{n+1} \cdot x_{n+1} \qquad \text{(IV.3.8)}$$

where

$$c_{n+1} = \frac{-\sum_{i=1}^{k}[(\alpha_i - h\lambda\beta_i)y_{n-i+1} - h\mu\beta_i x_{n-i+1}]}{(\alpha_0 - h\lambda\beta_0)} \qquad \text{(IV.3.9)}$$

and

$$m_{n+1} = \frac{-h\mu\beta_0}{(\alpha_0 - h\lambda\beta_0)}. \qquad \text{(IV.3.10)}$$

Although c_{n+1} and m_{n+1} are constant at a particular step, the solution is iterative using equations (IV.3.7) and (IV.3.8). Strictly in this simple case, x_{n+1} in equation (IV.3.8) could be removed using equation (IV.3.7) but in the general multivariable case this is not so.

The convergence of this method is now a function of the nonlinearity of the system. Provided that the step length is sufficiently small, a simple Jacobi form of iteration gives convergence in only a few iterations. It is equally possible to form a Jacobian matrix and obtain a solution by a Newton iterative process, although the storage necessary is much larger and as before, the step length must be sufficiently small to ensure convergence.

For a multivariable system, equation (IV.3.1) is coupled with

$$0 = G(Y, X) \qquad \text{(IV.3.11)}$$

and the solution of the integrable variables is given by the matrix equation

$$Y_{n+1} = C_{n+1} + M_{n+1} \cdot [Y_{n+1}, X_{n+1}]^t. \qquad \text{(IV.3.12)}$$

The elements of the vector C_{n+1} are as given in equation (IV.3.9) and the elements of the sparse M_{n+1} matrix are given in equation (IV.3.10).

The iterative solution may be started at any point in the loop, if Jacobi iteration is used. Because the number of algebraic variables (X) associated with equations (IV.3.1) or (IV.3.11) is small, it is most advantageous to extrapolate these algebraic variables and commence with a solution using equation (IV.3.11).

The disadvantage of any multistep method $(k > 2)$ is that is not self-starting. After a discontinuity $k - 1$, steps must be performed by some other self-starting method. Unfortunately, it is the period immediately after a step which is most critical as the largest negative eigenvalues are significant. As $k - 1$ is usually small, it is not essential to use an A-stable starting method. Accuracy over this period is of more importance.

IV.4   RUNGE–KUTTA METHODS

Runge–Kutta methods are able to achieve high accuracy while remaining single-step methods. This is done by making further evaluation of the functions within the step. The general form of the equation is

$$y_{n+1} = y_n + \sum_{i=1}^{v} w_i k_i \tag{IV.4.1}$$

where

$$k_i = hf\left(t_n + c_i h, y_n + \sum_{j=1}^{v} a_{ij} k_j\right) \qquad \text{for } i = 1, 2, \ldots, v \tag{IV.4.2}$$

$$\sum_{i=1}^{v} w_i = 1. \tag{IV.4.3}$$

Being single-step methods they are self-starting and the step length need not be constant. If j is restricted so that $j < i$, then the method is explicit and c_1 must be zero. When j is permitted to exceed i, then the method is implicit and an iterative solution is necessary.

Also of interest are the forms developed by Merson and Scraton. These are fourth-order methods ($p = 4$) but use five stages ($v = 5$). The extra degree of freedom obtained is used to give an estimate of the local truncation error at that step. This can be used to automatically control the step length.

Although they are accurate, the explicit Runge–Kutta methods are not A-stable. Stability is achieved by ensuring that the step length does not become large compared to any time constant. For a pth-order explicit method the characteristic root is

$$z_1 = 1 + \sum_{i=1}^{p} \frac{1}{i!} h^i \lambda^i + \sum_{i=p+1}^{v} \frac{a_i}{i!} h^i \lambda^i \tag{IV.4.4}$$

where the second summation term exists only when $v > p$ and where a_i are constant and dependent on the method.

For some implicit methods the characteristic root is equivalent to a Pade approximant to $e^{h\lambda}$.

The Padé approximant of a function $f(t)$ is given by

$$P_{MN}(f(t)) = \sum_{j=1}^{M} (a_j t^j) \bigg/ \sum_{j=1}^{N} (b_j t^j) \tag{IV.4.5}$$

and if

$$f(t) = \sum_{j=0}^{\infty} (c_j, t^j) \tag{IV.4.6}$$

then

$$f(t) - P_{MN}(f(t)) = \left(\sum_{j=0}^{\infty} (c_j t^j) \sum_{j=0}^{N} (b_j t^j) - \sum_{j=0}^{M} (a_j t^j)\right) \bigg/ \sum_{j=0}^{N} (b_j t^j). \tag{IV.4.7}$$

If the approximant is to have accuracy of order $M + N$ and if $f(0) = P_{MN}(f(0))$ then

$$\sum_{j=0}^{\infty} (c_j t^j) \sum_{j=0}^{N} (b_j t^j) - \sum_{j=0}^{M} (a_j t^j) = \sum_{j=M+N+1}^{\infty} (d_j t^j). \tag{IV.4.8}$$

It has been demonstrated that for approximations of λh where $M = N$, $M = N + 1$ and $M = N + 2$, the modulus is less than unity and thus a method with a characteristic root equivalent to these approximants is A-stable as well as having an order of accuracy of $M + N$.

IV.5   REFERENCES

[1] L. Lapidus and J. H. Seinfeld, 1971. *Numerical Solution of Ordinary Differential Equations*, Academic Press, New York.
[2] G. Dalquist, 1963. Stability questions for some numerical methods for ordinary differential equations, *Proc. Symposia in Applied Mathematics*.
[3] V. Zakian, 1975. Properties of I_{MN} and J_{MN} approximants and applications to numerical inversion of laplace transforms and initial value problems, *J. of Math. Analysis and Applications* **50** 191–222.

INDEX